# Disk Order Form

## TK!SOLVER FOR ENGINEERS
## Victor E. Wright
## ISBN 0-8359-7711-0

**Skip a time-consuming step . . .**

By ordering the companion diskette, you'll be able to maximize your learning time while minimizing the time normally taken to key in programs offered in this book. Developed by the author, this companion diskette is the fast, easy, and errorless way to jump right into programs that will help you learn and give you immediate applications.

**Partial Disk Contents:** Most of the sample programs that appear in this book; English-to-Metric Conversions; Properties of Liquids, Solids, and Vapors.

To order, use this handy coupon. Enclose a check or money order for $34.95, plus local sales tax. Remember to fill in your name, address, and other information *completely.*

NAME _____

ADDRESS _____

_____

CITY _____ STATE _____

ZIP _____

CHARGE TO MY CREDIT CARD INSTEAD

☐ VISA          ☐ MASTERCARD

Account No. _____

Expiration Date _____

_____

Signature as it appears on card

Mail orders to:

Prentice-Hall, Inc.
200 Old Tappan Road
Old Tappan, NJ 07675

*Attn.:* M.O.B.

V-0940-2G(0)

# TK!Solver for Engineers

VICTOR E. WRIGHT, PE

A Reston Computer Group Book
Reston Publishing Company, Inc.
A Prentice-Hall Company
Reston, Virginia

**Library of Congress Cataloging in Publication Data**

Wright, Victor E.
  TK!Solver for engineers.

  "A Reston Computer Group book."
  Bibliography: p.
  Includes index.
    1. TK!Solver (Computer program)   I. Title.
  TA345.W75  1985       620'.0028'5425       84–11477
  ISBN 0–8359–7711–0

# Contents

# Preface

I wrote this book for engineers—particularly for engineers who were not fortunate enough to grow up with computers, as engineering students do today.

I studied computer programming when I was in engineering school in the 1960's at the University of Cincinnati. It was a required course, and we all took it. The language we used in the course was FORTRAN II. Later, in a graduate course, I studied computer programming again, FORTRAN again, but by now it had become FORTRAN IV. Fortunately or unfortunately, I'm not sure which, I never had the opportunity to apply what little knowledge I absorbed in those courses—I have spent most of my working career with companies that had no use for the computers that ran the FORTRAN compilers I learned to program for.

Then came the microcomputer. I resisted the temptation to buy one for years, which I now see was a mistake. I should have tried to get in on the ground floor. When I finally did purchase a microcomputer, I found that I was computer illiterate, even after having taken no less than two courses in FORTRAN. Not only did I have to brush up on programming, but I had to learn operating systems as well. Of course, when I took those programming courses, the computers were off-limits to students, and I had never seen one up close.

I learned to program in BASIC, revived my knowledge of FOR-TRAN, and even learned some of the newer languages, such as Pascal. However, I also determined that, for all the power the micro has brought us, it is not a panacea. In the world of engineering, time is money, and no single job can support the time required for large scale development of engineering software, even with the micro.

Until the advent of tools like TK!Solver, there were only two ways to make use of the micro—develop your own software or buy it from someone else. Either way was expensive: a duct sizing program typically costs about $1000, a pipe sizing program about the same. So what do most engineers do? They continue to use tried and true methods—hydraulic tables, friction charts, systems sizer slide rules, etc. If the cost is not a sufficient deterrant to bringing computers into the organization, lack of familiarity with computers is. I know the feeling.

TK!Solver changes, or can change, all that. It is a program that allows the engineer (or the economist, financial analyst, chemist, etc.) to think in familiar terms about problem solving. TK!Solver provides an electronic replacement for the notebook of quadrille paper engineers use to write equations, variables, or units. It records the information just as well as the notebook does, but goes a step further: it solves the equations. If you can find an equation that will provide a solution to a problem, TK!Solver can probably solve it.

This book is for engineers who realize that the computer is a tool, not a gadget; for engineers who still want to practice engineering, not learn to be programmers; for engineers who want to do more engineering in less time.

# PART I

# THE TK!SOLVER PROGRAM

# CHAPTER 1

# What Is TK!Solver?

TK!Solver is a problem solver. It's a computer program that allows you to enter problems into a computer and solve them in a way that is natural to anyone who has taken a course in algebra, trigonometry, or even grade school arithmetic. TK!Solver is an application program designed for nonprogrammers. That means anyone can use it without having to learn programming. It's the latest development in the use of computers, specifically microcomputers, for solving a wide variety of problems.

TK!Solver was not designed for the practitioner in any particular field. The program is a general-purpose problem-solving tool. However, it is suitable for the solution of real problems, including those encountered in all fields of engineering.

TK!Solver is a significant development in engineering software because it allows the user to enter a problem into the computer in familiar terms. It is a logical and welcome extension of the trend started by VisiCalc® [1]. TK!Solver shares another characteristic with VisiCalc: even though it is powerful, and allows you to work in familiar terms and formats, it is *not easy* to use. Using TK!Solver to its full potential requires considerable time and effort.

TK!Solver is designed to be a practical tool, one that can be used day in and day out to solve real problems. It is designed to perform many functions quickly and efficiently and can accept commands almost as fast as a proficient typist can enter them. It spends much less time asking questions than many "user friendly" programs do.

At first, TK!Solver may seem confusing, and you may have to keep the manual open at all times. But if you make an effort to use

[1] VisiCalc is a registered trademark of VisiCorp.

TK!Solver to solve problems, even when you are tempted to use a pad of paper and a calculator, you will soon develop speed and proficiency. Then, you will not be tempted to return to the paper-and-pencil way of engineering.

## BACKGROUND

The first electronic computers allowed engineers and scientists to solve problems faster and more accurately than any known technique could do. These machines made practical the solution of problems that could not have been solved otherwise, or that could have been solved only at great expense and in long periods of time.

However, those early computers were not exactly easy to use. There was no software, no programmer, no computer department to call upon. Since the computer had to be programmed to solve the problem at hand, the first users had to invent programming. The method of solving a particular problem, the algorithm, had to be translated from the form that humans could comprehend to a form the computer could use. The translation process was so difficult and time-consuming, and so foreign to the way most people solve problems, that a new field of specialization soon appeared—computer programming.

The first computer programs were written directly in machine code. That is, instructions were converted to symbols consisting of strings of numbers on paper that represented patterns of "volts–no volts" in the machine. These patterns caused the computer to execute the program and produce a result that could then be interpreted. A typical early program might have looked like this:

```
1001000010111110000110000111101010100010101001010100100
1010010101010000101001010101010100001010010101010101010101
1001010101000001011111101011101010010100101010101010001
1010010101000001010100101010101010101010101010101011010101
1111100010101011110100101010101101000101010101101111101
1010010000011111111101010001001010101001010100101010010
1001001010100001110101011101010101010101010111111111000
```

Such programs were difficult to produce because the writing process was tedious and errors were a common occurrence. It must have been painfully obvious to those pioneer programmers that there had to be a better way.

The first milestone in the development of computer languages was the assembler, a computer program that allows the programmer to write programs in mnemonic form, that is, by using labels to represent the basic machine code instructions. For example, a programmer can remember that the label JCXZ means "Jump If CX Register is Zero" with much less effort than he can memorize the binary pattern 11100011. The assembler translates the mnemonics of the assembly language into the binary patterns of the machine code, thereby producing a machine code program.

The next major breakthrough in the effort to make computers easier to use was the development of the FORTRAN compiler (FORTRAN is an acronym for FORmula TRANslation, which is a programming language for scientific and mathematical use). As most engineers know, a FORTRAN compiler is a program that resembles a series of mathematical formulas. The compiler translates a series of FORTRAN statements, such as the ones shown in Figure 1-1, into a binary code that the computer can read and operate on, relieving the human users of the tedious task of translation and allowing them to concentrate instead on the problem to be solved. The end result of translating a FORTRAN program with a FORTRAN compiler is the same as translating an assembly language source program with an assembler—a series of binary instructions that tell the computer what function to perform. FORTRAN differs from assembly language in two significant ways: a FORTRAN program is not machine-specific, whereas an assembly language is; and a single FORTRAN statement is often the equivalent of several assembly language statements.

```
READ (IN, 106) N
M = N
IF (N .GT. MAXR) GOTO 5
IF (N .LT. 2) RETURN
DO 20 I = 1, N
    WRITE (OUT, 101) I
    K = 0
    L = 2 * I - 1
    DO 10 J = 1, N
      K = K + 1
```

Figure 1-1. A Partial FORTRAN Listing.

FORTRAN indeed helped to simplify the task of programming as its statements resemble familiar mathematical equations and simple English-like statements. Learning how to program a computer in its native machine language, or even to program it using its assembler, is far more difficult than learning FORTRAN. In the late 1950s and in the 1960s many an engineering student learned FORTRAN well enough to write small programs in a single semester. However, in using FORTRAN a programmer must state a problem in a form that follows the rules of this language. If a statement violates a FORTRAN rule, the compiler will not be able to translate the program properly.

Many other programming languages have been devised, but they all share a common goal: to allow the human operator to formulate the problems to be solved on the computer in terms that are easier for humans to process than the binary code used by the computer. The idea is always to allow the user, or programmer, to concentrate on the problem to be solved rather than on the details of a particular computer.

# A NEW TYPE OF PROGRAM

The reason that VisiCalc has been such a successful program is that it duplicates a familiar problem-solving tool electronically, and runs on a desktop computer, to boot. The VisiCalc display on the computer screen looks just like the columnar worksheet or spreadsheet that accountants, engineers, scientists, and others use to organize calculations and data. In a sense, the program can be said to have a language that is precisely the same as that of its users: to an accountant, VisiCalc is an accountant's worksheet; to an engineer, it is a calculation spreadsheet.

TK!Solver also allows its user to set up problems in familiar terms. In this case the terms are known as models, which take the form of an equation or a set of equations that can be solved. Figure 1-2 shows a typical TK!Solver display. This display will look familiar to anyone who has read an engineering textbook. It contains an area labeled RULE SHEET, where equations appear, and another labeled VARIABLE SHEET, where the variables are listed and explained.

At first glance, a rule in a TK!Solver model may appear somewhat similar to the statement in the partial FORTRAN listing in Figure 1-1, but there is a significant difference between the two. The FORTRAN statement

$$A = B * C + 4 * D$$

is not an equation in the same sense as the familiar expression below relating distance, time, and acceleration:

$$s = at^2/2$$

The FORTRAN statement is an assignment statement. That is, it instructs the computer to perform the operations described by the right side of the statement, and to store the result in the location named by the left side. Thus, the FORTRAN statement

$$A = A + 1$$

is a valid statement that is clearly not a mathematical equation. We won't go any further into the details of conventional computer programming at this point except to note that computer programming in any conventional programming language involves many concepts that are seldom formalized in the real world.

TK!Solver can, of course, work with equations. In the case of the equation relating distance, time, and acceleration, humans would solve for any one of these three variable factors by first

```
A>
(1r) Rule: rho*v1^2/2*gc+pa+pz1-pz1+p1+gamma*z1=rho*v2^2/2*gc+pa+pz2-pz2+p 201/!

=================== VARIABLE SHEET =============================================
St Input       Name    Output   Unit      Comment
-- -----       ----    ------   ----      -------
                                          ********************
                                          ***  STACKEFF  ***
                                          ********************
                                          CALCULATE THE STACK EFFECT
               rho               lbm/ft^3  mass density of fluid flowing in duct
               v1                ft/s      velocity at section 1
               gc                ft*lbm/lb dimensional constant
               pz1               lbf/in^2  atmospheric pressure at elev z1
=================== RULE SHEET ================================================
S Rule
- ----
* rho*v1^2/2*gc+pa+pz1-pz1+p1+gamma*z1=rho*v2^2/2*gc+pa+pz2-pz2+p2+gamma*z2+dPt
* g=32.174
* gc=32.174
* pz1=pa-(g/gc)*rho#a*z1
* pz2=pa-(g/gc)*rho#a*z2
* pse=(g/gc)*(rho#a-rho)*(z2-z1)
* gamma=rho*g/gc
* sv=1/rho
```

Figure 1–2. A TK!Solver Display.

substituting known numbers for the symbols representing two of these variables, and then rearranging and simplifying the equation until the symbol for the unknown number appeared on one side of the equals sign and a number on the other side. TK!Solver performs the same manipulations electronically, producing an output for any one of the variables once input values are assigned to all the others.

Within certain limitations, TK!Solver can solve a system of equations, working through one equation after another until a value is finally assigned to the last unknown, just as we would with a pad of paper and a calculator. As you examine TK!Solver's capabilities, you will see that its limitations are far outweighed by its power.

Like VisiCalc, TK!Solver is a "what-if" program—its models can be modified interactively. Like FORTRAN, it relies on statements in the form of equations. However, changing a FORTRAN program to solve a problem in a different way, or refining it to produce more accurate results, or even correcting errors, takes a considerable amount of time. The program must be edited (that is, the FORTRAN statements must be changed by means of an editor program), or new cards must be prepared on the keypunch machine. The revised program must then be compiled, or translated, with the FORTRAN compiler. Finally, it must be linked with a linker, which provides additional information required to transform the program you have written into one that will actually run on the computer. Because TK!Solver is an interactive program, its models can be changed and solved in a matter of seconds. If you wish to change, delete, or add a rule to the model, you can do so at any time, and then repeat the solution of the model.

What kind of problems can be solved with TK!Solver? Real ones—the kind of problems encountered in the practice of engineering. The program could be just as useful in any number of other fields, but we'll stick to engineering problems in this book.

Figures 1-3 through 1-6 illustrate the types of problems you can expect to solve with the TK!Solver program and the types of output you can produce. The basic method of output is shown in Figure 1-3: You enter input values in the input fields of the variable sheet, invoke the program's solver, and read the output values in the output fields of the variable sheet. Figure 1-4 shows an output list, produced by TK!Solver's list solver from input lists. Several lists, both input and output, can be combined to form a table, as shown in Figure 1-5. Finally, Figure 1-6 illustrates the graphing capability of the TK!Solver program.

```
(1c) Comment: PIPE SIZING MODEL                                    192/!

==================== VARIABLE SHEET ====================================
St Input      Name    Output     Unit     Comment
-- -----      ----    ------     ----     -------
                                          PIPE SIZING MODEL
                                          ****************
L             dP      .64093061  fth2o    pressure drop in L feet of pipe
              f       .01724504           friction factor, dimensionless
L    1        L                  ft       length of pipe
              rho     62.4       lbm/ft^3 fluid density
     5        V                  ft/s     fluid velocity
L             D       .28584123  in       inside diameter of pipe
              g       32.174     ft*lbm/lb acceleration of gravity
L    1        Q                  gal/min  volume flow rate of fluid in pipe
L             Re      9794.4502           Reynolds number
              v       .00001216  ft^2/s   kinematic viscosity
              epsilon .00015     ft       absolute roughness
L             Ds40    .364       in       standard sch 40 pipe id
L             dPs     .19139408  fth2o    pressure drop in sch40 pipe
L             Vs      3.0833086  ft/s     velocity in sch 40 pipe
L             Res     7691.3674           Reynolds number in sch 40 pipe
L             fs      .03920827           friction factor in sch 40 pipe
L             Dnom    'one_qtr            nominal pipe size
```

Figure 1-3. Output on the Variable Sheet.

TK!Solver enables people to use the power of the computer to advantage even though they don't have the time to learn how to program computers. It's also a boon to people who don't have the time to write computer programs to solve problems, but who know the algorithms to apply to the problems. The program makes the computer "transparent," as computer buffs are so fond of saying. In other words, you can use TK!Solver just as you would a self-

```
(1v) Value: 1.5350778293                                           192/!

==================== LIST: dP =========================================
Comment:
Display Unit:                    fth2o
Storage Unit:                    lbf/ft^2
Element Value
------- -----
1       1.535077829
2       3.719287795
3       .7041818425
4       .7578401749
5       1.115586548
6       .120054452
```

Figure 1-4. An Output List Produced by a TK!Solver Model.

```
                     Schedule 40 Pipe Sizes
   GPM      Size       I D, in.   Vel, fps   f          lgth, ft   delta P, f
   100      three_in    3.068     4.34019310 .020913416 45         1.07765071
   150      three_$_ha  3.548     4.86792411 .019931231 140        3.47851814
   225      five_in     5.047     3.60857446 .019114949 34          .311601105
   375      six_in      6.065     4.16475463 .018014536 50          .479872832
   475      eight_in    7.981     3.04648294 .017693275 85          .323277243
   550      eight_in    7.981     3.52750656 .017344158 10          .050190652
```

Figure 1–5. Several Lists Presented as a Table.

powered scratchpad or a tireless, obedient assistant—we write down the equations and data, and let the program do the dirty work.

A word of caution is in order here, however: an investment in time and effort is required. Although the TK!Solver manual will show you how to solve simple models in a single evening, you will spend many hours learning how to use all the features of TK!Solver. The difference between learning to use TK!Solver and learning to program in a conventional programming language lies not in the time required, but in the tasks you can accomplish while you are learning. It's powerful, to the point of performing many tasks not even mentioned in the manual, and it's designed for fast operation by an experienced operator.

By now, most engineering organizations have had to become acquainted with the computer. Consulting engineers who design heating, ventilating, and air conditioning systems for government buildings, for instance, are required to submit life cycle cost analyses of their proposed systems. The only way to prepare a comprehensive life cycle cost analysis is to use a computer; official design guidelines that have been established by state and federal agencies now specify several acceptable programs. Of course, these are large programs that run on large mainframe computers, and a single run can easily cost more than the TK!Solver program.

More and more engineers are being forced to rely on the computer because of economic factors. Since time is money, design work has to proceed at a quicker pace than it used to. In many cases the computer offers the only way to speed up the calculation part of engineering work.

Many an engineer who is forced into using a computer will select "canned" programs to accomplish the tasks at hand. This approach has two drawbacks: the programs are expensive; and the user has no control over the method of solution.

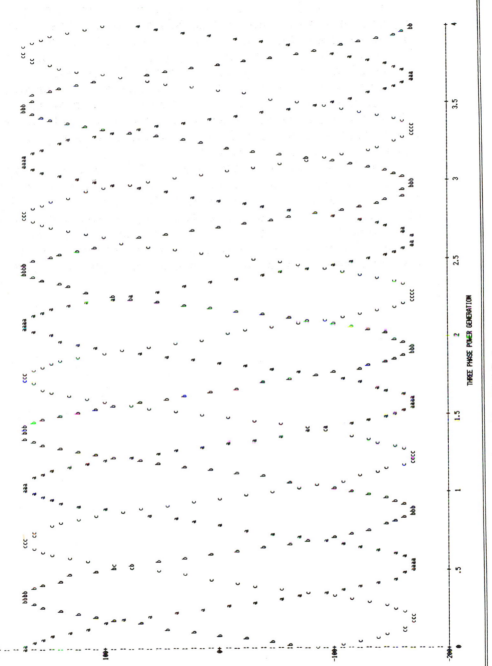

Figure 1-6. A Graph Produced by TK!Solver.

Again, the TK!Solver program differs from single-purpose programs in that the algorithm for solving a hydraulics problem or for modeling an electronic circuit, for example, is not built in. TK!Solver only solves equations: the engineer or user specifies the method of solving the engineering part of the problem.

How powerful is the TK!Solver approach? A single statistic may provide a hint. In later chapters showing how TK!Solver solves a list of equations and works with lists of variable values, we will see that we can enter 32,000 rules and lists with up to 32,000 values. That's a lot of information.

This book is about the organization of the TK!Solver program, the data you enter, and the commands you give it. It examines both the capabilities and the limitations of this program. The first part of the book describes how to use the program with simple models and problems. The second part looks at some TK!Solver models suitable for problems that engineers usually solve in the old-fashioned way, or by means of single-purpose computer programs.

# CHAPTER 2

# What Equipment Do You Need To Use TK!Solver?

TK!Solver is a computer program that runs on a number of personal computers. You must have the correct version for your computer. Some versions of the program are generic, and can be configured for any one of several computers by the user, but other versions are machine dependent and cannot be configured. The version used in this book is for the IBM® [1] Personal Computer (Figure 2–1), and it will not run on any other computer. The MS-DOS™ [2] version of the TK!Solver program will run on six different computers for which the MS-DOS operating system is available. Differences between the IBM PC version used in this book and other versions are presented in Appendix A.

## THE SOFTWARE PACKAGE

If you are familiar with computer software in general, you will know that some programs require other software to run. Many engineering programs require a BASIC interpreter, for instance. The TK!Solver program is a self-contained or stand-alone program that requires no additional software, other than the computer's operating system.

The program arrives in an attractive package similar to that of IBM's PC software. The documentation is assembled in a loose-leaf

[1] IBM Personal Computer is a registered trademark of the International Business Machines Corporation.
[2] MS-DOS is a trademark of Microsoft.

Figure 2–1. The IBM PC.

binder that is stored in a heavy-duty storage case. The printed text comprises two sections, a tutorial and a reference section. The package also includes a handy wall chart that illustrates the command syntax and the program organization, and a pocket guide that summarizes most of the commands.

The program resides on a single 5 1/4-inch floppy diskette. The program is copy protected; that is, backup copies cannot be made from the distribution disk. However, two copies are provided with the package in case one is ever damaged.

# THE IBM PERSONAL COMPUTER

In order to run the IBM PC version of TK!Solver, you need an IBM Personal Computer with one or two disk drives and 96K of user memory. (Other versions that run on other microcomputers may have different requirements.) Although TK!Solver can use up to 256K of memory if that much is present in the machine, all its capabilities will still be available when the program is run on a computer with the minimum amount of memory. If a minimial system is used, some operations may take a very long time. Nonetheless, you will be able to solve problems of the same type and complexity that can be handled by a computer with the maximum amount of memory.

Your IBM computer should have the IBM computer keyboard. The program checks the keyboard at various points in its operation, and the keyboard returns values to the program. Since nonstandard keyboards may not return the proper values or permit proper operation, the safest course of action is to use the standard keyboard.

The computer can use either the IBM monochrome monitor, a color monitor, or a television set and an RF adapter, but the display must be at least 40 characters wide. An 80-column display is preferable for TK!Solver. If you do use a 40-column or 64-column display, the displays on your computer will not match the illustrations in this book or the TK!Solver manual, but the problem-solving capabilities will still be available.

You must have the IBM PC-DOS® [1] (the Personal Computer Disk Operating System). Versions 1.0, 1.1, and 2.0 will do, but TK!Solver will not run under other operating systems that may be available for the computer. If you buy a version for another computer, you will have to make sure that you have the correct operating system for that computer.

You should also have a supply of blank diskettes so that you can store the models you will create. The TK!Solver program diskettes should not be used for that purpose. The distribution diskettes arrive with the write-protect notches covered so that you cannot write on them or accidentally erase them, and these protective tabs should be left in place. Always remember that you are working with diskettes that cannot be copied for backup purposes, so take every precaution possible to keep them intact.

---

[1] IBM PC-DOS is a registered trademark of the International Business Machines Corporation.

Although it is not absolutely necessary to have a printer, you will certainly want one for your system in order to print your results. The program allows you to print not only the answers to a problem, but also the rules, variable definitions, unit conversion factors, and functions used to reach the solution, all of which can add a great deal to an engineering report. Almost any printer will do as TK!Solver does not use graphics for any of its output. Of course, the IBM printer and the IBM graphics printer are excellent choices (installation and operation are covered in the IBM manuals).

The examples in this book were created on an IBM PC system having the following components:

- TK!Solver version TK-1 (2J)/PCDOS/IBM5150
- System unit with 256K user memory (RAM)
- IBM keyboard
- (2) Double-sided, double-density 5-1/4-inch disk drives
- IBM color/graphics board
- IBM color monitor
- Quadram QUADBOARD with 256K additional RAM and printer ports
- IBM printer (parallel) and Zenith Z-25 serial printer
- PC-DOS version 2.0

This system is not the minimal system for TK!Solver—the program will run with a less extensive system. Since TK!Solver does not take advantage of the color features of the PC or of the graphics features, the program will run just as well on a monochrome version.

## MEMORY EXPANSION

Because the original IBM PC accepts 64K of user RAM on the system board, a memory expansion board is needed to achieve the minimum 96K of user RAM that the program requires. An expansion board is not required for the later versions of the PC and the IBM® [1] PC-XT, which accept 256K of user RAM on the system

[1] IBM PC-XT is a registered trademark of the IBM Corporation.

board. In either case, one may as well install the maximum the program can use—256K. Even with 256K RAM in the computer, TK!Solver takes time to perform some tasks; thus operation of the program with a minimal system could be frustrating when a report is due and your TK!Solver model is its central feature.

The program will operate on a computer with only one disk drive, but practical considerations make a two-drive machine preferable. Although the TK!Solver disks are single-sided so that they will be compatible with the early PCs, the double-sided drives are more suitable for storing the models created in this book.

Three files on each of the distribution diskettes make up the TK!Solver program: TK.COM, TK.OVL, and TK.HLP. The first file, TK.COM, is the program per se—the executable file that is copy-protected. The other two files, TK.OVL and TK.HLP, can be copied onto another disk.

If you do have a PC-XT, or a PC with an external Winchester hard disk drive, you can copy the TK.OVL and TK.HLP files onto the Winchester drive, and you can store your models on it. Even though the TK!Solver program cannot be copied onto the Winchester, the program will operate faster if the supporting files are stored on it.

If you have a system with more than the maximum memory (256K) that TK!Solver can use, you should consider setting the excess memory up as a RAMdisk, and copying the TK.OVL and TK.HLP files onto the RAMdisk before starting a session. This arrangement is especially useful if you use the HELP facility because TK!Solver executes its commands much faster that it retrieves information about them from a floppy disk. The first version of TK!Solver had a very slow HELP function—it was almost as slow as flipping through the manual—but the current version has been greatly improved.

If you have a PC, and are familiar with its operation, you may wish to skim over the next chapter, or skip it and return to it later.

# CHAPTER 3

# Getting Started

If you already have an IBM Personal Computer and the TK!Solver program, you may wish to skim over this chapter. If you have neither one, reading it will make you a little more comfortable when you go shopping.

The IBM PC has three major components: the system unit, the keyboard, and the display. If the IBM printer is attached, it is a separate component. The IBM PC-XT includes a Winchester hard drive in the system unit, and one floppy disk drive. Expansion units are available for both the PC and the PC-XT; the expansion unit for the PC includes a Winchester hard drive controller, whereas the expansion unit for the PC-XT does not, as the controller is in the PC-XT system unit.

The TK!Solver promotional literature correctly states that you don't have to be a programmer to use the program. However, you do have to be your own computer operator. You should therefore have some idea of the use of the computer and the operating system as they relate to TK!Solver.

If your machine has two disk drives, the one on the left is drive A and the one on the right is drive B. Drive A is the default drive. When the computer is turned on, it will look for files on drive A. The TK!Solver manual always assumes that the DOS will look for files on drive A unless drive B is specified in the reference to the file.

## THE OPERATING SYSTEM

In order to run TK!Solver, you must first load PC-DOS into the computer, or "boot the system." PD-DOS is the operating system that helps you manage the information you use in running TK!Solver, or any other program, for that matter. The operating system is a program that stores information on the diskettes and retrieves it from the diskettes when you need it again. Throughout the book, you will be referred to "files" that are stored on disk. The term *file* signifies a program that performs some task, or data that will be used by a program. The operating system allows you to refer to these files using names of your own choosing; it will then create, copy, move, modify, or delete a file. You never know where it is stored on the disk.

In order to boot the system, you insert the system diskette (the one that contains PC-DOS) in drive A, close the drive door, and turn on the power. The computer then automatically locates the information it needs on the disk, transfers that information from the disk into its memory, and asks for the time and date. You then enter the information, and press the **RETURN** key. When the system is ready to accept a command, it will display this prompt:

A>

If you give the computer a command, it first tries to execute a built-in command—one that is in memory. If the command is not in memory, the computer will look for a file, or program, with that name on drive A. If you want to execute a command or program stored on the disk in Drive B, you must tell the computer to look for it there. You can either prefix the command B:, or type B: and press the **RETURN** key. In the first case, the computer will go to Drive B to find the command, but will still be logged in to Drive A; in the second case, the computer will log in Drive B, and look for all commands there, until A: RETURN is typed again.

Before you load TK!Solver, you should format some new, blank diskettes. You will need these extra disks to store the results of your TK!Solver sessions—you may wish to use your TK!Solver models over and over. The computer recognizes the information on a diskette by means of electronic marks that define storage areas, or sectors on the diskette. Since newly purchased diskettes are not always marked the way your computer wants, you must make sure that they are properly marked by formatting each one with the FORMAT program that is included with PC-DOS. Once you have TK!Solver running, you will not be able to format a disk without leaving the program and losing the information in memory. Therefore you must format the disks you need in advance.

**WARNING: Do NOT attempt to format the TK!Solver disk!**

To format a disk, insert the blank disk in drive B, and type the command,

FORMAT B: < cr >

The system will then type on the screen

Insert new diskette for drive B:
and strike any key when ready

Close the drive door and press any key; the drive light will come on, indicating that the system is writing the required information on the disk. When the disk has been formatted, the system will ask if you want to format another disk. At that point, you may choose to format another disk or exit from the format program.

Once you have DOS running and have formatted a few disks, you will be ready to load TK!Solver for the first time. Place the TK!Solver disk in drive B, and type

B:TK < cr >

Drive B will light, and the display shown in Figure 3–2 will appear on the screen.

Press the **ENTER** key. The display shown in Figure 3–3 will appear.

This display is discussed in more detail in the next chapter. For now, note that if it appears, the program has been loaded properly; if your disk has been damaged, or if you try to load the program from a copy of the distribution disk, you will receive a message informing you that the disk is bad and you will want to exit from the program. Thus, the first thing to learn about a program is how to exit from it. To exit from TK!Solver, you must know TK!Solver commands.

## TK!SOLVER COMMANDS

When TK!Solver presents its initial display, a white bar appears across the screen in the area labeled Rule sheet, as shown in Figure 3–3. At that point, you have two options: you can begin to enter a rule, or you can give TK!Solver a command to perform a task. TK!Solver has several commands that begin with the common character /. If you type the /, a list of these commands (see Figure

Version TK-1 (2J)/PCDOS/IBM5150
of the TK!Solver Program

This program is a product of

SOFTWARE ARTS, Inc.

Copyright (c) 1982 Software Arts, Inc.
All Rights Reserved

Except as specifically authorized in
the Software Arts License Agreement,
copying any part of this program
is prohibited.

***** Press ENTER to Start*****

Figure 3-1.  The TK!Solver Sign-on Message.

3-4) will appear on the second line of the display. Only the first letter of each command appears on the screen.

```
(1r) Rule:                                                                206/
For Help, type ?
===================== VARIABLE SHEET ========================================
St Input      Name    Output    Unit    Comment
-- -----      ----    ------    ----    -------




==================== RULE SHEET ============================================
S Rule
- ----
```

Figure 3-2.  The Initial Display.

```
B    Blank              P    Print
C    Copy               Q    Quit
D    Delete             R    Reset
E    Edit Field         S    Storage
I    Insert             W    Window
L    List               !    Solve
M    Move
```

Figure 3–3. TK!Solver Commands.


All the commands are discussed in the chapters that follow. For now, you need only look at the Q (Quit) option. If after typing /, you type Q, the program will write the following prompt on the second line (called the Prompt/Error line) of the screen:

Quit:   Y N

TK!Solver requires that you confirm the Quit command because everything stored in memory will be lost when the command is executed. The confirmation ensures that you will not quit the program accidentally and lose your work. If you type N, the Quit command is canceled; if you type Y, you will be returned to PC-DOS.

There is no need to wait for TK!Solver to write out each prompt before typing the next part of a command. You can type as fast as you wish, and TK!Solver will execute the command as soon as the last part is entered. Thus, the command to quit can be entered as

without hesitation.


Later you will discover an important feature of TK!Solver, the Help facility. Notice the second line of the display shown in Figure 3–3. Typing ? at any time will bring the Help facility on line. You can then enter a topic, and the Help facility will search its files for information on the topic. Typing another ? will display instructions for using the Help facility.

# TK!SOLVER PROGRAM FILES

When you type TK, followed by a carriage return, PC-DOS loads the file TK.COM from the TK!Solver diskette. This file is the executable file—that is, the actual program. The diskette contains some other files, which can be listed by name on the screen by typing DIR (carriage return). The contents of the diskette are listed below:

```
Volume in drive B has no label
Directory of B: \
```

| | | | | |
|---|---|---|---|---|
| TK | OVL | 49062 | 12–10–82 | 12:00p |
| TK | HLP | 29693 | 12–10–82 | 12:00p |
| GRAVITY | TK | 780 | 12–10–82 | 12:00p |
| COST | TK | 444 | 12–10–82 | 12:00p |
| TUNITS | TK | 208 | 12–10–82 | 12:00p |
| LINEAR | TK | 481 | 12–10–82 | 12:00p |
| MORTGAGE | TK | 931 | 12–10–82 | 12:00p |
| NPOWER | TK | 511 | 12–10–82 | 12:00p |
| FUNC | TK | 592 | 12–10–82 | 12:00p |
| TFUNC | TK | 479 | 12–10–82 | 12:00p |
| INSTRUCT | BAT | 128 | 12–10–82 | 12:00p |
| TK | COM | 4369 | 12–10–82 | 12:00p |

```
        12 File(s)                    12288 bytes free
```

Notice that several types of files are on the TK!Solver diskette, as indicated by the last two or three letters (the file extension) of each name. Although TK.COM is the only executable file, two other files must be present to run the program: TK.OVL and TK.HLP. TK.OVL is the Overlay file; it is not automatically loaded with TK.COM, and is only called into memory when the information it contains is needed. TK.HLP contains the text that is displayed by the Help facility, and it is loaded into memory only when it is needed.

The remaining files on the TK!Solver diskette are designed to be used with the instructional manual. These disks contain three types of files, all of which are created by TK!Solver. The type of file is indicated by the characters following the period in a complete file specification; those characters are known as the extension of the file name. TK!Solver creates files with three extensions: .TK, .DIF, and .PRF.

Files with a .TK extension make up a TK!Solver model, or a portion of a model. As you will see in more detail later, a TK!Solver model is organized into segments called sheets and subsheets. A .TK file can hold anywhere from one to a great number of these sheets.

Files with an extension of .DIF are special files that allow TK!Solver to exchange data with other programs, such as VisiCalc, which can also read from and write to the same type of file. The .DIF files are discussed in detail later.

The .PRF files contain information in the same format that would be sent to a printer. This type of file is used to store the output of a TK!Solver model on disk so that it can be printed later, or it can be merged with other files by means of a word processor. Thus TK!Solver models can be incorporated into engineering reports.

You can view the contents of .PRF and .DIF files from PC-DOS with the TYPE command, and you can manipulate them with edit and word processing programs. Files with the .TK extension cannot be displayed on the screen.

As noted earlier, the TK!Solver program is copy protected. Although you cannot make backup copies of the diskette, the program is furnished with two copies of the diskette; however, you must be extremely careful in handling the TK!Solver diskette. Whatever you do, do not remove the write-protect tab. The two auxiliary files, TK.OVL and TK.HLP, can and should be copied to the storage diskette—that is, the diskette used to store the TK!Solver models. You are allowed to make as many copies of these files as you wish.

The TK!Solver diskette can be removed from the disk drive once the program is loaded into memory; doing so will reduce the wear on the diskette. TK!Solver can operate properly as long as a diskette containing the TK.OVL file is in one of the drives. It doesn't matter which drive that diskette is in—TK!Solver can find the file when it needs to be loaded into memory.

You don't need the Help file on the storage diskette unless you want to use the Help facility. At first it is wise to have both the overlay file and the help file on each storage diskette; the Help facility saves flipping through the manual.

Remember that you can remove the PC-DOS system disk after the system is booted. This has been made possible because programs such as TK!Solver do not write over the PC-DOS operating system, as can happen with other operating systems. PC-DOS remains in memory until the computer is turned off or is reset.

It is wise to go through the tutorial portion of the TK!Solver manual as soon as possible after receiving the program. Doing so won't take more than an afternoon or an evening, and will introduce each sheet, if only briefly.

Now that you have seen how to prepare the computer and load the TK!Solver program, you are ready to examine the structure of TK!Solver in the next chapter.

# CHAPTER 4

# Models and Basic Commands

An engineer uses models to solve problems. Specifically, he or she uses models (that is, representations) of real processes or systems to predict or analyze the performance of those processes or systems. The representation is completely predictable, whereas the real process or system being modeled is usually not entirely predictable. The real process may be too complex to understand, let alone predict. Furthermore, a real process may not be readily observed. For instance, it is not possible to observe the shock waves that are formed on the various parts of a supersonic plane like the SR-/1 Blackbird. Therefore scale models of planes are tested in wind tunnels—then the data gathered from the wind tunnel tests are used to predict or analyze the performance of the real thing.

Models of processes and systems can also be built on paper, in the form of mathematical equations. Here's a familiar model:

$$V = A \times T$$
$$S = (A \times T^2)/2$$

Given a time period $T$ and an acceleration rate $A$, you can determine the velocity of a body at the end of the time period, and the distance traveled by the body during that period. If you are given the other values, you can also determine the acceleration rate or the time period required to reach a certain velocity or to travel a certain distance by rearranging the equations of the model. Note that although this model can be relied on to determine one of the variables if the other two are known, it is not an accurate model of the motion of a body, but a simplification. Many other equations and variables are required to accurately describe the motion of a ballistic missile, for instance.

TK!Solver is a modeling tool that allows you to construct models of processes in the form of equations, or rules. But instead of writing the equations on paper, you enter them into the computer via the keyboard, and the computer displays them on the screen. If you use paper and pencil to work with your equations, you must manipulate the equations mentally and record intermediate results on the paper, keeping track of both the terms of the equations and their various units. TK!Solver performs those details for you.

## THE TK!SOLVER SHEETS

The TK!Solver program is organized just like an engineer's notebook of calculations—it's organized in sheets (all of which are listed in Figure 4–1). It will take several chapters to cover the use of all the sheets in detail.

The first two sheets that you should become familiar with are the Rule sheet and the Variable sheet. When TK!Solver is loaded, it displays the copyright message, as you saw in the last chapter, and instructs you to press the **ENTER** key to start the program. When you press **ENTER**, TK!Solver erases the copyright message and displays the Rule sheet and the Variable sheet. The Rule sheet will be displayed on the lower half of the screen, and the Variable sheet will be displayed on the upper half, as shown in Figure 4–2. These two sheets are required for all models.

Equations, which are the heart of the model, are entered in the Rule sheet. A model can comprise a single equation, or a system of equations. These equations contain variables, and once they are entered on the Rule sheet, the variables they name will appear on the Variable sheet, except in certain cases, which will be examined later. As mentioned earlier, these two sheets correspond to the familiar format of equations presented in engineering textbooks: one or more equations followed by a list of the symbols appearing in the equations, with explanations and units for each.

## FIELDS

Each of these sheets is divided into fields. The Rule sheet has rows, each of which is divided into two fields, arranged in columns: the S or Status field, and the Rule field. (Throughout the book, the terms

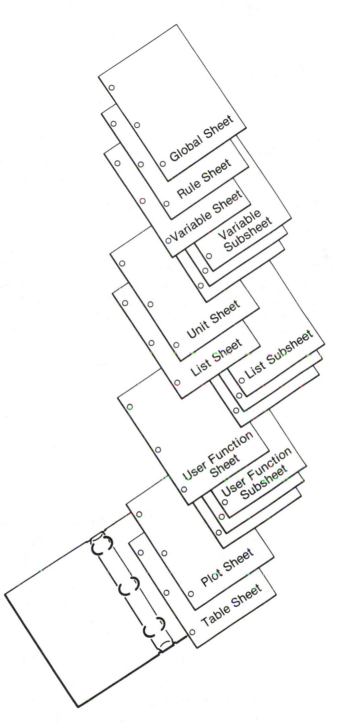

Figure 4–1.  The TK!Solver Sheets.

```
(1r) Rule:                                                          206/

=================== RULE SHEET =========================================
S Rule
- ----


(1i) Input:                                                        206/

==================== VARIABLE SHEET ====================================
St Input      Name      Output     Unit       Comment
-- -----      ----      ------     ----       -------
```

Figure 4-2. The Rule and Variable Sheets.

*field* and *column* will be used interchangeably; some sheets contain columns of fields and some contain single fields.) The Rule field is an entry field; you can make entries in an entry field. Specifically, you can enter rules, or equations, in the Rule field of the Rule sheet.

There are two other types of fields: output fields, and option fields. You cannot make entries in the output fields as they are reserved for information generated by TK!Solver. You can make entries in the option fields, but only from a limited set of characters. Because the entries that can be made in option fields and the meanings of those entries are different for each sheet, we will consider the option fields for each sheet as we study the sheet.

## THE CURSOR

When the first display appears, you will notice a green bar (or a white bar on a color monitor) extending from the first column of the Rule field of the Rule sheet to the right edge of the screen. That bar, which is known as the cursor, indicates where the next entry will be made. Notice that the cursor fills the field it is in: when in the Rule field of the Rule sheet, it is almost as wide as the display.

The length of the cursor does not, however, reflect the actual length of the field. The Rule field display on the screen is only 77 characters long, whereas the Rule field in memory will hold a rule 200 characters long. Various other fields are displayed in a similar manner.

TK!Solver uses the Status line, which is the first line on the screen, to indicate where the cursor is positioned. The number and letter in parentheses here are called the position indicator. The number tells you which line the cursor is on, and the letter tells you the name of the field the cursor is in by displaying the first letter of the field name. To the right of the position indicator is the field label, which is the name at the head of the column the cursor is in. The field label is followed by a colon. When the program is first loaded, the space to the right of the colon is blank. When an entry is made, the contents of the field where the cursor is positioned are displayed in this space.

## Moving the Cursor

TK!Solver allows you to move the cursor on the sheet with the arrow keys. The ← key moves the cursor to the previous field—that is, the one to the left of the cursor. If you want to move to the line above or below the current line, but remain in the same column, you use the ↑ key or the ↓ key. You can move the cursor to the extreme left field, the extreme right field, or the top line of a sheet by using the arrow keys. If you try to move past these boundaries, the cursor will not move and the computer will beep.

If you try to move past the last line on either the Rule sheet display or the Variable sheet display, the sheet will scroll up and the cursor will remain on the same line of the screen, allowing you to enter more lines that can be displayed on the screen at once. Not all the sheets are so cooperative; some will scroll for only a limited distance, and some not at all.

## THE GOTO COMMAND

If you want to move from one field to another without using the arrow keys to step through the intervening fields or lines, you can do so with the Goto command. When you type

```
:
```

TK!Solver displays the prompt

Goto: Destination or search:

on the second line of the screen (the Prompt/Error line), which TK!Solver uses to display error messages, prompts, and status messages.

You can respond to the Goto prompt in either of two ways.

You can enter the line number and the field name of the desired location on the sheet. The format of the response is the same as that of the position indicator in the Status line—that is, the line number followed by the first letter of the field label. If you want to move the cursor to the last line of a field that contains an entry, you type the Goto command and an asterisk,

followed by the field name.

Notice that TK!Solver reacts immediately when you type the Goto command, prompting for the destination. However, you must inform the program when you have entered the destination by pressing **ENTER** at the end of the entry, since the row number could be anywhere in the range of 1 to 32000. To move from the first row in the Input column (field) to the last possible row of the Output column, you would type

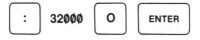

If you type a number without a field name, the cursor will move to the new line in the same field; if you type a field name without a line number, the cursor will move to that field in the same line. You can issue the Goto command with an asterisk and no field name in order to move to the bottom of the current column.

You can also have TK!Solver search for a location within the current column. You tell the program that you want it to search for an item by typing a double quotation mark (") *after* typing the Goto command, followed by the item as it appears on the screen. If the entry in a field is longer than the width of the screen display of that field, you do not have to enter the portion that is not displayed, but you must enter all the characters that are displayed.

## WINDOWS AND THE SWITCH COMMAND

One more cursor movement command should be mentioned before we leave the initial display: the Switch command. You will recall that two sheets appeared on the screen when you first loaded TK!Solver: the Rule sheet, and the Variable sheet. Remember that TK!Solver is organized as a series of sheets, and that there are far

too many to display on the screen at once. You can look at this array of sheets through windows. If you have used VisiCalc, the concept is a familiar one—the computer screen can be thought of as a
window through which you can view a portion of a large worksheet.
In TK!Solver, the windows look into the sheets that make up the
program, as illustrated in Figure 4-3.

    You can have one or two windows on the screen at a time. The
program initially loads with two windows: one into the Rule sheet,
and one into the Variable sheet. You can't build and solve a model
with less than these two sheets, and it is handy to have them both
on the screen when you are building a model.

    The arrow keys move the cursor around on a sheet; the sheet
will  scroll, in effect moving the window around on the sheet. But,
the arrow keys won't move the cursor off the sheet. To move the
cursor to another sheet, use the Switch command ;. Each time you
type

> [ ; ]

the cursor switches to the other window, if two windows are on the
screen.

## THE SELECT COMMAND

You can build and solve some simple models using just the Rule
sheet and the Variable sheet, but to build the type of models used
in the application section you will need all the sheets. To display
other sheets, use the Select command =. With the cursor in either
window, type

> [ = ]

to change the sheet displayed in that window. When you type =,
TK!Solver will prompt you on the Prompt/Error line with this
message:

       Sheet:  V  R  U  G  L  F  P  T

Each letter represents a sheet:

V   Variable Sheet

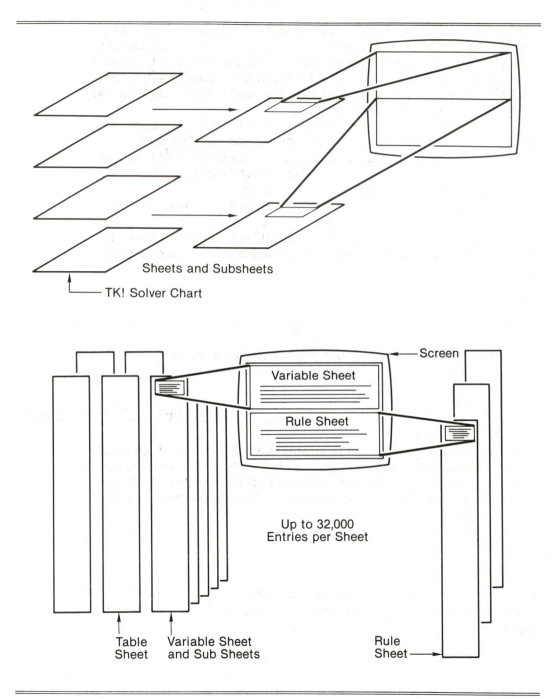

Sheets and Subsheets

TK! Solver Chart

Screen

Variable Sheet

Rule Sheet

Up to 32,000
Entries per Sheet

Table
Sheet

Variable Sheet
and Sub Sheets

Rule
Sheet

Figure 4–3.  Windows into the Sheets.

R   Rule Sheet

U   Unit Sheet

G   Global Sheet

L   List Sheet

F   Function Sheet

P   Plot Sheet

T   Table Sheet

When you enter one of the letters in response to the prompt, TK!Solver will change the sheet displayed in the current window (the window containing the cursor) to the one you have selected.

## THE WINDOW COMMAND

The last basic command you must know before proceeding to the subject of building TK!Solver models is the Window command. This command allows you to change the display from one window to two windows, or vice versa. The command is

When you type that command, TK!Solver displays the prompt

Window:  1 V R U G L F P T

The options are the same as those for the Select command, except for the initial number. If the current display is split into two windows, the 1 option is the only option available. Selecting the 1 will change the display to a single window into the sheet containing the cursor.

　　If you enter the /W command when only one window is on the screen, the screen will be split into two windows. For instance, if only the Variable sheet is on the screen, type

```
/   W   R
```

to display the Rule sheet in the lower half of the screen. The top window will always continue to show the current sheet, and the bottom window will show the sheet selected from the command options. Using the two commands, the Select command and the Window command, you can display any of the sheets or subsheets in either window.

## SOLVING A MODEL

Just for a preview of things to come, let's look at a model called GRAVITY, which comes with the program. It can be loaded from the TK!Solver disk with the Load option of the Storage command, which we enter with the command

<br>

| / |    | S |    | L |    **GRAVITY**    | ENTER |

When the model is loaded, the display will look like Figure 4-4.

This model includes a Unit sheet, but the Rule sheet and Variable Sheet will suffice for now. It's time to discuss how to enter input values and solve the model.

The model is loaded with the cursor in the lower window, where the Rule sheet is displayed. You move the cursor to the upper window by typing

| ; |

```
(1i) Input:                                                         204/!

===================== VARIABLE SHEET =====================================
St Input          Name    Output      Unit       Comment
-- -----          ----    ------      ----       -------
                  v                   ft/s       velocity
                  t                   s          time
                  s                   ft         distance
                  g                   ft/s^2     grav accel



================== RULE SHEET =============================================
S Rule
- ----
* v=g*t
* s=.5*g*t^2
* g=980.665     "in cm per second squared
```

Figure 4-4. The GRAVITY Model.

Next, you move the cursor to the Input field, second line, using the arrow keys. Type a value for time,

| 2 | | 0 |
|:-:|:-:|:-:|

and press **ENTER (carriage return)**

| ← |
|:-:|

The number you have typed will appear in the field, in black against the light background of the cursor, but the data will not be entered until you press the **ENTER** key, or one of the arrow keys. Before the entry or arrow keys have been pressed, you can back-space and change the value, if required.

To solve the model when the correct inputs have been entered, press

| / | | ! |
|:-:|:-:|:-:|

TK!Solver will solve the model and present the display shown in Figure 4–5 below.

Now that you have seen how TK!Solver solves the model, let's look at these two sheets more closely. The Rule sheet has two col-

```
  (2i) Input: 20                                                    204/

  ==================== VARIABLE SHEET ===================================
  St Input      Name    Output     Unit     Comment
  -- -----      ----    ------     ----     -------
                v       643.47970  ft/s     velocity
     20         t                  s        time
                s       6434.7970  ft       distance
                g       32.173985  ft/s^2   grav accel




  ==================== RULE SHEET ======================================
  S Rule
  - ----
    v=g*t
    s=.5*g*t^2
    g=980.665    "in cm per second squared
```

Figure 4–5. The GRAVITY Model, Solved.

umns of fields: the Status fields and the Rule fields. If you insert comments in the Rule field by entering a double quotation mark, followed by the comment, you can obtain a third field in each row. Because TK!Solver ignores everything in a line following a double quotation mark, the comments serve only as reminders, explanations, or the like. You will insert equations in the Rule field; the equations are the basis of the model. As you will see, our models are not restricted to the equations of the Rule sheet, but most models will be based on one or more equations.

The other column on the Rule sheet is the Status column. The fields of the Status column are output fields. You cannot enter information in these fields; TK!Solver uses them to display its messages. When you loaded GRAVITY, you saw an asterisk in the Status column for each equation. The asterisk means that the rule is not satisfied—that is, it hasn't been solved. If you move the cursor to one of the asterisks, TK!Solver will tell you the equation is not satisfied by means of a message on the second line (the Prompt/Error line) of the display:

Status: * Unsatisfied

When the model has been solved, the asterisks disappear, and if the cursor is moved to the status field for a rule, the message on the Prompt/Error line will read "Satisfied." When a model has been solved, all the asterisks will disappear. If one or more asterisks remain after you have attempted to solve a model, you will know that something has gone awry. (Later you will see that a model can provide useful results, even if all the rules are not satisfied.)

All the entries in the Rule column must be equations. Like any mathematical equation, each rule must have an equals sign and a valid expression on each side of the equals sign. TK!Solver will not accept an invalid expression in an equation, such as an expression containing more right than left parentheses, expressions ending with operators, and so on. Unlike FORTRAN or BASIC assignment statements, however, you can place an arbitrary number of terms on both sides of the equals sign. Remember that these rules are true equations, not assignment statements.

## OPERATORS

TK!Solver rules are like FORTRAN and BASIC statements in one important respect: the program does not recognize implied operators. You can use $ABC$ to represent "$A$ times $B$ times $C$" on paper, but not in computer programs, and TK!Solver is no exception. Nor can you enter an exponent as a superscript; you must use an explicit operator for exponentiation. The same rule applies to expressions in parentheses; for example, $a(b+c)(e-f)$ must be written as a*(b+c)*(e−f). Since computer terminals do not have the standard symbols for operators, you must use the following symbols:

| Operator | TK!Solver symbol | Meaning |
|----------|------------------|---------|
| + | + | Addition |
| − | − | Subtraction |
| x | * | Multiplication |
| ÷ | / | Division |
| exponent | ^ (or E for scientific notation) | Exponentiation |

## VARIABLES, CONSTANTS, AND FUNCTIONS

Besides operators, you can use variable names, constants, and functions in your rules. Variable names must begin with a letter of the alphabet or one of the characters @ # $ % __. The remaining characters must be letters, numbers, or the special characters listed above. TK!Solver distinguishes between upper and lower case letters in variable names, so you must be careful—Velocity and velocity, for example, would be treated as two different variables. For those who have struggled with some early BASIC interpreters, which allowed only one- and two-character variable names, TK!Solver's limit of 200 characters per variable name will be a sure sign of progress.

You can use two types of constants in your rules. The first type of constant is a numeric value. You can enter these numbers in decimal form, or in scientific notation, using the form 4.5E4 rather than the form $4.5 \times 10^4$. All numbers must be real numbers or integers; you cannot work with imaginary or complex numbers.

The second type of constant is the symbolic value, or string constant, as BASIC programmers would say. Symbolic values must be preceded by an apostrophe, cannot contain spaces, and cannot begin with numbers. Since the same special characters allowed for variable names are also allowed for symbolic values, they can be made more readable by using one of the special characters instead of spaces; for example, flow rate can be written as 'flow__rate or 'flow$rate. Symbolic values are limited to 200 characters, as are variable names.

Lastly, you can use functions in your rules. For example, you can model the relationship of the length of a side of a square to the area with the rule

SIDE = SQRT(AREA)

as well as with the rule

AREA = SIDE^2

TK!Solver has more than thirty built-in functions. The list includes mathematical functions and special TK!Solver functions. We'll study these functions in detail in a later chapter, and we'll also study user-defined functions. For now, rest assured that TK!Solver will do all that nasty trig for you.

# THE VARIABLE SHEET

Now let's look at the Variable sheet. It has six columns, or groups of fields. The Variable sheet keeps track of your variables and their characteristics. Unless you tell TK!Solver to do otherwise, it will record each variable on the Variable sheet as you enter it on the Rule sheet. You'll see how to change that arrangement later. First, let's look at the fields of the Variable sheet.

The first field is the Status field, which is both an output field and an option field. As mentioned earlier, an option field is an entry field with only a limited number of possible entries. As an output field, the Status field will tell you if a variable has caused an error during the solution of the model. If the value assigned to the variable causes an error, the program will place a > in the Status field for that variable. You'll learn how to use the Status field as an option field after we look at the other fields.

The next column is the Input column. These are the fields where you enter the input values for the variables for which you have known values. You can enter numbers or expressions that TK!Solver can evaluate to numbers. For example, if you have entered the rule

VOLUME = AREA * LENGTH

to find the volume of cylinders, you can enter a value for the cross-sectional area of the cylinder using an expression for the area, rather than entering another rule, or calculating the value of the expression

$$PI(\ )*4\hat{}2 <ENTER>$$

When you enter this expression, it will appear in the field; when you press **ENTER**, TK!Solver evaluates the expression, and a single value appears in the field.

You can also enter symbolic values in the Input fields, following the rules mentioned above for symbolic values. Again, a symbolic value must always be preceded with an apostrophe—otherwise, TK!Solver will attempt to evaluate the value as an expression, and will return an error.

The third column is the Name column. Each variable has a name, constructed as outlined above. Every variable used in a rule must appear in the Name column of the Variable sheet. If you don't tell the program otherwise, it will make these entries for you as you enter the rules on the Rule sheet.

The fourth column is the Output column. TK!Solver uses these fields to display its output when a model is solved. As with the Input field, your models can display numeric or symbolic values as outputs.

The fifth column, which is the Unit column, displays the units associated with the variables. As you will see in the next chapter, TK!Solver keeps track of units for you. For now, it is interesting to note that you can define the units in your own terminology. TK!Solver does not work with predefined units; it simply converts units according to rules that you define for each model, and it performs its calculations in a consistent set of units, as you define them.

The last column is the Comment column. The program ignores the comments, just as it ignores the comments in the Rule field of the Rule sheet, but allows you to keep notes, enter instructions, and so forth. Unlike the comments you enter in the Rule field of the Rule sheet, you do not have to precede these comments with double quotation marks. Similar comment fields are provided on several sheets, and all follow the same convention.

Let's return to the Status column and look at the options you can enter there. Options are single letters, and you don't have to press **ENTER** to make the entry. Just place the cursor in the desired field, and type the first letter of the option. The options are listed and explained in Figure 4-6.

Before going on to the next chapter, let's look at two more commands, the Dive command and the Return command. You saw

| OPTION | RESULT |
|--------|--------|
| Input (I) | Moves the output value to the Input field. Deletes the Guess option. |
| Output (O) | Moves the input value to the Output field. Deletes the Guess option |
| List (L) | Associates the variable with a List of values. Can coexist with the Guess option. |
| Guess (G) | Designates the input value as the first guess for an iterative solution. |
| Blank (B) | Blanks the variable value. |

Figure 4-6. Options for the Variable Sheet.

earlier that TK!Solver is organized into sheets and subsheets. You can look at a subsheet using the Dive command. Let's look at one of the Variable subsheets of the GRAVITY model that is furnished on the TK!Solver diskette. First, position the cursor in the window showing the Variable sheet; then move it to one of the rows containing a variable. If you place the cursor in the row containing v and then type

> 

TK!Solver will display the subsheet shown in Figure 4-7.
    Note that some of the entries in the Variable subsheet correspond to the columns of the Variable sheet. In fact, the Variable

```
(s) Status:                                                            201/!

==================== VARIABLE: v1 ==========================================
Status:
First Guess:
Associated List:
Input Value:
Output Value:
Display Unit:            ft/s
Calculation Unit:        ft/s
Comment:                 velocity at section 1
```

Figure 4-7. The Variable Subsheet.

sheet is a summary of all the Variable subsheets in a model, and the Variable subsheets actually define the variables. Note also that some entries on the subsheet are not shown on the Variable sheet: First guess, Associated list, Display unit, and Calculation unit. We'll discuss these entries in more detail later as they are not required for the first models we will build.

To return to the main sheet, the Variable sheet in this case, type the Return command,

```
<
```

Or, you can return to the Variable sheet by typing

```
=    V
```

You can't move directly from one Variable subsheet to another without returning to the Variable sheet, but you can leave the Variable subsheet for any other sheet.

So far, a TK!Solver model looks almost like an equation in a textbook: the equation appears on one line, and the variables and constants, and their units, appear in a list below the equation, just as we noted above. If you enter equations with consistent units, the Rule sheet and the Variable sheet are all that you need for a complete model. You can even construct a workable model without defining units—but that would be a waste of TK!Solver's power.

# CHAPTER 5

# Building a TK!Solver Model: More Sheets, Commands

The variable sheet and the Rule sheet discussed in the last chapter correspond to textbook equations and to the definitions of the variables used in those equations. These two sheets are used in almost all TK!Solver models, and some powerful models can be constructed with just the two. But TK!Solver has even more to offer.

You can construct a theoretical model of a process or system with a few equations, but you will not be able to use that model in solving actual problems unless the units of the variables are consistent. Back in the dark ages when slide rules were still used, you had to keep track of both the units and the decimal point mentally. Although the calculator and the personal computer subsequently took over the job of keeping track of the decimal point, the units of the variables remained the operator's responsibility. TK!Solver has now changed that, as it will also watch the units.

## THE UNIT SHEET

Engineering problems are rarely presented with consistent units. If you are calculating the size of an air conditioning system, for example, you need to know how much heat is generated by the lights in the room. The electrical engineer, however, tells you how much power they draw in watts. You must then convert the number he gives you in watts to heat generated in British themal units by multiplying the number of watts by 3.413. If you are solving a heat radiation problem, you will be given the temperature of the radiating body in degrees Fahrenheit, but to solve the equation you must convert that temperature to absolute temperature in

```
(1f) From: ft                                                                191/!

===================== UNIT SHEET =============================================
From        To          Multiply By  Add Offset
----        --          -----------  ----------
ft          in          12
ft^2        in^2        144
```

Figure 5-1. The Unit Sheet.

degrees Rankine by adding 460 to it. In fact, you must examine each term in an equation to ensure that the data are expressed in the proper units. When you build a TK!Solver model, on the other hand, the program watches the units with the Unit sheet (Figure 5-1).

Before that is possible, however, you must first determine the rules and the variables when you build a TK!Solver model. Next, you must choose units for each of the variables, and enter the units on the Variable sheet, or directly in the Variable subsheets (see Figure 5-2). These units can be defined consistently to begin with, or they can be defined according to some other scheme—in an air conditioning load calculation, you can use watts for the units of the lighting variable. Then you enter conversion factors in the Unit sheet for each pair of units. The Unit sheet performs all conversions with the two basic operations of multiplying by a factor and then adding an offset, if necessary, as the column headings indicate. Thus you can enter all the possible conversion factors and let TK!Solver watch the units from then on.

```
(1i) Input:                                                                  205/!

===================== VARIABLE SHEET =========================================
St Input      Name    Output    Unit      Comment
-- -----      ----    ------    ----      -------
              VP                INCHES_WG  velocity pressure
              VEL               FT/MIN     air velocity
              TP                INCHES_WG  total pressure
              SP                INCHES_WG  static pressure



===================== RULE SHEET =============================================
S Rule
- ----
* VP = (VEL / 4005)^2          "The Toricelli equation
* TP = SP + VP
```

Figure 5-2. The Variable Subsheet.

In Chapter 4 you saw that TK!Solver accommodates two kinds of units for each variable: Calculation units and Display units. The Calculation unit is the one the program uses in its calculations, and the Display unit is the one you see on the screen.

When you first enter a unit in the unit column of the Variable sheet, that unit becomes both the Display unit and the Calculation unit, and it is automatically entered in both the Display unit field and the Calculation unit field. If you later change the Display unit on the Variable sheet, the Calculation unit remains unchanged. You can build a model that works properly by entering units in the Unit column of the Variable sheet as long as those units are consistent with the rules of the models. For instance, if you construct a model of the distance traveled by a body moving at a constant velocity,

$$s = v * t$$

*Where:*   s = distance in miles

v = velocity in miles per hour

You must define t in hours, and you must enter input values for t in hours. If you mistakenly enter a value in kilometers per hour for the variable v, you must interpret the answer as distance in kilometers, or the answer will be wrong.

By using the Unit sheet, you can define the conversion factors that allow TK!Solver to solve the model in its Calculation units and display the results in different units. You can enter values in English units, for instance, and read the output in metric units.

If you define units inconsistently, TK!Solver will display a question mark next to the numbers in the Calculation units that are in doubt.

## THE TORRICELLI MODEL

Now that you have looked at the GRAVITY model that comes with the program, it is time to build a model of your own and to learn the commands that can be used with it. This is a simple model that relates air velocity, total pressure, and static pressure in an air duct connected to a centrifugal fan.

The model is composed of two equations. The first equation is a form of the Torricelli equation, which is familiar to anyone who has had to select fans for moving air in ducts. A fan produces an increase in air pressure, which in turn causes the air to flow from the fan housing in the direction of lower pressure. If a centrifugal fan discharges into a closed vessel, it increases the pressure in the vessel, up to a certain limit. When you measure the pressure exerted on the walls of such a vessel, you are measuring the total pressure created by the fan.

However, if the fan discharges into a duct with outlets, the pressure exerted by the air stream on the walls of the duct is less than the total pressure measured in the closed vessel. The pressure exerted on the walls, perpendicular to the direction of air flow, is referred to as static pressure. The difference between the total pressure produced by the fan and the static pressure measured in a duct connected to the fan is the velocity pressure—which can be measured with a pitot tube aligned with the direction of air flow, connected to a manometer (Figure 5-3).

The Torricelli equation relates air velocity to velocity pressure, and the second equation asserts that total pressure is the sum of static pressure and velocity pressure. Therefore, if you know static pressure and velocity pressure, you can calculate velocity and total pressure.

Here are the two entries to make on the Rule sheet, preceded by the cell pointer:

(1r)   VP = (VEL / 4005)^2 "The Torricelli equation < ENTER >
(2r)   TP = SP + VP < ENTER >

The model will be accurate if you enter the pressures in inches of water and the velocity in feet per minute, common units for this type of problem. You won't have to make any unit entries in the Variable sheet if you always remember the units. If, you plan to expand the model, however, you should enter the units. TK!Solver makes the entries in the Name field of the Variable sheet automatically, and so all you have to do is add the units and comments. Move the cursor to the Variable sheet by typing the Switch command

```
;
```

and then make these entries:

(1i)    : 1u                  < ENTER >
(1u)    INCHES__WG        ↓
(2u)    FT/MIN            ↓

| (3u) | INCHES__WG | ↓ |
|------|------------|---|
| (4u) | INCHES__WG | <ENTER> |
| (4u) | : 1c | <ENTER> |
| (1c) | velocity pressure | ↓ |
| (2c) | air velocity | ↓ |
| (3c) | total pressure | ↓ |
| (4c) | static pressure | <ENTER> |

That's all there is to it. The display of the Variable sheet and the Rule sheet is shown in Figure 4-2, the Unit sheet is shown in Figure 5-1, and the Variable subsheets are shown in Figure 5-2. Now you can enter input values for two of the variables and solve for the other two, since you have two equations.

## TWO WAYS TO SOLVE A MODEL

Earlier you used one of the commands beginning with the / to solve a model. The command

is called the Solve command. The Solve command works from any point in the model—that is, the cursor can be in any field of any sheet, and TK!Solver will solve the model.

You can use one other command to solve the model, but only from the Rule sheet or the Variable sheet. It's called the Action command, and it is the exclamation point

```
!
```

without the slash (/). The Action command also generates action when entered from other sheets, but may not necessarily be used to solve the model. The function of the Action command in connection with the other sheets is discussed later.

If you enter values for the velocity pressure (VP) and static pressure (SP) and then give the Action command from the Rule sheet of the model you have built above, the display shown in Figure 5-4 will appear.

While TK!Solver is working on the solution of the model, it will tell you which method it is using—the Direct Solver or the

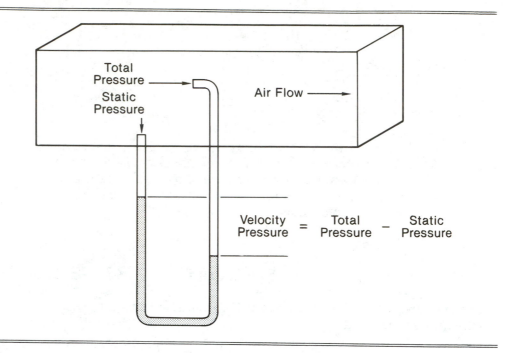

Figure 5–3. Diagram of Pressure Measurements in a Duct.

```
(s) Status:                                                           205/!

===================== VARIABLE: VP ================================================
Status:
First Guess:
Associated List:
Input Value:
Output Value:
Display Unit:               INCHES_WG
Calculation Unit:           INCHES_WG
Comment:                    velocity pressure

(s) Status:                                                           205/!

===================== VARIABLE: VEL ================================================
Status:
First Guess:
Associated List:
Input Value:
Output Value:
Display Unit:               FT/MIN
Calculation Unit:           FT/MIN
Comment:                    air velocity
```

Figure 5–4. Solution of the Torricelli Model.

```
(s) Status:                                                           205/!

==================== VARIABLE: TP =========================================
Status:
First Guess:
Associated List:
Input Value:
Output Value:
Display Unit:            INCHES_WG
Calculation Unit:       INCHES_WG
Comment:                total pressure



   (s) Status:                                                        205/!

==================== VARIABLE: SP =========================================
Status:
First Guess:
Associated List:
Input Value:
Output Value:
Display Unit:           INCHES_WG
Calculation Unit:       INCHES_WG
Comment:                static pressure



(4i) Input: .08                                                       205/

==================== VARIABLE SHEET =======================================
St Input      Name     Output     Unit       Comment
-- -----      ----     ------     ----       -------
   .1         VP                  INCHES_WG  velocity pressure
              VEL      1266.4922  FT/MIN     air velocity
              TP       .18        INCHES_WG  total pressure
   .08        SP                  INCHES_WG  static pressure



==================== RULE SHEET ===========================================
S Rule
- ----
  VP = (VEL / 4005)^2             "The Toricelli equation
  TP = SP + VP
```

Figure 5–4. (continued).

Iterative Solver. The model just discussed was solved with the Direct Solver because it could be solved in one step, just as you would have solved the equations manually by substituting values and simplifying. The two solvers are discussed in detail later.

## CHANGING VALUES

It would be a rare model that needed to be solved only once. Usually a model is solved many times, with different values. TK!Solver enables you to change the values in entry fields in three ways.

First, you can place the cursor in the field you want to change, and simply enter a new value in place of the existing value. As soon as you type the first character of the new value, the field will be blanked out, and the first character of the new value will appear in the field. As you type the remainder of the entry, it will appear in the cursor, although the Status line will still display the previous contents of the field. Then, when you press **ENTER** or move the cursor to another field with the arrow keys, the new value will be entered. If you decide to leave the value as it is, you can press the break key before pressing **ENTER,** and TK!Solver will ignore the attempt to change the value. When you change the model by entering a new value after the model has been solved, TK!Solver does not notify you that the output values are no longer valid by replacing the solution indicator on the right end of the Rule sheet Status line, nor does the program automatically solve the model each time a value is changed—you have to instruct it to do so either with the Solve command or the Action command.

A second way to change the values in entry fields is to blank the entries, or to edit them. Earlier you saw that you could blank an item on the Variable sheet with the Blank option by placing the cursor in the Status field of the Variable to be blanked, and by entering B. You can also use the Blank command, which is a B preceded by a slash (/B). The Blank command will blank a single field or a column of fields. When you type

| / | | B |

TK!Solver gives the prompt:

Blank: Point to the last field (1n)

The number in parentheses is the position indicator, which tells you where the cursor is. You can blank a single field by typing a carriage return immediately after typing /B, or you can move the cursor down the column in which the cursor is located before typing the carriage return. The program will blank all the fields between the initial and final locations of the cursor. Since the Blank command assumes the first field to be blanked is the current location of the cursor, it is imperative to locate the cursor in the desired field before entering the command.

You can abort the Blank command, or any other command that is terminated with **ENTER**, with the Break key. It's always wise to double-check before hitting **ENTER** when entering commands that delete previous entries, or that can overwrite previous entries.

A third way to change the value in an entry field is to edit the field. Here you would use the Edit Field command,

with the cursor located in the field you want to edit. TK!Solver tells you the editor is in use by writing a message on the Prompt/Error line, and placing a black underline and a space (called the cue) in the field. The message is

(Edit)

In the Edit mode, you still use the arrow keys, but now they move the cue—the black underline—in the field being edited. The right and left arrows move the cue one character to the right and left, respectively, and the up and down arrows move the cue to the beginning or end of the field, respectively.

The first time you use the editor, you may think it is creating blanks in the field, but the blank actually marks the position of the cue in the field. Whenever you type a character on the keyboard, it will appear to the left of the blank space; if you press the Backspace (←) key, the character to the left of the blank space will be deleted.

If you have used other editors, you might conclude that TK!Solver's editor is somewhat primitive. You would be correct. TK!Solver's edit mode enables you to perform only six functions:

Move ahead one character.

Move back one character.

Move to the beginning of the field.

Move to the end of the field.

Delete a character.

Insert a character.

But, when we are editing a single field with a maximum length of

200 characters per field, those six commands are sufficient, and they are easy to learn.

You can use the editor to change the entry in any entry field, including the Rule field of the Rule sheet. To exit the editor, press the **ENTER** key. The cue will thus disappear from the field, and the edited value will be entered.

As an example, suppose you had entered the Torricelli model as follows:

(1r)   VP = (VEL / 4005)*2 "The Torricelli equation < ENTER >
(2r)   TP = SP + VT < ENTER >

You can correct the mistake in the second rule by retyping the entry, but since only one character is wrong, it is simpler to use the editor. (When you are working with longer rules, it will obviously be easier to correct mistakes with the editor.) In this case, after ensuring that the cursor is positioned in field 2r, you would inovke the editor with the command

producing the display

$$\_TP = SP + VT$$

Next, you would move the cue to the end of the field by typing

producing the display

$$TP = SP + VT\_$$

Then you would erase the T with the **BACKSPACE** key and type

producing the correct rule,

$$TP = SP + VP$$

Typographical errors made when rules are entered have an irritating side effect: they produce unwanted entries in the Variable sheet. Thus, had you entered the Torricelli model with VT instead of VP in the second rule, you would find VT appearing in the Variable sheet, and correcting the rule with the editor would not remove the extraneous variable from the Variable sheet. The presence of an unused variable usually doesn't affect the solution of a model, but for the sake of neatness, you should remove the variable VT by moving to the Variable sheet with the Switch command

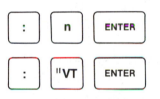

and then to the Name column with the Goto command

or

Then, you can blank the field containing VT with the Blank command

## THE RESET COMMAND

If you wish to abandon a model or one sheet of a model and start over, you can use the Reset command

The Reset command gives you several options with its prompt

Reset: V S A

The Reset Variables (V) option clears the values from the Input and Output fields of the Variable sheet and all its subsheets. The model will still be intact and ready for new input values and a new solution.

The Reset Sheet (S) option clears all fields on the sheet where the cursor is currently located. Since this command can ruin your model by erasing an essential sheet, TK!Solver asks for confirmation with the prompt

Reset current sheet:  Y  N

Only a response of Y will clear the sheet; pressing **N**, or any other key besides **Y**, will cancel the command.

The Reset All (A) option clears all information from all the sheets and subsheets. This command is used to completely abandon a model. TK!Solver will then be returned to the same condition as it would be in if we had exited and reloaded the program.

# THE STORAGE COMMAND

Let's take a look at the commands that begin with the slash (/) again (see Figure 5-5). We have discussed the Blank, Edit Field, Quit, Reset, Window, and Solve (!) commands, and we have mentioned the Storage command. Now let's take a closer look at the Storage command.

When you enter the Storage command

```
B    Blank              P    Print
C    Copy               Q    Quit
D    Delete             R    Reset
E    Edit Field         S    Storage
I    Insert             W    Window
L    List               !    Solve
M    Move
```

Figure 5-5. The Commands.

the following prompt appears on the Prompt/Error line:

Storage:  L  S  V  U  F  #  D

As with the Window (/W) command, the letters stand for options of the command. The options are listed below:

L   Load File
S   Save Model
V   Save Variables
U   Save Units
F   Save Functions
#   Save/Load DIF™[1] File (lists only)
D   Delete File

In Chapter 4 you learned that the Load File option is the command used to load the GRAVITY model furnished with TK!Solver. If you enter the Load File option, TK!Solver prompts you for a file name with a message on the Prompt/Error line. When you enter the file name, including the disk drive letter if necessary, the program will load the file. The file to be loaded does not have to be a complete model but can be a portion of a model, such as a predefined Unit sheet or User Function sheet. Now that we have discussed several sheets, the six options following the Load option should be self-explanatory.

You would use the Save Model option (S) to save the entire model as a disk file; that is, all the sheets in memory are written to a disk file. When you type

the following prompt appears:

Save: filename:

You would then enter a file name according to the rules of PC-DOS for naming files and press the **ENTER** key.

If you wish to save each version of a model as you refine it, you must give each version a different name, perhaps by assigning a

[1] DIF is a trademark of Software Arts Products Corp.

different number as the eighth character in the file name. If the older versions are of no interest, you can use the same name each time you save the model. When you use the same name, TK!Solver writes the new version over the old version. The program will protect you from mistakes, however. If you enter the name of an existing file, TK!Solver will prompt you as follows:

Save: Overwrite filename: thisfile:  Y  N

If you answer Y to this prompt, the program will write the new data in place of the exising file; if you answer N, the Storage command will be cancelled.

TK!Solver offers assistance in another form called Directory Scrolling. If you press the right arrow key while the "Save filename:" prompt is on the Prompt/Error line, another file name will be displayed. By repeatedly pressing the right arrow key, you can review the names of all the TK!Solver storage files on the disk. When you have selected a file by using Directory Scrolling, or by typing the name in from the keyboard, you tell TK!Solver to save the file by pressing the **ENTER** key. The program tells you that it is saving the model with the prompt

Saving to filename: <filename>

on the Prompt/Error line.

You can use Directory Scrolling with any options of the Storage command, which eases the strain of remembering the name of models created weeks previously. If you scroll through the directory without selecting a file name by pressing **ENTER,** the Storage command will not be executed. Thus, you can check the contents of a disk without leaving the program, or without actually saving or loading a model.

If you want to save only the Variable information, you must use the Save Variable (V) option

This option operates just like the Save Model option, except that it saves only the Variable sheet and Variable subsheets. If you plan to build several models of a system by using the same variables, you can save the variables from the first model and reuse that file for all the following models, thereby eliminating the need to retype all the variables.

The Save Units (U) option also works like the Save Model option, except that it saves only the Unit sheet. The Save Function (F) option works like the Save Units and Save Variable options, saving the User Function sheet, the List sheet, and any related subsheets. We'll look at the Save DIF (#) option later when we discuss the List sheet.

Saving portions of models with the Save options allows you to construct models from modules. You can enter a set of units and save those units for later incorporation in various models. You can also build modules of User Functions, variable names, even sets of rules that are common to many different rules.

The last option is the Delete File (D) option. When you enter the D option, TK!Solver prompts you on the Prompt/Error line as follows:

Delete filename:

If you then enter "thisfile" in response, the program will ask for confirmation with the message

Delete filename: thisfile: Y N

To delete the file, you must enter Y, just as you did when you wanted to overwrite an existing file.

# CHAPTER 6

# The Solvers

You now know how to build TK!Solver models and solve them by using three of the available sheets: the Rule sheet, the Variable sheet, and the Unit sheet. Before introducing TK!Solver's other sheets, which will allow you to build more complex models, let's look at the two ways TK!Solver arrives at solutions to its models: the Direct Solver, and the Iterative Solver.

## THE DIRECT SOLVER

The Direct Solver solves equations in the same way you do with pencil and paper. It substitutes known values for all the variables except the one you want to solve for. It then reduces the equation by performing the arithmetic operations contained in the equation. When all the operations have been completed and one value remains, the equation is solved. This method of solution corresponds exactly to the methods traditionally used to solve algebraic equations. However, this method has some limitations.

First, given a single-equation model, you must assign a value to every variable except the one you wish to solve for. If you leave two or more variables undefined, the Direct Solver cannot solve the equation. Of course, this requirement is not unusual—the same rule would hold if you were solving the equation manually.

Second, the variable for which you wish to solve must appear only once in the equation. Other variables may appear more than once, but they must be given input values. This is not an insurmountable limitation because you can often write equations so that the unknown variable appears only once. For instance, the Direct Solver cannot solve for $x$ in the equation:

$$x * x = 2 * (a + b)$$

61

but it can solve for $x$ if the equation is written this way:

$$x*2 = 2 * (a + b)$$

Engineering problems often contain equations that cannot be arranged so that the unknown variable appears only once, but we will look at some techniques for dealing with that situation later.

Third, the unknown variable cannot be the argument of a function that does not have a unique inverse, such as int(x), which rounds a real number off to an integer. As you can imagine, an infinite number of real numbers between 5 and 6 would result in the same integer, 5, when used as the argument of that function. Nor is the integer function the only function to produce that situation—many trigonometric functions present similar constraints.

You can solve models for more than one unknown if you have a set of equations. If each equation meets the three conditions at some time during the solution process, the Direct Solver can solve the model. If you build the model shown in Figure 6-1, TK!Solver will solve it using the Direct Solver with only two values as inputs.

You can solve this model by entering values for the two variables, a and b. The fourth equation is solved first, because only one variable in it is unknown. The second equation is solved next, because now only one of its variables is unknown; the value for d was obtained from the solution of the fourth equation. The third equation is solved for c, since we now have another value from the

```
(5s) Status:                                                         204/!

===================== VARIABLE  SHEET ====================================
St Input        Name    Output    Unit     Comment
-- -----        ----    ------    ----     -------
                a
                b
                c
                e
                f
                d


================== RULE  SHEET ===========================================
S Rule
- ----
*  a+b*c-e=f
*  a+f=d
*  c+1=f/2
*  b-d=4
```

Figure 6-1. Simultaneous Equations.

solution of the second equation. Finally, the first equation is solved for e, since only one unknown remains, and the entire model is solved.

Now consider what happens if you build and try to solve a model that doesn't meet the conditions listed above. If you use the same model but enter only one value (see Figure 6-2) and type the Solve or the Action command, the asterisks will remain in the status column of the first, second, and fourth equations, and there will be no output for the variables a, b, e, and d. However, TK!Solver does solve the third equation, because it has enough information to do so.

If you move the cursor to one of the fields containing an asterisk, the Prompt/Error line will contain the message * Unsatisfied. The model was not completely solved because it was underdefined—that is, there were not enough known values.

Now, if you enter input values for the variables a, b, c, and f, and then attempt to solve the model, you will get an error symbol in the Status field for the third equation, as well as in the Status fields of the variables, c and f. (The display is shown in Figure 6-3). If you place the cursor over that error symbol for variable f, TK!Solver will give you an error message in the Prompt/Error line: > Overdefined. You have given TK!Solver conflicting values.

Note that it is not always necessary to satisfy every rule in a model. You may wish to include several redundant rules, knowing that only the appropriate rules will be used. You can construct models with control structures by including a set of mutually exclusive rules—only one of which can be satisfied at any given time.

```
(1u) Unit: hr*ft^2*oF/Btu                                          204/

==================== VARIABLE SHEET ================================
St Input      Name    Output    Unit    Comment
-- -----      ----    ------    ----    -------
              R       45.814537 hr*ft^2*o resistance to heat transfer
    5         D                 inches   outside diameter of insulation
    2         d                 inches   inside diameter of insulation
   .05        K                 Btu/hr*ft thermal conductivity of insulation
              Q                 Btu/hr   heat loss
              deltaT            oF       temperature difference across insulati

==================== RULE SHEET ===================================
S Rule
- ----
  R = D*LN(D/d)/(2*K)
* Q = deltaT/(d*LN(D/d)*2*K)
```

Figure 6-2. An Underdefined Model.

```
(4s) Status:  > Overdefined                                        204/!

=================== VARIABLE SHEET ============================================
St Input        Name     Output     Unit       Comment
-- -----        ----     ------     ----       -------
  > 50          R                   hr*ft^2*o  resistance to heat transfer
  > 5           D                   inches     outside diameter of insulation
  > 2           d                   inches     inside diameter of insulation
  > .05         K                   Btu/hr*ft  thermal conductivity of insulation
    500         Q                   Btu/hr     heat loss
              deltaT                oF         temperature difference across insulati


=================== RULE SHEET ================================================
S Rule
- ----
  > R = D*LN(D/d)/(2*K)
  * Q = deltaT/(d*LN(D/d)*2*K)
```

Figure 6-3. An Overdefined Model.

For instance, you can construct a fluid flow model that accommodates both laminar flow and turbulent flow by including a variable that allows one set of rules to be solved if the flow is laminar, and another to be solved if the flow is turbulent. You will see exactly how to accomplish this objective in a later chapter. For now, you only need to note that the notion of overdefinition can be used to your advantage.

## THE ITERATIVE SOLVER

Because TK!Solver always tries to use the Direct Solver first, it is worthwhile to meet its conditions if possible—it works faster than the other method available, the Iterative Solver. But if you have to solve the heat transfer model shown in Figure 6-4, where the heat loss, emissivity, and ambient temperature are known and the surface temperature is unknown, the Direct Solver will not work, and it's necessary to rely on the Iterative Solver.

Iteration is commonly used to solve engineering problems. Many physical processes of interest to engineers, such as heat transfer by radiation, are too complicated for direct solution. Here is the equation of Figure 6-4 as it would be presented in a textbook or a handbook:

$$Q = 0.296(T_s - T_a)^{1.25} + 0.174e[T_s/100)^4 - (T_a/100)^4]$$

*Where:*    $Q$ = *heat loss in Btu/(hr)(SF)*

$T_s$ = surface temperature in degrees Rankine

$T_a$ = ambient temperature in degrees Rankine

$e$ = surface emittance ratio to black-body

This equation relates heat loss by radiation and convection from a hot surface to its surroundings. If you know the temperatures, it is a simple matter to solve for the heat loss. But if you know the heat loss you want to achieve and the ambient temperature, and want to find the allowable temperature of the hot surface, the situation is entirely different. In that case, you would resort to trial and error, or iterative solution. The reason, as mentioned above, is that the variable $T_s$ appears in more than one location in the equation.

It is important to remember that TK!Solver does not perform symbolic manipulation, or simplification, when solving models, as is often the case when manual methods are used. If a model can be stated in a way that allows direct solution, you must enter it that way; TK!Solver works with equations as they are entered, and if you enter them in a manner that does not meet the restrictions of the Direct Solver, the program uses iteration to reach a solution.

All iterative methods involve guessing a value for an unknown variable, making a calculation, and using the result of the calculation to make a more intelligent guess for the unknown variable. This process is repeated until successive guesses are within an acceptable range—that is, until the difference between successive guesses is small. In the problem above, you would guess a value for

```
(2s) Status:                                                           204/!

==================== VARIABLE SHEET ====================================
St Input       Name    Output    Unit       Comment
-- -----       ----    ------    ----       -------
               Q                  Btu/ft^2/  heat loss
               Ts                 oR         surface temperature, degrees R
               Ta                 oR         ambient temperature, degrees R
               e

=================== RULE SHEET ==========================================
S Rule
- ----
* Q = 0.296*(Ts-Ta)^1.25+0.174*e*((Ts/100)^4-(Ta/100)^4)
```

Figure 6–4. A Model Requiring Iteration.

$T_s$, calculate $Q$, and then compare the calculated value of $Q$ to the given value of $Q$. If the calculated value was higher than the given value, you would reduce the value of $T_s$, and repeat the calculation.

In order to make the process efficient, you must have not only a rule, or model, of the process such as the one given above, but also a rule for generating the successive guesses. You try to obtain a procedure that converges on the correct answer. A good example of an iterative process is the one taught in grade school for extracting square roots by hand (Figure 6–5). This method incorporates rules that ensure that the correct answer will be produced.

When you use the TK!Solver's Iterative Solver, you usually don't have to worry about the rule for producing the next guess. The method for generating the next guess is built into the Solver. All you have to do is make the first guess, enter it, and type the Solve or the Action command. You enter a guess by entering a value in the input field of the unknown variable, and then setting it to a guess with the Guess option. You place the cursor in the Status field of the unknown, and type

G

Figures 6–6(a) and 6–6(b) illustrate the use of the Iterative Solver with the heat transfer equation listed above.

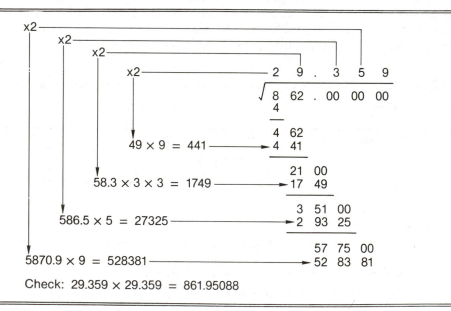

Figure 6–5. Square Root Extraction—An Iterative Process.

```
(2s) Status:  Guess                                             204/!

=================== VARIABLE SHEET ================================
St Input       Name     Output    Unit    Comment
-- -----       ----     ------    ----    -------
   50          Q                  Btu/ft^2/ heat loss
 G 600         Ts                 oR       surface temperature, degrees R
   540         Ta                 oR       ambient temperature, degrees R
   .9          e                           surface emittance ratio (to black body



=================== RULE SHEET ===================================
S Rule
- ----
   Q = 0.296*(Ts-Ta)^1.25+0.174*e*((Ts/100)^4-(Ta/100)^4)


(2s) Status:  Guess                                             204/!
Iterative Solver: 2 out of 10
=================== VARIABLE SHEET ================================
St Input       Name     Output    Unit    Comment
-- -----       ----     ------    ----    -------
   50          Q                  Btu/ft^2/ heat loss
 G 570.94853   Ts                 oR       surface temperature, degrees R
   540         Ta                 oR       ambient temperature, degrees R
   .9          e                           surface emittance ratio (to black body



=================== RULE SHEET ===================================
S Rule
- ----
 * Q = 0.296*(Ts-Ta)^1.25+0.174*e*((Ts/100)^4-(Ta/100)^4)
```

Figure 6-6(a).  The Heat Transfer Model with a Guess.

```
(2s) Status:                                                    204/

=================== VARIABLE SHEET ================================
St Input       Name     Output    Unit    Comment
-- -----       ----     ------    ----    -------
   50          Q                  Btu/ft^2/ heat loss
               Ts       568.54610 oR       surface temperature, degrees R
   540         Ta                 oR       ambient temperature, degrees R
   .9          e                           surface emittance ratio (to black body



=================== RULE SHEET ===================================
S Rule
- ----
   Q = 0.296*(Ts-Ta)^1.25+0.174*e*((Ts/100)^4-(Ta/100)^4)
```

Figure 6-6(b).  The Heat Transfer Model Solved.

Iterative solutions with TK!Solver are not always successful, at least on the first attempt. As with numerical methods used on mainframe computers or manual numerical calculations, the first guess affects the result. The desired result is, of course, a solution. However, you may obtain one of three unsuccessful results.

The first type of unsuccessful result is slow convergence, which means that the Solver is approaching the correct solution, but that it will stop before reaching it. TK!Solver has two parameters that govern when it will stop working on an iterative solution: the Maximum Iteration Count, and the Comparison Tolerance. If two successive guesses differ by the Comparison Tolerance or less, TK!Solver stops and announces a solution. TK!Solver also counts the steps or iterations it has made, and if the number of steps reaches the Maximum Iteration Count value, it stops. If the model is converging but stops before reaching a solution, you can reach the solution by typing the Action command or the Solve command a second, or perhaps a third time. Each time you reenter the Action or Solve command, TK!Solver resumes, not with the initial guess, but with the guess used for the last iteration. Slow convergence is usually the result of a bad first guess, but can be a consequence of the model itself. Nonetheless, a solution is possible.

The second type of unsuccessful result is called divergence, which means that each successive iteration is further removed from the answer instead of being closer to it. In this case, you cannot arrive at a correct answer, even by repeating the Action or Solve command. Divergence can be the result of a bad first guess, but it can also be the result of a faulty model. Inconsistent equations will produce divergence, and can be corrected to obtain a model that converges. However, if the solution to the model includes imaginary or complex numbers, TK!Solver will not be able to cope with them, and divergence will be the inevitable result.

The third type of unsuccessful result is oscillation, which means that successive guesses alternate between two numbers. As with divergence, the Iterative Solver will stop at the Maximum Iteration Count. The causes for oscillation are generally the same as those for divergence.

In its standard configuration, TK!Solver displays the resuls of each iteration; this display is controlled by the Global sheet discussed in the next section. Once a model is proved, there is no need to watch the display of intermediate results, as you are interested only in the final result. Until the model is proved, including initial guesses, it is wise to watch the display of intermediate results, as they will reveal whether a model is oscillating, diverging, or merely converging slowly.

# THE GLOBAL SHEET

This is an appropriate time to look at the Global sheet (Figure 6-7) since it contains values that determine how TK!Solver will perform various tasks. When the program is first loaded, the Global sheet contains the default values shown in Figure 6-7. However, these values can be changed whenever necessary.

Starting at the top field, you will see the phrase Variable Insert ON. This item tells you whether TK!Solver will automatically enter the variables in the Variable sheet as you enter rules in the Rule sheet. So far, we have assumed that this variable was set to Yes. You can change the value to No by placing the cursor in the field and typing N. Subsequently TK!Solver will not enter the variables automatically.

The next field is the Intermediate Redisplay ON: field. This value is associated with the Iterative Solver. If it is set to Yes, the values associated with each step of an iterative solution are displayed on the screen. It's useful to have this feature on, as it may help you to determine why the solution to a model is not converging quickly. Howver, if you are working with a model that you know is accurate and will produce convergent solutions, turning the feature off may allow the solver to work faster.

```
(v) Variable Insert ON: Yes                                         204/

=================== GLOBAL SHEET ====================================
Variable Insert ON:          Yes

Intermediate Redisplay ON:   Yes
Automatic Iteration ON:      Yes
Comparison Tolerance:        .000001
Typical Value:               1
Maximum Iteration Count:     10

Page Breaks ON:              Yes
Page Numbers ON:             Yes
Form Length:                 66
Printed Page Length:         60
Printed Page Width:          80
Left Margin:                 0
Printer Device or Filename:  PRN
Printer Setup String:
Line End:                    CR&LF
```

Figure 6-7. The Global Sheet.

The third field is Automatic Iteration ON:, which has a default value of Yes. If Automatic Iteration is set to Yes, you do not have to instruct TK!Solver to use the Iterative Solver. It will attempt to use the Direct Solver, and if it is unsuccessful, it will then call on the Iterative Solver. However, if Automatic Iteration is to work, you must also make an entry in the First Guess field of the Variable subsheet for the variable you want to solve for.

The Iterative Solver is available whether or not the Automatic Iteration field is set to Yes, as it is invoked whenever the Action command (given from the Rule sheet or the Variable sheet) or the Solve command is entered and the Guess option is entered in the Status field of a variable on the Variable sheet. However, if the field is set to No, guess values must be entered for each solution of the model. For single solutions of a model, the field need not be set to Yes, but if the List Solver is to be used, that setting allows Automatic Iteration to take place. Then, the program retrieves the initial guess from the First Guess field of the Variable subsheet at the beginning of each solution of the model. Automatic Iteration works best with proven models in which a best value has been determined for each of the initial guesses required for solution of the model.

The fourth field is the Comparison Tolerance field, the purpose of which has already been explained. Each time TK!Solver performs an iteration, it compares the current value of the variable being solved for to the previous value of the same variable. If the absolute value of the difference between the two divided by the larger of the two values is greater than the Comparison Tolerance, the Solver will make another iteration; if smaller, TK!Solver will announce a solution. The default value of the Comparison Tolerance is 0.000001, which means that TK!Solver announces a solution when the difference between two successive guesses is one millionth of the larger of the two guesses. Thus if you were working on problems involving dimensions of metal parts on the order of 1 inch in size, you could expect your answers to be accurate to a millionth of an inch or so. On the other hand, if you were working on problems involving distances on the order of the distance from the earth to the sun, your answers would be accurate to about a hundred miles.

You can change the Comparison Tolerance to suit the model at hand by calling up the Global sheet, moving the cursor to that field, and typing in the new value. Choosing the Comparison Tolerance to suit the requirements of the model allows faster solution when accuracy of less than one part in a million is not required.

You can enter values in the Comparison Tolerance field as numeric constants, or as expressions that can be evaluated to valid numeric values. Valid numeric values range from 1E–120 to 1E120. As with the entry of variable values, entering a Comparison Tolerance as an expression can save considerable hand calculation.

The next field is the Typical Value field, which is used by the Iterative Solver in conjunction with the Comparison Tolerance field. TK!Solver uses the product of the value in the Comparison Tolerance field and the value in the Typical Value field to determine the point at which values should be treated as approaching zero. When the program must compare values that are less than the product of the Comparison Tolerance and the Typical Value, it divides the difference of the two values by the Typical Value rather than the greater of the two values. The default value of the Typical Value field is 1.

The value for both the Comparison Tolerance field and the Typical Value field affect the precision of model solutions. An increase in either value will decrease the accuracy of equality determination.

The last field related to the Iterative Solver is the Maximum Iteration Count field, which contains a default value of 10. The field must contain an entry, and the value must be an integer—between 1 and 10000. The value stored in this field determines how many iterations TK!Solver will attempt before "giving up."

Although it is handy to have a computer program that can use iteration to solve problems, it is generally better to be able to solve them directly. The Direct Solver is faster than the Iterative Solver and does not require a good first guess.

You can avoid using the Iterative Solver by providing a sufficient number of independent equations, as you would try to do with manual methods. Now and then you can manipulate the equations of the model to provide one or more redundant equations, which will allow the Direct Solver to solve the model. This subject is discussed further in Part II of the book.

Since the remaining fields of the Global sheet are related to the printer, they will be discussed together with the printer.

# CHAPTER 7

# Lists

You should now be able to create a model and then solve it to obtain values for one or more unknown variables, given values for a sufficient number of known variables. That type of solution to a model is all that you may need in many cases. If you are calculating the design heat load for a building, or the size of a tank, for example, a single answer or set of answers is enough.

For some models, however, you will need a series of answers for each unknown variable, so that you can see how it varies with the known variables. A good example is a control problem in which you need to see how the output of a system varies with the input. Given a system that has become stabilized and is then disturbed, you may need to know how long it will take for the system to become stabilized again. To determine that period of time, you will need to see a series of values for the system output plotted or tabulated versus time.

With what you now know, you could construct a table of values by repeatedly solving the model, increasing the time variable by the same increment for each successive solution, and recording the results on paper. However, TK!Solver can easily do that mundane work for you by using three other sheets it is equipped with: the List sheet, the Plot sheet, and the Table sheet. What you want to do, and what you can do, is create a model and then enter a list of input values from which TK!Solver will produce a list of output values.

## THE LIST SHEET

You can build a model that uses lists in exactly the same manner as the earlier models used variables. In fact, a list is a form of variable.

Lists are defined with the List sheet and List subsheets, in conjunction with the Variable sheet and Variable subsheets. As you construct a model by entering rules in the Rule sheet, TK!Solver sets up the Variable sheet and the Variable subsheets; with the aid of the List sheet and the List subsheets, you can associate these variables with lists.

The List sheet (Figure 7–1) is a summary of all the lists in a model, just as the Variable sheet is a summary of all the variables in a model. The List sheet contains four fields: the Name, Elements, Unit, and Comment fields.

List names must follow the same rules as Variable names. They must begin with letters, and may contain letters, numbers, and the characters @, #, $, %, and __. The Name field is an entry field.

The Elements field is an output field that tells you how many elements are in the list. You'll see how the Elements field is filled when we generate a list.

The Unit field is analogous to the Unit field in the Variable sheet except for a slight difference. The two types of units associated with lists are called Storage units and Display units, whereas those associated with variables are called Calculation units and Display units.

Your first entry in the Unit field for a List becomes the Storage unit and the Display unit. The Storage unit is the unit that is assumed internally, just as the Calculation unit for Variables is the unit that is used in calculations. You can change the Display unit for a list from the List sheet, but to change the Storage unit, you must go to the List subsheet. If you change the Display unit

```
(10n) Name: opcost                                                      192/!
==================== LIST SHEET ===========================================
Name           Elements   Unit      Comment
----           --------   ----      -------
des_kw         39         kw
kwperton       39         kw/ton
evap_pd        39         ft_wg
cond_pd        39         ft_wg
max_tons       39         ton
kw@max         39         kw
price          39         $
$perton        39         $/ton
line           39
opcost         39
```

Figure 7–1. The List Sheet.

without changing the Storage unit, you must define a conversion between the two units on the Unit sheet, just as you did for Variables.

The last field in the List sheet is the Comment field, which you can use for instructions or notes. As with comment fields in other sheets, TK!Solver ignores comments in its calculations.

## THE LIST SUBSHEET

If you want to see the complete definition of a list, you must go to the List subsheet by means of the Dive command (ⓒ), just as you did to access the Variable subsheet from the Variable sheet. You place the cursor in the Name field for the List you want to examine, and type

```
>
```

The List subsheet is shown in Figure 7-2.

There is a List subsheet for each list named in the List sheet Name column. The name of the list appears on the third line to the

```
(1v) Value: .696                                                      192/!

==================== LIST: kwperton =========================================
Comment:
Display Unit:           kw/ton
Storage Unit:           kw/ton
Element Value
------- ------
1       .696
2       .6906666667
3       .688
4       .6826666667
5       .6746666667
6       .6746666667
7       .672
8       .6693333333
9       .6586666667
10      .6586666667
11      .6533333333
12      .648
13      .648
14      .648
15      .6426666667
16      .6426666667
```

Figure 7-2. The List Subsheet.

right of the word List. There are three single fields, and two columns of fields.

The first field is the Comment field, which is again a space for notes and for a longer title than the one used for the List Name. This comment is the same comment as the one that appears on the List sheet comment field for the list.

The next two fields are the Display Unit field and the Storage unit field. The Display unit field is used in the same manner as the Display unit field in the Variable subsheet, and the Storage unit field corresponds to the Calculation unit field of the Variable subsheet.

The remainder of the List subsheet differs from the Variable subsheet. Instead of single entry fields, here you will find two columns of fields, the Element column and the Value column.

The Element column is your index into the list. Element 1 is the position marker of the first Value in the list. The Element field is an output field; when you place a value in a list, the program generates an Element number for it.

The Value column is the list itself. These values are the ones you are interested in, whether they are inputs or outputs. The Value column can be either input or output.

If you want to use a list as an input, you must first define the list from the List sheet, as you would with all the lists in the model. Then you Dive to the List subsheet you want to use as an input. Next, place the cursor in the first row of the Value column, and make an entry. The simplest way to make the entries is to type the value and then move to the next field by typing the Down arrow, which will move the cursor to the next row in the Value column. Each time you make an entry in the Value column, TK!Solver will make a corresponding entry in the Element column.

# THE ACTION COMMAND FROM THE LIST SUBSHEET

If you want to create a list of values that increase or decrease by a constant difference, TK!Solver will do part of the work for you. After creating the list from the List sheet, you Dive to the subsheet and enter the first value in the first row of the Value column. Then you move the cursor to the last element with the Goto command (:), and enter the last value. Next, you use the Action command (!). Remember that the effect of the Action command varies from sheet to sheet.

When you use the Action command from the List subsheet, you invoke the Fill List command. The message on the Prompt/Error line is

Fill List:  Y  N

As usual, TK!Solver is watching for possible unintentional errros. If you were to use the Fill List command carelessly, you could overwrite an existing list, so TK!Solver wants a confirmation before it fills the list.

The sequence of commands needed to create a list of 20 elements from any sheet is

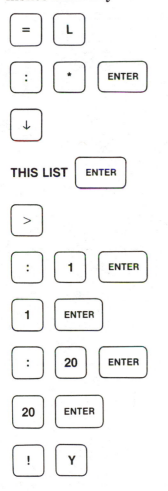

After the Action command prompt is answered with the Y, TK!Solver fills the list with 20 equally spaced values, as shown in Figure 7–3.

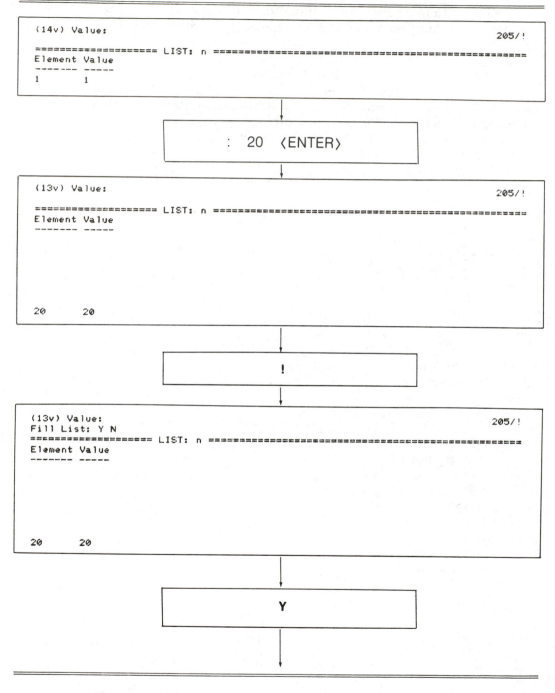

```
(14v) Value:                                                                205/!
==================== LIST: n =================================================
Element Value
------- -----
1       1
```

```
:  20  ⟨ENTER⟩
```

```
(13v) Value:                                                                205/!
==================== LIST: n =================================================
Element Value
------- -----




20      20
```

```
!
```

```
(13v) Value:                                                                205/!
Fill List: Y N
==================== LIST: n =================================================
Element Value
------- -----




20      20
```

```
Y
```

Figure 7-3.  List Created with the Action Command, Fill List Option.

```
(2v) Value: 2                                                        205/!
===================== LIST: n =============================================
Element Value
-------  -----
2       2
3       3
4       4
5       5
6       6
7       7
8       8
9       9
10      10
11      11
12      12
13      13
14      14
15      15
16      16
17      17
18      18
19      19
20      20
```

Figure 7-3. (continued).

So far, the entries and outputs discussed here have been assumed to be numbers. However, you can also create and use lists of symbolic values, as mentioned earlier in the discussion of Variables. Symbolic values in a list must be preceded by an apostrophe, or single quotation mark (')—TK!Solver beeps at unmarked symbolic values.

## ASSOCIATING VARIABLES AND LISTS

Lists, as noted earlier, can be considered variables. However, you must tell TK!Solver that a variable is represented by a list before it can use the list to solve a model. You have already seen the mechanism used to associate a variable with a list, but let's take another look at the Variable subsheet (Figure 7-4) and examine the third field there—the Associated List field. When you tell TK!Solver that a list is associated with a variable, the name of the list will appear in this field.

```
(1c) Comment: TORRICELLI MODEL                                          205/!

==================== VARIABLE SHEET ============================================
St Input          Name    Output    Unit    Comment
-- -----          ----    ------    ----    -------
                                            TORRICELLI MODEL
                                            ****************
                  VP                INCHES_WG velocity pressure
                  VEL               FT/MIN    air velocity
                  TP                INCHES_WG total pressure
                  SP                INCHES_WG static pressure
```

### Variables Not Associated with Lists

```
(6i) Input:                                                             204/!

==================== VARIABLE SHEET ============================================
St Input          Name    Output    Unit    Comment
-- -----          ----    ------    ----    -------
                                            TORRICELLI MODEL
                                            ****************
L                 VP                INCHES_WG velocity pressure
L                 VEL               FT/MIN    air velocity
L                 TP                INCHES_WG total pressure
L                 SP                INCHES_WG static pressure
```

### Variables Associated with Lists

```
(1n) Name: VP                                                           204/!

==================== LIST SHEET ===============================================
Name      Elements  Unit    Comment
----      --------  ----    -------
VP                  INCHES_WG
VEL                 FT/MIN
TP                  INCHES_WG
SP                  INCHES_WG
```

### Lists Created by Setting Variable Status to L

Figure 7-4. The Variable Sheet.

You will have to make two entries to associate a list with a variable. You make one of the entries on the Variable sheet by placing the cursor in the Status field of the variable and typing the List option (L). The List option is a toggle or a switch. If you type L in a blank Status field, the variable will be associated with a list, but if you type L in a Status field that already contains an L, the variable and its associated list will be disassociated, and the L will disappear from the Status field.

You make the other entry on the Variable subsheet, in the Associated List field. You can Dive (>) from the Variable sheet and make the entry directly, or you can let TK!Solver do it for you. If you type L in the Status field of a variable on the Variable sheet without first making an entry in the Associated List field of the Variable subsheet, TK!Solver will enter the name of the variable in the Associated List field. Notice that you can associate any list with a variable; the name of the list does not have to be the same as the name of the variable, but TK!Solver will use the name of the variable for the name of the list if the Associated List field has not been previously filled with another name.

Once a list has been entered in the Associated List field of the Variable subsheet, you can Dive to the List subsheet from the Variable subsheet, as well as from the List sheet. The Return command (<) will return you to the previous sheet. If you Dive to the List subsheet from the Variable subsheet, the Return command will take you back to that subsheet, not the List sheet.

There is no need to use the Return command several times to return to the Variable sheet. You must use the Dive command to reach a Variable or List subsheet, but you can return to the Variable sheet with the Select command (=). In the example in Figure 7-4, you could reach the t List subsheet and return to the variable sheet with the following sequence of commands:

| : |

| "t" |

| > |

| > |

Since you will use lists to obtain a succession of solutions to a model, you need both input lists and output lists. You associate both types of lists with variables—input lists with input variables, and output lists with output variables. When you enter a name of a list in the Associated List field of a Variable subsheet, an empty list is created if a list does not already exist.

Models that use lists for input and output require particular attention to units. You must make sure that three conversion factors are included in the Units sheet for each variable and associated list:

- Variable Calculation units to Variable Display units.
- Variable Calculation units to List Storage units.
- List Storage units to List Display units.

TK!Solver will usually indicate the absence of a conversion factor between Variable Calculation units and Variable Display units, but it will not always indicate the absence of a conversion factor between Variable Calculation units and List Storage units. You must therefore pay particular attention to including the factors.

## THE LIST COMMAND

Once your lists have been defined and associated with variables, you can solve the model using the List command.

The List command can be entered from any sheet, as can the other commands. When you type /L, you will get a message showing the options to the List command:

List:  !  B  P  G

The first option is the Solve option (!). The Solve option of the

List command calls on one of the two solvers, the Direct Solver or the Iterative Solver, to solve the model. When you invoke one of the Solvers with the List command, it will solve the model using the first element of each input list, and will produce the first element of each output list. Then it will solve the model again, using the second element of each input list and produce the second element of each output list, and continue until the input lists are exhausted and the output lists are filled.

Unless you are sure that a model is bug-free, you will probably want to see each element of the output as it is generated. You can observe the process if the Global sheet has the Intermediate Redisplay ON: field set to Yes. TK!Solver will keep you posted with a message such as this one:

List Solver: < element number > , Direct Solver

If the Intermediate Redisplay ON: field is set to Yes, TK!Solver will also display the inputs and outputs for each solution. When you are sure a model functions correctly, you can allow it to work faster by turning off the Intermediate Redisplay option.

The second option of the List command is the Block option (B), which works in the same manner as the Solve option. The difference between the two is that the Block option will solve a model for a portion of the associated lists. When you solve a model using the Block option of the List command, you receive two additional prompts. The first prompt asks for the first element of the block we want to solve for:

List Block: First Element:

You answer that prompt with the number of the first element, followed by the **ENTER** key. TK!Solver will then prompt you again, this time for the last element:

List Block: Last Element:

After you enter the number of the last element of the block, and press **ENTER**, TK!Solver will begin to solve the model and will display the same message as it would with the Solve option.

You will often want to examine the set of data being used as the input, and perhaps the set of data produced as output as well. You can do so with the Get option (G) of the List command. When you invoke the Get option, TK!Solver will display all the inputs and outputs of a particular instance of the model on the Variable sheet. To use the option, type the List command

followed by the Get option

from any location on the Variable sheet. The program will display this prompt:

Get from element:

If you answer the prompt with the number 10, TK!Solver will get the tenth element from each list associated with a variable in the model, and will display those values in the Input and Output fields of the Variable sheet.

Figure 7–5 illustrates a model that calculates the future value (FV) of a principal amount (PV) after n periods at an interest rate i. Both the number of periods and the future value variables are associated with lists; n is an input list, and FV is an output list. The lists resulting from a solution of the model, using the List Solver, are also shown in Figure 7–5.

The first display of the Variable and Rule sheets shows the fourth element of both lists. If you wish to examine the tenth element of both lists along with the other variables, you give the following sequence of commands, with the cursor positioned anywhere on the Variable sheet, or the Rule sheet:

Actually, you can use the Get Option of the List command whether or not the Variable sheet is on the screen. TK!Solver will insert the selected set of list elements in the Variable sheet, and will display them when the Variable sheet is selected.

The last option of the List command is the Put option. This option is the converse of the Get command. If you have a model that uses lists, you can solve the model and then place all the values shown on the Variable sheet in the associated lists. To use the Put option, type P from the List command prompt

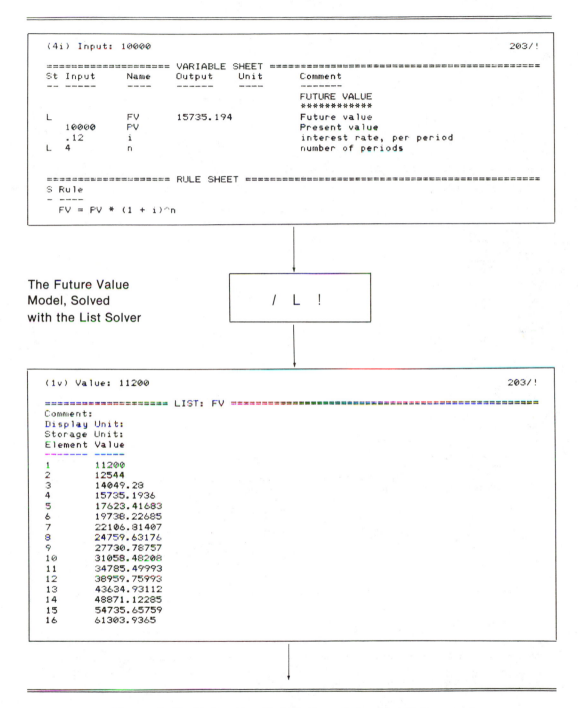

```
   (4i) Input: 10000                                               203/!

   ==================== VARIABLE SHEET ==================================
   St Input       Name    Output     Unit     Comment
   -- -----       ----    ------     ----     -------
                                              FUTURE VALUE
                                              ************
   L              FV      15735.194           Future value
       10000      PV                          Present value
        .12       i                           interest rate, per period
   L   4          n                           number of periods


   ==================== RULE SHEET =====================================
   S Rule
   - ----
     FV = PV * (1 + i)^n
```

The Future Value
Model, Solved
with the List Solver

```
                              /  L  !
```

```
   (1v) Value: 11200                                               203/!

   ================== LIST: FV =========================================
   Comment:
   Display Unit:
   Storage Unit:
   Element Value
   ------- -----
   1        11200
   2        12544
   3        14049.28
   4        15735.1936
   5        17623.41683
   6        19738.22685
   7        22106.81407
   8        24759.63176
   9        27730.78757
   10       31058.48208
   11       34785.49993
   12       38959.75993
   13       43634.93112
   14       48871.12285
   15       54735.65759
   16       61303.9365
```

Figure 7-5. Using the Get Option of the List Command.

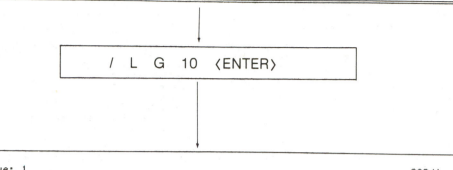

```
(1v) Value: 1                                                               203/!

=================== LIST: n ==================================================
Comment:
Display Unit:
Storage Unit:
Element Value
------- -----
1       1
2       2
3       3
4       4
5       5
6       6
7       7
8       8
9       9
10      10
11      11
12      12
13      13
14      14
15      15
16      16


(1r) Rule: FV = PV * (1 + i)^n                                              203/!

=================== VARIABLE SHEET ===========================================
St Input       Name    Output      Unit    Comment
-- -----       ----    ------      ----    -------
                                           FUTURE VALUE
                                           ************
L              FV      31058.482           Future value
   10000       PV                          Present value
   .12         i                           interest rate, per period
L  10          n                           number of periods


=================== RULE SHEET ===============================================
S Rule
- ----
   FV = PV * (1 + i)^n
```

Figure 7–5. (continued).

TK!Solver will prompt you for the element you wish to fill:

Put to element:

    You will then respond with an element number, and TK!Solver will fill the corresponding element of all associated lists from the variables on the Variable sheet. You can use this option to construct a set of lists with values that do not vary in a uniform manner, or to store the results of a series of solutions of a model.

    Figure 7-6 illustrates the use of the Put option with the Toricelli model. Each of the variables has been associated with a list, but you should solve the model using the Action command, rather than the List command. After each solution, use the command

to store the variable values, both input and output, in the associated lists. The resulting lists are shown in the lower part of Figure 7-6.

    The Put option is useful for storing repeated solutions of a given model, particularly when the model is solved in several different sessions, as in the case of a model used to store experimental values from tensile stress tests and values calculated from the measured values. After you have spent weeks solving the model and storing the values, the lists stored in the model will still be available for analysis.

    Figure 7-7 illustrates the screen display that makes the Put option easy to use. Since no command is needed to locate the "next" element to use when appending an element to a list with the Put Option, it is handy to have the List sheet, rather than the Rule sheet, displayed with the Variable sheet, so that the current number of elements can easily be determined.

    Notice that the Get and Put options are commands, not functions that can be included in the rules of a model. There is a function analogous to the Get command—it allows you to retrieve any element of a list and use it as a value—but there is no function analogous to the Put command.

    You can use the options of the List command to speed the development and verification of your models. In an engineering application, a model may have to be solved for several hundred sets of data to produce useful results. It was noted earlier that TK!Solver solves models faster than you could ever hope to do manually, and

```
(4i) Input: .08                                                           203/!

===================== VARIABLE SHEET =====================================
St Input   Name    Output      Unit        Comment
-- -----   ----    ------      ----        -------
                                           the TORICELLI MODEL
                                           *******************
L  .05     VP                  INCHES_WG   velocity pressure
L  .08     SP                  INCHES_WG   static pressure
L          VEL   895.54523     FT/MIN      air velocity
L          TP        .13       INCHES_WG   total pressure


===================== RULE SHEET =========================================
S Rule
-  ----
-  VP = (VEL / 4005)^2
   TP = SP + VP                "The Toricelli equation"
```

```
. 
```

```
/ L P 1  <ENTER>
```

```
(3i) Input: .06                                                           203/

===================== VARIABLE SHEET =====================================
St Input   Name    Output      Unit        Comment
-- -----   ----    ------      ----        -------
                                           the TORICELLI MODEL
                                           *******************
L  .06     VP                  INCHES_WG   velocity pressure
L  .09     SP                  INCHES_WG   static pressure
L          VEL   981.02064     FT/MIN      air velocity
L          TP        .15       INCHES_WG   total pressure


===================== RULE SHEET =========================================
S Rule
-  ----
-  VP = (VEL / 4005)^2
   TP = SP + VP                "The Toricelli equation"
```

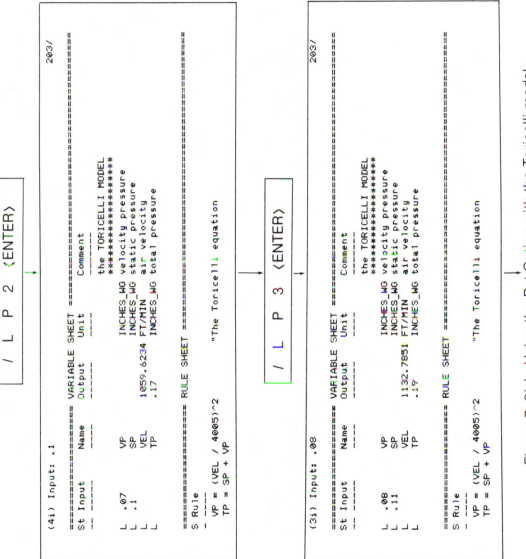

Figure 7-6(a). Using the Put Option with the Toricelli model.

89

```
(1v) Value: .05

================== LIST: VP ==================================

Comment:
Display Unit:      INCHES_WG
Storage Unit:      INCHES_WG
Element Value
-------
1      .05
2      .06
3      .07
4      .08
```

```
(1v) Value: 895.54522497

================== LIST: VEL ==================================

Comment:
Display Unit:      FT/MIN
Storage Unit:      FT/MIN
Element Value
-------
1      895.545225
2      981.020642
3      1059.6234
4      1132.785063
```

```
(1v) Value: .13

================== LIST: TP ==================================

Comment:
Display Unit:      INCHES_WG
Storage Unit:      INCHES_WG
Element Value
-------
1      .13
2      .15
3      .17
4      .19
```

(1v) Value: .08                          203/

=================== LIST: SP ====================================

Comment:
Display Unit:            INCHES_WG
Storage Unit:            INCHES_WG
Element Value
------- -----
1    .08
2    .09
3    .1
4    .11

(4n) Name: SP                            203/

=================== LIST SHEET ====================================
Name   Elements   Unit         Comment
----   --------   ----         -------
VP     4          INCHES_WG
VEL    4          FT/MIN
TP     4          INCHES_WG
SP     4          INCHES_WG

Figure 7-6(b). Lists Resulting from Use of the Put Option.

91

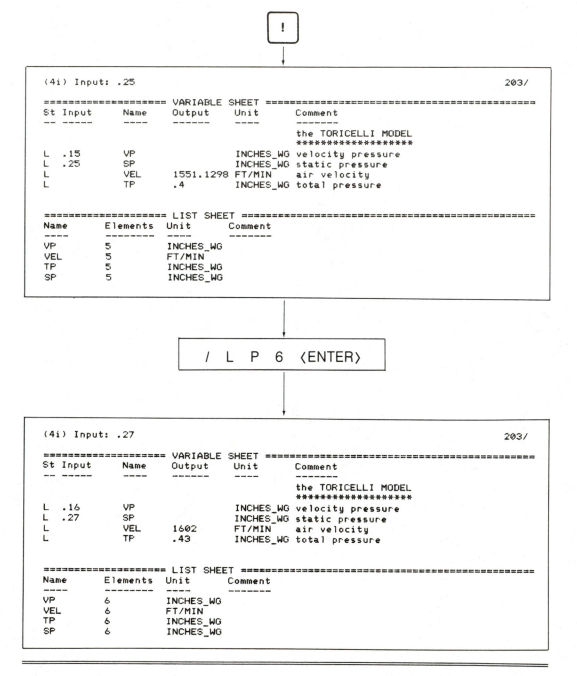

```
            ┌─────┐
            │  !  │
            └──┬──┘
               │
               ▼
┌──────────────────────────────────────────────────────────────────────────┐
│ (4i) Input: .25                                                      203/  │
│                                                                            │
│ ==================== VARIABLE SHEET ====================================== │
│ St Input        Name      Output     Unit      Comment                     │
│ -- -----        ----      ------     ----      -------                     │
│                                                 the TORICELLI MODEL        │
│                                                 ******************         │
│ L  .15          VP                   INCHES_WG velocity pressure           │
│ L  .25          SP                   INCHES_WG static pressure             │
│ L               VEL      1551.1298   FT/MIN    air velocity                │
│ L               TP        .4         INCHES_WG total pressure              │
│                                                                            │
│                                                                            │
│ ==================== LIST SHEET ========================================== │
│ Name           Elements  Unit       Comment                               │
│ ----           --------  ----       -------                               │
│ VP             5         INCHES_WG                                        │
│ VEL            5         FT/MIN                                           │
│ TP             5         INCHES_WG                                        │
│ SP             5         INCHES_WG                                        │
│                                                                            │
└──────────────────────────────────────────────────────────────────────────┘
               │
               ▼
      ┌──────────────────────────────┐
      │  /  L  P  6  〈ENTER〉         │
      └──────────────────────────────┘
               │
               ▼
┌──────────────────────────────────────────────────────────────────────────┐
│ (4i) Input: .27                                                      203/  │
│                                                                            │
│ ==================== VARIABLE SHEET ====================================== │
│ St Input        Name      Output     Unit      Comment                     │
│ -- -----        ----      ------     ----      -------                     │
│                                                 the TORICELLI MODEL        │
│                                                 ******************         │
│ L  .16          VP                   INCHES_WG velocity pressure           │
│ L  .27          SP                   INCHES_WG static pressure             │
│ L               VEL      1602        FT/MIN    air velocity                │
│ L               TP        .43        INCHES_WG total pressure              │
│                                                                            │
│                                                                            │
│ ==================== LIST SHEET ========================================== │
│ Name           Elements  Unit       Comment                               │
│ ----           --------  ----       -------                               │
│ VP             6         INCHES_WG                                        │
│ VEL            6         FT/MIN                                           │
│ TP             6         INCHES_WG                                        │
│ SP             6         INCHES_WG                                        │
│                                                                            │
└──────────────────────────────────────────────────────────────────────────┘
```

Figure 7-7. Using the Put Option of the List Command.

it does so even faster than a computer department would, considering the time that the job might have to wait its turn on the computer. However, TK!Solver does not solve large models instantaneously, particularly if the List processor is invoked. With the Block option, you can develop and verify a model using only a few sets of input data. When you know everything is working correctly, you invoke the List solver with the Solve option

and then do something else while TK!Solver is working.

The Get option allows you to spot-check a model that has been solved for many sets of data. After you have solved a model with the List solver, you can recall any given instance and check the calculation, or compare that instance with experimental measurements.

You can use the Put option to record the results of a series of solutions reached without the List solver. For instance, you may wish to use a model to check the results of experiments performed over an extended period of time. You would load the same model on each occasion, solve the model, and use the Put option to record the results. Then you can save the model, and recall it later. At any time, you can review the individual lists, or Get the results of a previous solution. Again, TK!Solver is your electronic notebook.

## SAVING LISTS

When you give the command to save a model, TK!Solver saves the entire model, including the lists it contains. When you load the same model for another session, the lists are loaded just as they were saved. But suppose you wish to use the same input lists for several models. Or, suppose you wish to use the output lists of one model as the input lists of another model. You certainly would not want to reenter the lists for each occasion.

You can save a model's lists without saving the rest of the model, in either of two formats, by using options of the Storage command (/S). You can also load a file containing only lists with other options of the Storage command. The two file formats are TK!Solver's normal internal format (which has a file extension of TK), and the DIF file (which has an extension of DIF). Either type of file can be used to transfer lists from one TK!Solver model to another, but DIF files must be used to transfer lists from TK!Solver to other programs, or vice versa.

You can save lists in the TK format with the User Function Save option of the Storage command with this entry:

TK!Solver prompts you for the name to be assigned to the file, with the message

Function Save: Filename:

TK!Solver then saves the User Function sheet, the List sheet, and the List and User Function subsheets in a file with the specified name. Of course, you must take care to assign a name other than that of the model from which you are excerpting the lists, so that TK!Solver will not overwrite the file containing the complete model.

When you construct another model so that it will use with the same lists, you load the file containing the lists with the Load option of the Storage command. Note that there are no suboptions to the Load option; you simply respond with the name of the file to be loaded.

You can load a file saved with the User Function Save option at any time during the construction of a model. TK!Solver does not overwrite a model in the computer's memory—it adds the contents of the new file to the existing model. Of course, the file of lists still exists on the disk, so it can be loaded into any number of different models.

DIF files are discussed in a later chapter, in which we will also examine the structure of the file. Nor have you seen the last of lists. The List Solver is one of the most powerful features of TK!Solver, and you will use it in almost every model you consider.

# CHAPTER 8

# Plots, Tables, and the Print Command

You now have the tools required to create useful models: you know how to enter the rules and variables that are the basis of any engineering model; and you can enter units and conversion factors and let TK!Solver keep track of the units for you. In addition, you can create models and solve them for a number of instances, and store the input and output data as part of the model. The next step is to learn how to present models visually by means of graphs and tables.

## THE PRINTER AREA OF THE GLOBAL SHEET

You can display graphs (which the TK!Solver manual calls plots) or tables either on the printer or the screen. TK!Solver normally puts the display on the screen. You have to tell the program to print the display on the printer. First, however, you need to know a little more about the Global sheet (Figure 8–1).

The fields of the Global sheet are divided into three groups, the first two of which have already been discussed. The first group comprises the single field, Automatic Variable Insert ON:. The second group comprises the parameters used by the Solvers. The third group contains parameters TK!Solver needs to communicate with the printer. You must be familiar with these parameters in order to prepare tables and plots.

The first parameters in the last group is the Page Breaks ON: field. If this field is set to Yes, TK!Solver will treat the printout as a series of sheets of a specified length. In this case, the program will print a preset number of lines on each sheet and then send a Form

```
(v) Variable Insert ON: Yes                                          192/!

==================== GLOBAL SHEET =========================================
Variable Insert ON:           Yes

Intermediate Redisplay ON:    Yes
Automatic Iteration ON:       Yes
Comparison Tolerance:         .000001
Typical Value:                1
Maximum Iteration Count:      10

Page Breaks ON:               Yes
Page Numbers ON:              Yes
Form Length:                  66
Printed Page Length:          60
Printed Page Width:           132
Left Margin:                  0
Printer Device or Filename:   PRN
Printer Setup String:         ^Ec^R
Line End:                     CR&LF
```

Figure 8-1. The Global Sheet.

Feed to the printer so that it will advance to the next sheet. If the display calls for titles, they will be repeated at the top of each page.

If the Page Breaks ON: field is set to No, the program will ignore page and form lengths; the printout will be one continuous listing. Titles and headers will appear once at the top of the listing and will not be repeated.

The second parameter in this group is the Page Numbers ON: field. If this field is set to Yes, TK!Solver will number each page of the printout at the bottom. This field is ignored if the Page Breaks ON: field is set to No. When the setting for this field, and for the Page Breaks On field, is Yes:, the printer will skip lines at each page break, and number the pages.

The third parameter is the Form Length: field. This field specifies the length of a page in lines of print. The number entered must be an integer representing the actual length of the sheet of paper. If you are using continuous fan-fold paper, you will enter the length between perforations. TK!Solver will accept a form length of 6 to 1000 lines. The program ignores this field if the Page Breaks ON: field is set to No. The default setting for this field is 66 lines; that value corresponds to a length of 11 inches and a line spacing of 6 lines per inch.

The fourth parameter here is the Printed Page Length: field. This field is used to enter the number of lines of printing desired on each page of the printout. This number should be less than or equal to the form length; if you enter a printed page length that is greater than the form length, the program will ignore the form length and leave no space between printed pages. These settings are important because some of the tables you can generate are too large to be printed on a single sheet and must be pieced together from the separate sheets TK!Solver prints. The default setting here is 60 lines, as shown in Figure 8–1.

The fifth parameter is the Printed Page Width: field. This field specifies the width of each printout page in characters. The program will accept any integer value between 6 and 200 in this field.

As a point of comparison, if you use standard computer paper that is 14 7/8 inches wide and your printer prints 16.5 characters per inch, you could enter the maximum value of 200 and produce a line of print 12 inches long. If you use the same paper with the printer set for a pitch of 10 characters per inch, you could only enter 135 characters, which would produce a 13.5-inch line of print. The IBM Printer uses 8 1/2-inch wide paper, and will print at pitches of 5, 8.25, 10, and 16.5 characters per inch. The maximum number you can enter in the Printed Page Width field with the IBM printer attached is 132, which is the maximum number of characters per line at 16.5 characters per inch.

The sixth parameter is the Left Margin: field. This field specifies the number of spaces the computer will leave to the left of the first character of each line. You can enter any number between 0 and 200. A value of 0 will produce a left margin width of approximately 1/4 inch. It is wise to check that the sum of the Left Margin value and the Printed Page Width value is not greater than the physical page width (in characters); if that sum is greater than the physical page width, the printed line may be truncated at the right end.

The default setting for the Left Margin: field is 0, and the default setting for the Page width: field is 80. These values correspond to the normal settings for the IBM 80 CPS printer (an Epson MX-80 in disguise) of 10 characters per inch, and 80 characters per line.

The seventh parameter in this group is the Printer Device or Filename: field. The Logical Device Name of the printer is entered in this field. The Logical Device Name for the printer under IBM PC-DOS is either LST: or PRN:. Notice that a file name is acceptable for this field. PC-DOS is a device-independent operating system; the operating system treats hardware devices in the same manner as it treats disk files. Because of this device independence, you can use the Print command to write the program output to disk, and then print the disk file later.

If the field is filled with a file name other than LPT or PRN when the Plot or Table sheet Screen or Printer fields are set to Printer, the Print command (/P) or the Action command (!) will send the output to that file. If the Printer Device or file name is left blank and the Screen or Printer field is set to Printer, the same commands will send the output to the screen one page at a time formatted according to the other printer parameters.

The normal printer output for the IBM PC is LPT1, which is the parallel printer port (the IBM Matrix and Graphics printers are both parallel printers). If you wish to use a serial printer instead, you must connect the printer to a COM port, and enter COM1 or COM2 in the Printer Device or Filename: field. As an alternative, you may enter LPT in this field, and redirect the printer output to a COMx port through the operating system. (See the PC-DOS manual for more information on redirection of output.)

You will shortly see that both the Plot and Table sheets allow you to specify whether a plot or a table is displayed on the screen or the printer. The Printer Device or Filename field has no effect if the Plot and Table sheets are set to send output to the screen. If one or the other is set to send output to the printer, the Printer Device or Filename field affects only that sheet.

If "SCREEN" is entered in the Printer Device or Filename field, and the Plot and Table sheets are set to send output to the printer, the output will be sent to a disk file with the name SCREEN.PRF, and will be displayed on the screen while it is being sent to disk.

The eighth parameter is the Printer Setup String: field. It is used to give commands to the printer when the actual text is about to be printed. All printers recognize certain character sequences as commands. When a printer receives a sequence that it recognizes, it will take the appropriate action rather than print the sequence. For example, the IBM Printer interprets the ASCII character 0D(hex) as a carriage return command, and returns the print head to the left side of the paper.

Two types of character sequences are recognized as commands by printers: control characters and escape sequences. You will recall that you can enter control characters from the keyboard by holding the **CTRL** key down and pressing another key at the same time. Holding the **CTRL** key down changes the code sent by the other key. A computer program can also send a control character to a device—the printer in this case—by sending the correct ASCII code for the desired control character.

Escape sequences are formed by sending the ASCII character 1B(hex) and a series of other characters. The escape character is normally written as ESC. If you send ESC E to the IBM printer, it will begin printing emphasized characters; if you send ESC F, it will stop printing bold face characters.

The Printer Setup String: field is used to store the characters that will tell the printer how to print for the specific model. You may want the tables for one model printed in small letters, for example, and those for another model printed in bold characters. TK!Solver lets you choose the format for each model.

As noted above, some printers recognize control characters, some recognize escape sequences, and some recognize both. TK!Solver accepts both types of commands. The following format is used to enter the commands in the Printer Setup String field:

^          Marks the beginning of a control character sequence

^^         Used to represent the caret (^) character

^R         Represents a carriage return (ASCII 0D HEX)

^L         Represents a line feed (ASCII 0A HEX)

^C         The next character is a control character

^E         Represents the ESC character

^H         The next two digits are the value of an ASCII character in hexadecimal

As an example, the Heath/Zenith line printer used to prepare the illustrations for this book accepts escape sequences for some actions, and control characters for others. (Most line printers accept control characters for the basic teletype-like commands.) Entering the string

^E[4w^R^E[;200s^R

in the Printer Setup String field produces a character pitch of 16.5 characters per inch, and a right margin of 200 characters; both commands are required because that particular printer does not change its right margin automatically. Each ^E tells TK!Solver to send an ESCAPE character, which tells the printer that the following characters are commands, and not text to be printed. Each escape sequence must be terminated with a carriage return (for this printer) by ^R.

The last parameter of this group is the Line End (CR/LF or CR): field. You can enter any one of three characters in this field. If you enter a C, the program will display CR in the field, and will send a carriage return character to the printer after printing each line. If you enter an L in the field, the program will display LF in the field, and will send a line feed character after each line. If you place the cursor in the Line End field, and enter &, the program will display CR&LF in the field, and will send both the carriage return character and the line feed character at the end of each line. Some printers automatically execute a line feed for each carriage return, and others require that both characters be sent. The IBM printer can be set for either mode of operation.

A fourth option for the Line End field is available on some types of computer, but does not apply to the IBM PC. In certain systems, the TK!Solver program will set the Line End field.

# THE TABLE SHEET

Now that you are familiar with all the fields on the Global sheet, you are ready to work with graphs and tables. Let's look at tables first, beginning with the Table sheet (Figure 8-2).

A TK!Solver table is a group of lists in which the corresponding elements are aligned either vertically or horizontally. A typical table is shown in Figure 8-3. The Table sheet tells TK!Solver what lists to use and how to arrange them.

The first field on the Table sheet is the Screen or Printer field. This field tells the program whether to display the table on the computer screen or the printer. You enter the desired value by placing the cursor in the field and typing either P or S; TK!Solver spells the word out for you. Remember, the effect of the Screen or Printer

```
(10h) Header:                                                    192/!

================= TABLE SHEET ==========================================
Screen or Printer:          Printer
Title:                      Trane Centravac Selection Summary
Vertical or Horizontal:     Vertical
List        Width First Header
----        ----- ----- ------
line        10    1     SELECTION
des_kw      10    1     DESIGN KW
kwperton    10    1     KW/TON
evap_pd     10    1     EVAP P.D.
cond_pd     10    1     COND P.D.
max_tons    10    1     MAX TONS
kw@max      10    1     KW@ MAX
price       10    1     PRICE
$perton     10    1     PRICE/TON
```

Figure 8-2. The Table Sheet.

Trane Centravac Selection Summary

| SELECTION | DESIGN KW | KW/TON | EVAP P.D. | COND P.D. | MAX TONS | KW@ MAX | PRICE | PRICE/TON |
|---|---|---|---|---|---|---|---|---|
| 1 | 261 | .696 | 19.5 | 16 | 436 | 322 | 80800 | 215.466667 |
| 2 | 259 | .690666667 | 19.5 | 16 | 416 | 319 | 80800 | 215.466667 |
| 3 | 258 | .688 | 19.5 | 16 | 437 | 338 | 80800 | 215.466667 |
| 4 | 256 | .682666667 | 19.5 | 16 | 435 | 322 | 80800 | 215.466667 |
| 5 | 253 | .674666667 | 19.5 | 13 | 416 | 312 | 80100 | 213.6 |
| 6 | 253 | .674666667 | 19.5 | 13 | 437 | 330 | 80100 | 213.6 |
| 7 | 252 | .672 | 19.5 | 13 | 436 | 315 | 80100 | 213.6 |
| 8 | 251 | .669333333 | 19.5 | 10.6 | 420 | 313 | 82400 | 219.733333 |
| 9 | 247 | .658666667 | 19.5 | 19.8 | 417 | 304 | 81150 | 216.4 |
| 10 | 247 | .658666667 | 19.5 | 19.8 | 435 | 302 | 81150 | 216.4 |
| 11 | 245 | .653333333 | 19.5 | 19.7 | 436 | 315 | 81150 | 216.4 |
| 12 | 243 | .648 | 19.5 | 19.7 | 435 | 304 | 81150 | 216.4 |
| 13 | 243 | .648 | 16.7 | 19.5 | 491 | 377 | 88850 | 236.933333 |
| 14 | 243 | .648 | 16.7 | 19.5 | 511 | 380 | 88850 | 236.933333 |
| 15 | 241 | .642666667 | 16.7 | 19.5 | 510 | 381 | 88850 | 236.933333 |
| 16 | 241 | .642666667 | 19.5 | 13.1 | 435 | 295 | 87550 | 233.466667 |
| 17 | 241 | .642666667 | 19.5 | 11 | 420 | 297 | 88850 | 236.933333 |
| 18 | 240 | .64 | 16.8 | 19.5 | 463 | 345 | 88850 | 236.933333 |
| 19 | 240 | .64 | 19.5 | 16 | 435 | 298 | 83750 | 223.333333 |
| 20 | 240 | .64 | 19.5 | 13.1 | 434 | 296 | 87550 | 233.466667 |
| 21 | 237 | .632 | 16.8 | 15.5 | 460 | 337 | 91800 | 244.8 |
| 22 | 237 | .632 | 16.7 | 15.5 | 496 | 378 | 91800 | 244.8 |
| 23 | 237 | .632 | 16.7 | 15.5 | 516 | 381 | 91800 | 244.8 |
| 24 | 234 | .624 | 16.8 | 12.7 | 460 | 333 | 94150 | 251.066667 |
| 25 | 227 | .605333333 | 16.7 | 19.2 | 544 | 381 | 95200 | 253.866667 |
| 26 | 224 | .597333333 | 16.8 | 19.2 | 461 | 325 | 95200 | 253.866667 |
| 27 | 224 | .597333333 | 16.7 | 19.2 | 509 | 377 | 95200 | 253.866667 |
| 28 | 224 | .597333333 | 16.7 | 19.1 | 534 | 381 | 95200 | 253.866667 |
| 29 | 222 | .592 | 16.8 | 15.6 | 458 | 320 | 97800 | 260.8 |
| 30 | 222 | .592 | 16.7 | 15.6 | 512 | 376 | 97800 | 260.8 |
| 31 | 221 | .589333333 | 16.7 | 15.6 | 537 | 381 | 97800 | 260.8 |
| 32 | 220 | .586666667 | 16.8 | 12.8 | 455 | 316 | 101650 | 271.066667 |
| 33 | 218 | .581333333 | 16.7 | 10.7 | 513 | 373 | 102900 | 274.4 |
| 34 | 218 | .581333333 | 16.7 | 10.7 | 543 | 381 | 102900 | 274.4 |
| 35 | 217 | .578666667 | 16.8 | 10.7 | 449 | 306 | 102900 | 274.4 |
| 36 | 217 | .578666667 | 13.5 | 15.6 | 453 | 310 | 103500 | 276 |
| 37 | 216 | .576 | 13.5 | 10.7 | 456 | 310 | 108600 | 289.6 |
| 38 | 216 | .576 | 11.2 | 12.7 | 460 | 313 | 108900 | 290.4 |
| 39 | 214 | .570666667 | 11.2 | 10.7 | 461 | 313 | 110150 | 293.733333 |

Figure 8-3. A TK!Solver Table.

option depends on the entry in the Printer Device or Filename: field in the Global sheet.

The second field on the Table sheet is the Title field. This field specifies the title that will appear on the tables you create. This field is optional; if no title is entered in this field, the program will print an extra two lines of data. If a title is entered, it will be printed as the first line on the table, and will be followed by a blank line.

The title can be up to 200 characters long, but it should not be any longer than the Printed Page Width value entered on the Global sheet. Any title that is longer will be truncated on the right end. If you set the Page Breaks ON: to Yes, the title will be printed at the top of each page of the printout.

The third field is the Vertical or Horizontal: field. This field specifies whether successive elements of each list are to be arranged in columns or rows. If the value is set to Vertical, each list will consist of a column of values with a column heading at the top of each column. If the field is set to Horizontal, each list will be a row of values with a row heading at the left side of each row. Set the field by placing the cursor in the field and typing either V or H (in upper or lower case)—it is not necessary to press **ENTER**.

The next four items on the Table sheet are columns of fields. The first column is the List column, which is used to specify the lists that will make up your table. Use the list names as they appear on the List sheet. If you enter a name in this field that does not appear on the List sheet, the program will add the name to the last row of the List sheet, which will then create a new, but empty, list. If you print the table before filling the empty list, the program will print a blank column in the table.

The second column is the Width column. This field specifies the width of each column in characters. It must have an entry, but there is a default value of 10 characters. You can specify widths from 3 to 128 characters. If a number is too long to fit the width that has been entered, TK!Solver will display the number in scientific notation, or it will round it off.

The third column contains the First field for each list. This field specifies which element of a list is to be printed as the first element. You must have an entry in the field, but again there is a default value of 1.

The fourth column is the Header column. This field specifies the heading for each column in a vertical table, or for each row in a horizontal table. This field is an optional field—you do not have to specify headings. The headings for a vertical table will be repeated at the top of each page of the printout if the Page Breaks ON: field is set to Yes. You should make certain that the first few characters of each heading are unique, as the program will print only the number of characters specified in the Width field, and will truncate any longer titles on the right. If no heading is specified, the program will head the columns or rows with the list name as specified in the List column.

The table is actually printed, or displayed on the screen, with the Action command

!

given from any field on the Table sheet. The Table sheet can be displayed as a single sheet, or in either window of a two-sheet display. The cursor, by the way, must be on the Table sheet when the Action command is given. If the Action command is given with the Screen or Printer field set to Screen, the table will be displayed on the screen, and it will remain until a key is struck; if the display requires more than one screen, striking a key will cause successive screens to be displayed. Striking any key when the last screenful is displayed will return the Table sheet to the screen, or the Table sheet and any other sheet that may have been displayed along with the Table sheet.

If the Action command is given with the Screen or Printer field set to Printer, the table will be sent to the device or file named in the Printer Device or Filename field of the Global sheet. At the same time the table is being sent to the printer or disk file, it will be displayed on the screen in the same format. If the Printed Page Width field of the Global sheet is set to the same width as the computer screen (80, 64, or 40 characters are typical), the entire table will be displayed on the screen. If the value in the Printed Page Width field is greater than 80 characters, only the first 80 characters of each line will be displayed on the screen. The output sent to the printer (or disk file) or to the screen will not pause at the ends of pages, and the previous display will return to the screen as soon as the last page is sent.

Notice that there are several constraints on the printing of tables. For one thing, a single list cannot be wider than 128 characters or the maximum number of characters the printer can print on a line, whichever is smaller. Second, the printer can print only so many characters on a line, regardless of the number of fields or lists involved. However, you can print tables that are much wider than one page. If you define a table that is wider than one page, TK!Solver will print the lists that will fit on the first page up to the number of characters the printer can print on a line, from the first element to the last. Then the program will start at the top of the next group of lists, and print as many as will fit the width of the page. When the program has printed all the lists, you can tear the pages apart and align them so that the elements of all lists are aligned vertically. When these pages are taped together, you will have a table that is as wide as you need. Figure 8–4 shows a large table that has been pieced together.

CAPITAL RECOVERY FACTORS

| n | i = 1% | i = 2% | i = 3% | i = 4% | i = 5% | i = 6% | i = 7% |
|---|--------|--------|--------|--------|--------|--------|--------|
| 1 | 1.01 | 1.02 | 1.03 | 1.04 | 1.05 | 1.06 | 1.07 |
| 2 | .507512438 | .515049505 | .522610837 | .530196078 | .537804878 | .545436893 | .553091787 |
| 3 | .340022112 | .346754673 | .353530363 | .360348539 | .367208565 | .374109813 | .381051666 |
| 4 | .256281094 | .262623753 | .269027045 | .275490045 | .282011833 | .288591492 | .295228117 |
| 5 | .206039800 | .212158394 | .218354571 | .224627114 | .230974798 | .237396400 | .243890694 |
| 6 | .172548367 | .178525812 | .184597501 | .190761903 | .197017468 | .203362629 | .209795800 |
| 7 | .148628283 | .154511956 | .160506354 | .166609612 | .172819818 | .179135018 | .185553220 |
| 8 | .130690292 | .136509799 | .142456389 | .148527832 | .154721814 | .161035943 | .167467763 |
| 9 | .116740363 | .122515437 | .128433857 | .134492993 | .14069008 | .147022235 | .153486470 |
| 10 | .105582077 | .111326528 | .117230507 | .123290944 | .129504575 | .135867958 | .142377503 |
| 11 | .096454076 | .102177943 | .108077448 | .114149039 | .120388892 | .126792938 | .133356905 |
| 12 | .088848789 | .094559597 | .100462086 | .106552173 | .11282541 | .119277029 | .125901989 |
| 13 | .082414820 | .088118353 | .094029544 | .100143728 | .106455765 | .112960105 | .119650848 |
| 14 | .076901172 | .082601970 | .088526339 | .094668973 | .101023970 | .107584909 | .114344939 |
| 15 | .072123780 | .077825472 | .083766588 | .089941100 | .096342288 | .102962764 | .109794625 |
| 16 | .067944597 | .073650126 | .079610849 | .085819999 | .092269908 | .098952144 | .105857648 |
| 17 | .064258055 | .069969841 | .075952529 | .082198522 | .088699142 | .095444804 | .102425193 |
| 18 | .060982048 | .066702102 | .072708696 | .078993328 | .085546222 | .092356541 | .099412602 |
| 19 | .058051754 | .063781766 | .069813881 | .076138618 | .082745010 | .089620860 | .096753015 |
| 20 | .055415315 | .061156718 | .067215708 | .073581750 | .080242587 | .087184557 | .094392926 |
| 21 | .053030752 | .058784769 | .064871776 | .071280105 | .077996107 | .085004547 | .092289002 |
| 22 | .050863718 | .056631401 | .062747395 | .069198811 | .075970509 | .083045569 | .090405773 |
| 23 | .048885840 | .054668098 | .060813903 | .067309057 | .074136822 | .081278485 | .088713926 |
| 24 | .047073472 | .052871097 | .059047416 | .065586831 | .072470901 | .079679005 | .087189021 |
| 25 | .045406753 | .051220438 | .057427871 | .064011963 | .070952457 | .078226718 | .085810517 |
| 26 | .043868878 | .049699231 | .055938290 | .062567380 | .069564321 | .076904347 | .084561028 |
| 27 | .042445529 | .048293086 | .054564210 | .061238541 | .068291860 | .075697166 | .083425734 |
| 28 | .041124436 | .046989672 | .053293233 | .060012975 | .067122530 | .074592552 | .082391928 |
| 29 | .039895020 | .045778355 | .052114671 | .058879934 | .066045515 | .073579614 | .081448652 |
| 30 | .038748113 | .044649922 | .051019209 | .057830009 | .065051435 | .072648911 | .080586404 |
| 31 | .037675731 | .043596347 | .049998929 | .056855352 | .064132120 | .071792220 | .079796906 |
| 32 | .036670886 | .042610607 | .049046618 | .055948590 | .063280419 | .071002337 | .079072916 |
| 33 | .035727438 | .041686531 | .048156122 | .055103567 | .062490044 | .070272935 | .078408065 |
| 34 | .034839969 | .040818673 | .047321963 | .054314772 | .061755445 | .069598425 | .077796738 |
| 35 | .034003682 | .040002209 | .046539292 | .053577322 | .061071707 | .068973859 | .077233960 |
| 36 | .033214310 | .039232853 | .045803794 | .052886878 | .060434457 | .068394835 | .076715310 |
| 37 | .032468049 | .038506779 | .045111624 | .052239566 | .059839794 | .067857427 | .076236848 |
| 38 | .031761496 | .037820566 | .044459340 | .051631919 | .059284228 | .067358124 | .075795052 |
| 39 | .031091595 | .037171144 | .043843852 | .051060827 | .058764624 | .066893772 | .075386762 |
| 40 | .030455598 | .036555748 | .043262378 | .050523489 | .058278161 | .066461536 | .075009139 |
| 41 | .029851023 | .035971884 | .042712409 | .050017377 | .057822292 | .066058855 | .074659624 |
| 42 | .029275626 | .035417295 | .042191673 | .049540201 | .057394713 | .065683415 | .074335907 |
| 43 | .028727371 | .034889933 | .041698110 | .049089886 | .056993333 | .065333118 | .074035895 |
| 44 | .028204406 | .034387939 | .041229847 | .048664544 | .056616251 | .065006057 | .073757691 |
| 45 | .027705046 | .033909616 | .040785176 | .048262456 | .056261735 | .064700496 | .073499571 |
| 46 | .027227750 | .033453416 | .040362538 | .047882049 | .055928204 | .064414853 | .073259965 |
| 47 | .026771110 | .033017922 | .039960507 | .047521885 | .055614211 | .064147680 | .073037442 |
| 48 | .026333835 | .032601836 | .039577774 | .047180648 | .055318431 | .063897655 | .072830695 |

Figure 8–4. A Large Printed Table in Sections.

| 49 | .025914739 | .032203964 | .039213138 | .046857124 | .055039645 | .063663562 | .072638529 |
| 50 | .025512731 | .031823210 | .038865494 | .046550200 | .054776735 | .063444286 | .072459850 |
| 51 | .025126805 | .031458562 | .038533823 | .046258850 | .054528670 | .063238803 | .072293652 |
| 52 | .024756033 | .031109086 | .038217184 | .045982124 | .054294497 | .063046167 | .072139015 |
| 53 | .024399557 | .030773919 | .037914706 | .045719145 | .054073337 | .062865508 | .071995091 |
| 54 | .024056583 | .030452262 | .037625584 | .045469102 | .053864377 | .062696021 | .071861101 |

Page 1


CAPITAL RECOVERY FACTORS

| n | i = 1% | i = 2% | i = 3% | i = 4% | i = 5% | i = 6% | i = 7% |
|---|--------|--------|--------|--------|--------|--------|--------|
| 55 | .023726373 | .030143373 | .037349071 | .045231243 | .053666864 | .062536963 | .071736326 |
| 56 | .023408244 | .029846565 | .037084473 | .045004866 | .053480098 | .062387647 | .071620106 |
| 57 | .023101559 | .029561196 | .036831143 | .044789323 | .053303430 | .062247435 | .071511829 |
| 58 | .022805727 | .029286671 | .036588482 | .044584009 | .053136257 | .062115736 | .071410930 |
| 59 | .022520195 | .029022434 | .036355928 | .044388358 | .052978016 | .061992001 | .071316890 |
| 60 | .022244448 | .028767966 | .036132959 | .044201845 | .052828185 | .061875722 | .071229226 |
| 61 | .021978004 | .028522783 | .035919085 | .044023978 | .052686274 | .061766423 | .071147491 |
| 62 | .021720412 | .028286431 | .035713848 | .043854296 | .052551827 | .061663664 | .071071272 |
| 63 | .021471252 | .028058485 | .035516822 | .043692370 | .052424420 | .061567035 | .071000188 |
| 64 | .021230127 | .027838547 | .035327602 | .043537795 | .052303652 | .061476153 | .070933882 |
| 65 | .020996667 | .027626244 | .035145813 | .043390194 | .052189151 | .061390660 | .070872026 |
| 66 | .020770521 | .027421223 | .034971100 | .043249210 | .052080568 | .061310225 | .070814314 |
| 67 | .020551364 | .027223155 | .034803129 | .043114510 | .051977575 | .061234535 | .070760462 |
| 68 | .020338886 | .027031729 | .034641587 | .042985780 | .051879864 | .061163301 | .070710207 |
| 69 | .020132796 | .026846653 | .034486179 | .042862723 | .051787147 | .061096251 | .070663305 |
| 70 | .019932821 | .026667649 | .034336625 | .042745062 | .051699173 | .061033130 | .070619527 |
| 71 | .019738701 | .026494457 | .034192663 | .042632534 | .051615627 | .060973702 | .070578662 |
| 72 | .019550193 | .026326831 | .034054045 | .042524892 | .051536328 | .060917744 | .070540514 |
| 73 | .019367065 | .026164538 | .033920535 | .042421901 | .051461032 | .060865047 | .070504898 |
| 74 | .019189099 | .026007358 | .033791911 | .042323340 | .051389525 | .060815417 | .070471645 |
| 75 | .019016088 | .025855083 | .033667963 | .042229002 | .051321609 | .060768670 | .070440595 |
| 76 | .018847837 | .025707515 | .033548493 | .042138687 | .051257093 | .060724635 | .070411602 |
| 77 | .018684159 | .025564466 | .033433310 | .042052210 | .051195799 | .060683151 | .070384527 |
| 78 | .018524879 | .025425759 | .033322237 | .041969392 | .051137561 | .060644067 | .070359241 |
| 79 | .018369829 | .025291226 | .033215103 | .041890067 | .051082219 | .060607241 | .070335627 |
| 80 | .018218850 | .025160705 | .033111746 | .041814075 | .051029623 | .060572541 | .070313572 |
| 81 | .018071791 | .025034045 | .033012013 | .041741266 | .050979633 | .060539841 | .070292972 |
| 82 | .017928509 | .024911101 | .032915758 | .041671496 | .050932114 | .060509025 | .070273731 |
| 83 | .017788866 | .024791733 | .032822842 | .041604628 | .050886941 | .060479782 | .070255758 |
| 84 | .017652733 | .024675812 | .032733133 | .041540535 | .050843992 | .060452608 | .070238969 |
| 85 | .017519985 | .024563211 | .032646504 | .041479093 | .050803157 | .060426307 | .070223285 |
| 86 | .017390503 | .024453811 | .032562837 | .041420185 | .050764326 | .060402486 | .070208634 |
| 87 | .017264175 | .024347498 | .032482015 | .041363700 | .050727400 | .060379559 | .070194947 |
| 88 | .017140894 | .024244163 | .032403931 | .041309533 | .050692283 | .060357947 | .070182161 |
| 89 | .017020555 | .024143703 | .032328479 | .041257583 | .050658883 | .060337572 | .070170215 |
| 90 | .016903061 | .024046017 | .032255560 | .041207754 | .050627114 | .060318362 | .070159054 |
| 91 | .016788318 | .023951011 | .032185079 | .041159955 | .050596895 | .060300252 | .070148626 |
| 92 | .016676235 | .023858594 | .032116945 | .041114098 | .050568148 | .060283176 | .070138884 |
| 93 | .016566727 | .023768678 | .032051071 | .041070102 | .050540801 | .060267076 | .070129781 |
| 94 | .016459711 | .023681181 | .031987373 | .041027887 | .050514783 | .060251895 | .070121276 |
| 95 | .016355107 | .023596023 | .031925773 | .040987377 | .050490029 | .060237580 | .070113329 |
| 96 | .016252841 | .023513127 | .031866193 | .040948500 | .050466477 | .060224082 | .070105904 |
| 97 | .016152840 | .023432421 | .031808561 | .040911188 | .050444067 | .060211354 | .070098966 |
| 98 | .016055034 | .023353832 | .031752807 | .040875376 | .050422742 | .060199350 | .070092483 |
| 99 | .015959357 | .023277295 | .031698863 | .040841000 | .050402449 | .060188031 | .070086425 |
| 100 | .015865743 | .023202744 | .031646666 | .040808000 | .050383138 | .060177356 | .070080765 |

Page 2


Figure 8–4. (continued).

## CAPITAL RECOVERY FACTORS

| i = 8% | i = 9% | i = 10% | i = 11% | i = 12% | i = 13% | i = 14% |
|---|---|---|---|---|---|---|
| 1.08 | 1.09 | 1.1 | 1.11 | 1.12 | 1.13 | 1.14 |
| .560769231 | .568468900 | .576190476 | .583933649 | .591698113 | .599483568 | .607289720 |
| .388033514 | .395054757 | .402114804 | .409213070 | .416348981 | .423521970 | .430731480 |
| .301920805 | .308668662 | .315470804 | .322326352 | .329234436 | .336194197 | .343204783 |
| .250456455 | .257092457 | .263797481 | .270570310 | .277409732 | .284314543 | .291283547 |
| .216315386 | .222919783 | .229607380 | .236376564 | .243225718 | .250153232 | .257157496 |
| .192072401 | .198690517 | .205405500 | .212215270 | .219117736 | .226110804 | .233192377 |
| .174014761 | .180674378 | .187444018 | .194321054 | .201302841 | .208386720 | .215570024 |
| .160079709 | .166798802 | .173640539 | .180601664 | .187678889 | .194868902 | .202168384 |
| .149029489 | .155820090 | .162745395 | .169801427 | .176984164 | .184289556 | .191713541 |
| .140076342 | .146946657 | .153963142 | .161121007 | .168415404 | .175841455 | .183394271 |
| .132695017 | .139650659 | .146763315 | .154027286 | .161436808 | .168986085 | .176669327 |
| .126521805 | .133566560 | .140778524 | .148150993 | .155677195 | .163350341 | .171163664 |
| .121296853 | .128433173 | .135746223 | .143228202 | .150871246 | .158667496 | .166609145 |
| .116829545 | .124058883 | .131473777 | .139065240 | .146824240 | .154741780 | .162808963 |
| .112976872 | .120299910 | .127816621 | .135516747 | .143390018 | .151426245 | .1596154 |
| .109629432 | .117046249 | .124664134 | .132471485 | .140456728 | .148608439 | .156915436 |
| .106702096 | .114212291 | .121930222 | .129842870 | .137937311 | .146200855 | .154621152 |
| .104127628 | .111730411 | .119546868 | .127562504 | .135763005 | .144134394 | .152663159 |
| .101852209 | .109546475 | .117459625 | .125575637 | .13387878 | .142353788 | .150986002 |
| .099832250 | .107616635 | .115624390 | .12383793 | .132240092 | .140814328 | .149544861 |
| .098032068 | .105904993 | .114005063 | .122313101 | .130810509 | .139479481 | .148303165 |
| .096422169 | .10438188 | .112571813 | .120971182 | .129559965 | .138319133 | .147230813 |
| .094977962 | .103022561 | .111299776 | .119787211 | .128463442 | .137308261 | .146302841 |
| .093678779 | .101806251 | .110168072 | .118740242 | .127499970 | .136425928 | .145498408 |
| .092507127 | .100715360 | .109159039 | .117812575 | .126651858 | .135654506 | .144800014 |
| .091448096 | .099734905 | .108257642 | .116989164 | .125904094 | .134979073 | .144192884 |
| .090488906 | .098852047 | .107451013 | .116257145 | .125243869 | .134386929 | .143664491 |
| .089618535 | .098055723 | .106728075 | .115605470 | .124660207 | .133867225 | .143204166 |
| .088827433 | .097336351 | .106079248 | .115024599 | .124143658 | .133410650 | .142802794 |
| .088107284 | .096685600 | .105496214 | .114506267 | .123686057 | .133009192 | .142452561 |
| .087450813 | .096096186 | .104971717 | .114043285 | .123280326 | .132655929 | .142146751 |
| .086851632 | .095561726 | .104499406 | .113629379 | .122920310 | .132344868 | .141879576 |
| .086304110 | .095076597 | .104073706 | .113259055 | .122600638 | .132070808 | .141646037 |
| .085803265 | .094635837 | .103689705 | .11292749 | .122316619 | .131829221 | .141441810 |
| .085344674 | .094235050 | .103343064 | .112630441 | .122064141 | .131616163 | .141263148 |
| .084924403 | .093870329 | .103029941 | .112364164 | .121839592 | .131428190 | .141106798 |
| .084538936 | .093538198 | .102746925 | .112125351 | .121639800 | .131262290 | .140969934 |
| .084185130 | .093235550 | .102490984 | .111911071 | .121461967 | .131115824 | .140850096 |
| .083860162 | .092959609 | .102259414 | .111718727 | .121303626 | .130986481 | .140745143 |
| .083561494 | .092707885 | .102049803 | .111546009 | .121162598 | .130872231 | .140653207 |
| .083286841 | .092478142 | .101859991 | .111390863 | .121036958 | .130771290 | .140572660 |
| .083034137 | .092268368 | .101688047 | .111251462 | .120924899 | .130682092 | .140502081 |
| .082801516 | .092076749 | .101532237 | .111126174 | .120825210 | .130603257 | .140440228 |
| .082587285 | .091901651 | .101391005 | .111013542 | .120736252 | .130533571 | .140386016 |
| .082389908 | .091741596 | .101262953 | .110912268 | .120656936 | .130471964 | .140338496 |
| .082207992 | .091595246 | .101146822 | .110821188 | .120586206 | .130417493 | .140296838 |
| .082040266 | .091461389 | .101041480 | .110739262 | .120523125 | .130369326 | .140260317 |
| .081885573 | .091338929 | .100945904 | .110665559 | .120466858 | .130326731 | .140228296 |
| .081742858 | .091226868 | .100859174 | .110599243 | .120416664 | .130289059 | .140200219 |
| .081611158 | .091124302 | .100780458 | .110539568 | .120371883 | .130255739 | .140175600 |
| .081489590 | .091030407 | .100709004 | .110485861 | .120331928 | .130226266 | .140154012 |
| .081377351 | .090944434 | .100644134 | .110437521 | .120296276 | .130200195 | .140135080 |
| .081273700 | .090865703 | .100585234 | .110394008 | .120264463 | .130177133 | .140118477 |

Figure 8–4. (continued).

CAPITAL RECOVERY FACTORS

| i = 8% | i = 9% | i = 10% | i = 11% | i = 12% | i = 13% | i = 14% |
|---|---|---|---|---|---|---|
| .081177963 | .090793593 | .100531748 | .110354836 | .120236072 | .13015673 | .140103916 |
| .081089518 | .090727537 | .100483173 | .110319570 | .120210734 | .130138680 | .140091146 |
| .081007796 | .090667020 | .100439056 | .110287818 | .120188120 | .130122711 | .140079947 |
| .080932275 | .090611571 | .100398982 | .110259228 | .120167936 | .130108582 | .140070124 |
| .080862473 | .090560760 | .100362580 | .110233484 | .120149920 | .130096081 | .140061508 |
| .080797949 | .090514194 | .100329509 | .110210302 | .120133839 | .130085020 | .140053952 |
| .080738296 | .090471515 | .100299464 | .110189425 | .120119485 | .130075233 | .140047324 |
| .080683140 | .090432396 | .100272166 | .110170624 | .120106672 | .130066574 | .140041510 |
| .080632138 | .090396536 | .100247363 | .110153692 | .120095234 | .130058911 | .140036411 |
| .080584970 | .090363662 | .100224824 | .110138442 | .120085023 | .130052131 | .140031939 |
| .080541346 | .090333524 | .100204344 | .110124707 | .120075908 | .130046132 | .140028016 |
| .080500995 | .090305891 | .100185733 | .110112336 | .120067770 | .130040823 | .140024575 |
| .080463669 | .090280556 | .100168820 | .110101194 | .120060505 | .130036125 | .140021556 |
| .080429139 | .090257324 | .100153449 | .110091157 | .120054020 | .130031968 | .140018909 |
| .080397193 | .090236022 | .100139479 | .110082117 | .120048229 | .130028290 | .140016586 |
| .080367636 | .090216487 | .100126783 | .110073974 | .120043060 | .130025034 | .140014549 |
| .080340288 | .090198572 | .100115244 | .110066638 | .120038445 | .130022154 | .140012762 |
| .080314982 | .090182143 | .100104757 | .110060031 | .120034325 | .130019605 | .140011195 |
| .080291565 | .090167076 | .100095224 | .110054079 | .120030646 | .130017349 | .140009820 |
| .080269895 | .090153257 | .10008656 | .110048718 | .120027362 | .130015353 | .140008614 |
| .080249840 | .090140583 | .100078685 | .110043888 | .120024430 | .130013586 | .140007556 |
| .080231280 | .090128959 | .100071526 | .110039537 | .120021812 | .130012023 | .140006628 |
| .080214102 | .090118297 | .100065020 | .110035618 | .120019474 | .130010640 | .140005814 |
| .080198204 | .090108517 | .100059105 | .110032087 | .120017388 | .130009416 | .140005100 |
| .080183488 | .090099547 | .100053729 | .110028906 | .120015524 | .130008333 | .140004474 |
| .080169868 | .090091319 | .100048842 | .110026041 | .120013061 | .130007374 | .140003924 |
| .080157260 | .090083772 | .100044400 | .110023460 | .120012376 | .130006526 | .140003442 |
| .080145590 | .090076849 | .100040362 | .110021135 | .120011050 | .130005775 | .140003020 |
| .080134787 | .090070499 | .100036692 | .110019040 | .120009866 | .130005110 | .140002649 |
| .080124788 | .090064674 | .100033355 | .110017153 | .120008808 | .130004522 | .140002323 |
| .080115531 | .090059330 | .100030322 | .110015453 | .120007865 | .130004002 | .140002038 |
| .080106961 | .090054428 | .100027564 | .110013921 | .120007022 | .130003542 | .140001788 |
| .080099029 | .090049932 | .100025058 | .110012541 | .120006270 | .130003134 | .140001568 |
| .080091685 | .090045807 | .100022779 | .110011298 | .120005598 | .130002774 | .140001376 |
| .080084886 | .090042023 | .100020708 | .110010179 | .120004998 | .130002455 | .140001207 |
| .080078592 | .090038552 | .100018825 | .110009170 | .120004463 | .130002172 | .140001059 |
| .080072765 | .090035367 | .100017114 | .110008261 | .120003984 | .130001922 | .140000929 |
| .080067371 | .090032446 | .100015558 | .110007442 | .120003557 | .130001701 | .140000815 |
| .080062376 | .090029766 | .100014143 | .110006705 | .120003176 | .130001505 | .140000715 |
| .080057752 | .090027308 | .100012857 | .110006040 | .120002836 | .130001332 | .140000627 |
| .080053472 | .090025052 | .100011688 | .110005442 | .120002532 | .130001179 | .140000550 |
| .080049508 | .090022983 | .100010626 | .110004902 | .120002261 | .130001043 | .140000482 |
| .080045839 | .090021085 | .100009659 | .110004417 | .120002019 | .130000923 | .140000423 |
| .080042442 | .090019344 | .100008781 | .110003979 | .120001802 | .130000817 | .140000371 |
| .080039296 | .090017746 | .100007983 | .110003585 | .120001609 | .130000723 | .140000326 |
| .080036384 | .090016281 | .100007257 | .110003229 | .120001437 | .130000640 | .140000286 |

**Page 4**

CAPITAL RECOVERY FACTORS

| i = 15% | i = 16% | i = 17% | i = 18% | i = 19% | i = 20% |
|---|---|---|---|---|---|
| 1.15 | 1.16 | 1.17 | 1.18 | 1.19 | 1.2 |
| .615116279 | .622962963 | .630829493 | .638715596 | .646621005 | .654545455 |
| .437976962 | .445257873 | .452573681 | .459923861 | .467307895 | .474725275 |
| .350265352 | .357375070 | .364533114 | .371738671 | .378990938 | .386289121 |

Figure 8–4. (continued).

```
.298315553 .305409382 .312563864 .319777842 .327050167 .334379703
.264236907 .271389870 .278614802 .285910129 .293274292 .300705746
.240360364 .247612677 .254947243 .262361999 .269854902 .277423926
.222850090 .230224260 .237689892 .245244359 .252885060 .260609422
.209574015 .217082487 .224690510 .232394824 .240192202 .248079462
.199252063 .206901083 .214656597 .222514641 .230471309 .238522757
.191068983 .198860752 .206764792 .214776386 .222890901 .231103794
.184480776 .192414733 .200465582 .208627809 .216896022 .225264965
.179110457 .18718411  .195378139 .203686207 .212102153 .220620001
.174688490 .182897973 .191230218 .199678058 .208234563 .216893055
.171017053 .179357522 .187822095 .196402783 .205091906 .213882120
.167947691 .176413616 .185004010 .193710084 .202523448 .211436135
.165366862 .173952250 .182661569 .191485271 .200414307 .209440147
.163186287 .171884853 .180705995 .189639457 .198675594 .207805386
.161336350 .170141656 .179067452 .188102839 .197237650 .206462453
.159761470 .168667032 .177690359 .186819981 .196045291 .205356531
.158416791 .167416169 .176530035 .185746433 .195054399 .204443939
.157265771 .166352635 .175550249 .184846258 .194229430 .203689619
.156278395 .165446582 .174721405 .184090200 .193541556 .203065258
.155429830 .164673386 .174019170 .183454297 .192967267 .202547873
.154699402 .164012615 .173423428 .182918826 .192487299 .202118729
.154069806 .163447227 .172917471 .182467478 .192085808 .201762496
.153526482 .162962942 .172487362 .182086720 .191749713 .201466592
.153057131 .162547753 .172121440 .181765285 .191468188 .201220668
.152651327 .162191525 .171809915 .181493769 .191232251 .201011619
.152300198 .161885683 .171544547 .181264306 .191034434 .200846109
.151996180 .161622951 .171318385 .181070299 .190868517 .200704594
.151732801 .161397141 .171125557 .180906211 .190729314 .200586817
.151504516 .161202983 .170961090 .180767386 .190612494 .200488775
.151306566 .161035980 .170820770 .180649904 .190514436 .200407147
.151134855 .160892289 .170701021 .180550463 .190432112 .200339174
.150985857 .160768624 .170598804 .180466277 .190362988 .200282565
.150856533 .160662168 .170511537 .180394994 .190304939 .200235415
.150744257 .160570509 .170437020 .180334628 .190256185 .200196141
.150646761 .160491576 .170373382 .180283503 .190215236 .200163424
.150562085 .160423593 .170319028 .180240199 .190180837 .200136168
.150488531 .160365033 .170272599 .180203517 .190151941 .200113461
.150424629 .160314585 .170232936 .180172442 .190127665 .200094542
.150369106 .160271120 .170199051 .180146116 .190107270 .200078779
.150320859 .160233670 .170170100 .180123812 .190090135 .200065644
.15027893  .160201399 .170145364 .180104914 .190075738 .200054701
.150242489 .160173590 .170124227 .180088903 .190063641 .200045582
.150210816 .160149624 .170106166 .180075336 .190053477 .200037983
.150183284 .160128969 .170090732 .180064048 .190044937 .200031652
.150159352 .160111168 .170077543 .180054098 .190037761 .200026376
.150138548 .160095825 .170066271 .180045844 .190031731 .200021979
.150120462 .160082601 .170056639 .180038849 .190026664 .200018316
.150104739 .160071203 .170048407 .180032922 .190022406 .200015263
.150091069 .160061378 .170041372 .180027899 .190018828 .200012719
.150079184 .160052909 .170035359 .180023643 .190015822 .200010599
```

Page 5

CAPITAL RECOVERY FACTORS

| i = 15% | i = 16% | i = 17% | i = 18% | i = 19% | i = 20% |
|---|---|---|---|---|---|
| .150068851 | .160045609 | .170030221 | .180020036 | .190013295 | .200008832 |
| .150059867 | .160039317 | .170025829 | .180016979 | .190011173 | .200007360 |
| .150052055 | .160033893 | .170022076 | .180014389 | .190009389 | .200006134 |
| .150045263 | .160029217 | .170018868 | .180012194 | .190007890 | .200005111 |
| .150039358 | .160025187 | .170016126 | .180010334 | .190006630 | .200004259 |

Figure 8-4.  (continued).

```
.150034223  .160021712  .170013783  .180008757  .190005571  .200003550
.150029758  .160018717  .170011780  .180007421  .190004682  .200002958
.150025876  .160016135  .170010068  .180006289  .190003934  .200002465
.150022501  .160013909  .170008605  .180005330  .190003306  .200002054
.150019565  .160011991  .170007355  .180004517  .190002778  .200001712
.150017013  .160010337  .170006286  .180003828  .190002335  .200001426
.150014794  .160008911  .170005373  .180003244  .190001962  .200001189
.150012864  .160007682  .170004592  .180002749  .190001649  .200000991
.150011186  .160006622  .170003925  .180002330  .190001385  .200000826
.150009727  .160005709  .170003355  .180001974  .190001164  .200000688
.150008458  .160004921  .170002867  .180001673  .190000978  .200000573
.150007355  .160004242  .170002451  .180001418  .190000822  .200000478
.150006395  .160003657  .170002095  .180001202  .190000691  .200000398
.150005561  .160003153  .170001790  .180001018  .190000581  .200000332
.150004836  .160002718  .17000153   .180000863  .190000488  .200000277
.150004205  .160002343  .170001308  .180000731  .19000041   .200000230
.150003657  .160002020  .170001118  .180000620  .190000345  .200000192
.150003180  .160001741  .170000955  .180000525  .190000290  .20000016
.150002765  .160001501  .170000817  .180000445  .190000243  .200000133
.150002404  .160001294  .170000698  .180000377  .190000204  .200000111
.150002091  .160001116  .170000597  .180000320  .190000172  .200000093
.150001818  .160000962  .170000510  .180000271  .190000144  .200000077
.150001581  .160000829  .170000436  .180000230  .190000121  .200000064
.150001375  .160000715  .170000372  .180000195  .190000102  .200000054
.150001195  .160000616  .170000318  .180000165  .190000086  .200000045
.150001039  .160000531  .170000272  .180000140  .190000072  .200000037
.150000904  .160000458  .170000233  .180000118  .190000061  .200000031
.150000786  .160000395  .170000199  .180000100  .190000051  .200000026
.150000683  .160000340  .170000170  .180000085  .190000043  .200000022
.150000594  .160000293  .170000145  .190000072  .190000036  .200000018
.150000517  .160000253  .170000124  .180000061  .190000030  .200000015
.150000449  .160000218  .170000106  .180000052  .190000025  .200000013
.150000391  .160000188  .170000091  .180000044  .190000021  .200000010
.150000340  .160000162  .170000078  .180000037  .190000018  .200000009
.150000296  .160000140  .170000066  .180000032  .190000015  .200000007
.150000257  .160000120  .170000057  .180000027  .190000013  .200000006
.150000223  .160000104  .170000048  .180000023  .190000011  .200000005
.150000194  .160000090  .170000041  .180000019  .190000009  .200000004
.150000169  .160000077  .170000035  .180000016  .190000008  .200000004
.150000147  .160000067  .170000030  .180000014  .190000006  .200000003
.150000128  .160000057  .170000026  .180000012  .190000005  .200000002
```

Figure 8–4. (continued).

# THE PLOT SHEET

If you require a graphic representation, you must use the Plot sheet (Figure 8–5).

The first two fields are the same as the first two fields of the Table sheet. The Screen or Printer: field specifies where the graph defined by the sheet will be displayed. If the field is set to Screen, the graph is displayed on the screen in a format compatible with

```
(t) Title: Trane Centravac Selection Summary                        192/!

==================== PLOT SHEET ====================================================
Screen or Printer:         Printer
Title:                     Trane Centravac Selection Summary
Display Scale ON:          Yes
X-Axis:                    price
Y-Axis     Character
------     ---------
des_kw     *
max_tons   o
kw@max     x
$perton    +
```

Figure 8–5. The Plot Sheet.

the screen size, and the printer parameters are ignored. If the field is set to Printer, the graph will be sent to the logical device or file defined on the Global sheet, and the display will be defined by the printer parameters on the Global sheet.

The second field is the Title field, which is used to specify the title of the graph. This field is optional. If you make an entry in the field, TK!Solver will display it at the bottom of the screen or the printed page. As with the title for a table, you can enter up to 200 characters, but the title will be truncated if it is wider than the width of the printed page.

The next field is the Display Scale ON: field. Note in Figure 8–6 that a TK!Solver graph, or plot, is displayed with a pair of axes, one vertical axis on the left side of the graph, and one horizontal axis at the bottom. The axes are always displayed, but the numeric scales are optional.

TK!Solver calculates the scale numbers for you, interpolating between the smallest and largest numbers to be displayed. We'll return to this subject in a moment.

The fourth field is the X-Axis: field. This field is used to specify the name of the list that will supply the values for the horizontal coordinates of each data point. Only one list can be specified for this coordinate. If you specify the name of a nonexistent list, the program will add the name to the List sheet in the Name column.

The X-Axis field is followed by two columns. The first column is the Y-Axis column. This column is used to specify the list that will provide the vertical coordinates of the data points. You can enter a number of lists in this column and plot several lists as functions of the list named in the X-Axis field.

The second column is the Character: column. This column is used to specify what character will be used to plot the data points

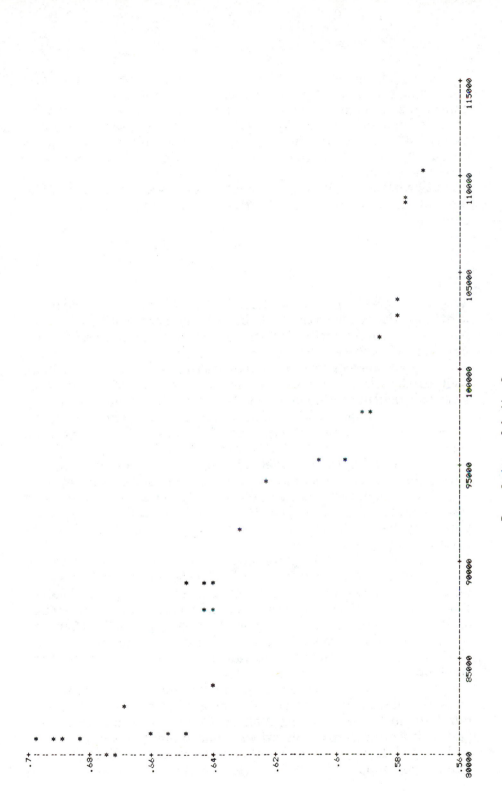

Trane Centravac Selection Summary

Figure 8–6.  A TK!Solver Graph.

for each list named in the Y-Axis column. You can use any character the computer or printer can display, although the characters used as command characters (?/=::;><!) must be entered by means of the Edit command. The program enters the asterisk (*) as a default in this field. To change the * to another character, you must place the cursor in the field to be changed and type the new character.

As with the Table sheet, you produce the actual plot by placing the cursor anywhere on the Plot sheet and typing the Action command

```
!
```

The entire screen will then be cleared, whether or not a second sheet is displayed along with the Plot sheet. If the Screen or Printer field is set to screen, the entire plot will appear on the screen, and remain until a key is struck.

If the Screen or Printer field is set to Printer, the plot will be sent to the device or disk file named in the Printer Device or Filename field of the Global sheet. The plot will be scaled according to the Printed Page Width field of the Global sheet, and the first 80 (or 64 or 40) characters of each line of the plot will appear on the screen as they are sent to the printer or disk file.

Notice in Figure 8–6 that TK!Solver does not plot smooth curves; in fact, it does not plot curves. Plots consist of isolated data points with no interconnecting lines. The program does not make use of the graphic mode of the IBM PC; the advantage here is that plots can be produced on any type of printer, whether it has graphics capabilities or not.

TK!Solver handles plots and tables differently. Whereas tables are printed on as many pages as are necessary to display the entire table, plots are always presented on a single page. TK!Solver scales the plot so that it will fit on a single page, even if that means omitting some of the values. If you attempt to plot a list that contains more values than your printer can print in a single line, including the Y-axis and the scale values, you can rest assured that some of the points will not be plotted.

If you select the screen as the display device, TK!Solver will scale the plot to fit in the 80 columns available. A list of several hundred closely spaced points can take on a ragged appearance, just as it would if only a few widely spaced points were calculated and plotted. If you need to plot more points than the display device allows, you must resort to a little subterfuge—solve the model for as many points as you wish, placing the output in several sequential lists, each as long as the printer can plot. Then plot each list separately, and piece the plot together (see Figure 8-7).

As you will see again in later chapters, TK!Solver has certain limits, but those limits will not prevent you from accomplishing your objectives if you take the same attitude toward those limits as you would toward the limits of any other tool—just use a little imagination when you meet an obstacle.

## THE PRINT COMMAND

Printing tables and graphs is only part of presenting an engineering model, whether it be a TK!Solver model or some other type. You must present the rules, descriptions of variables, units, and so on. Once you have constructed a TK!Solver model, you will have all the information you need to describe the model, and this information will be recorded on the sheets of the model.

The Print command allows you to print the sheets and subsheets that make up the model. When you give the Print command

TK!Solver will prompt you with the message

Print:   Point to last line

followed by the position indicator.

You must then instruct TK!Solver where to stop printing by moving the cursor to the last line to be printed. If you wish to print the entire sheet, be sure to give the Print command with the cursor on the first line of the sheet, because TK!Solver begins printing where the cursor was positioned when the command was entered.

You do not have to move the cursor to the last line with the arrow key, although that method will work. You can enter the Goto command after entering the Print command—that is, you can nest the commands. The quick way to print an entire sheet is to enter the command

| / | P | : | * | ENTER | ENTER |

with the cursor positioned on the first line.

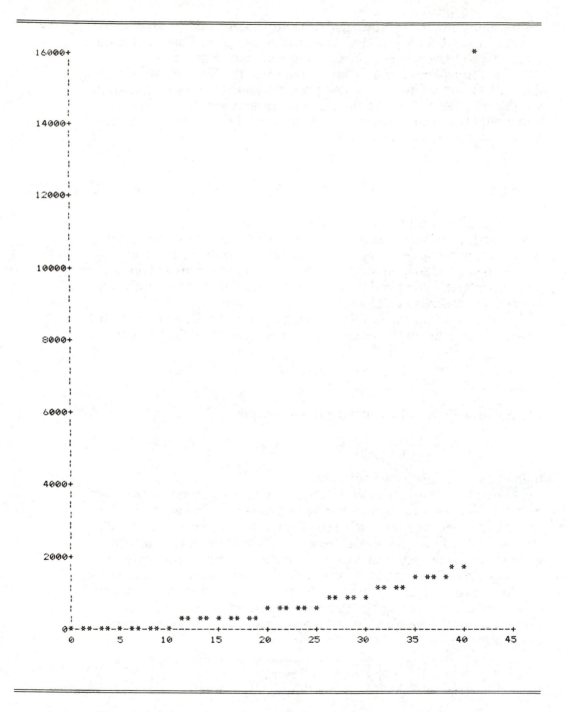

Figure 8–7. Segmented Plots.

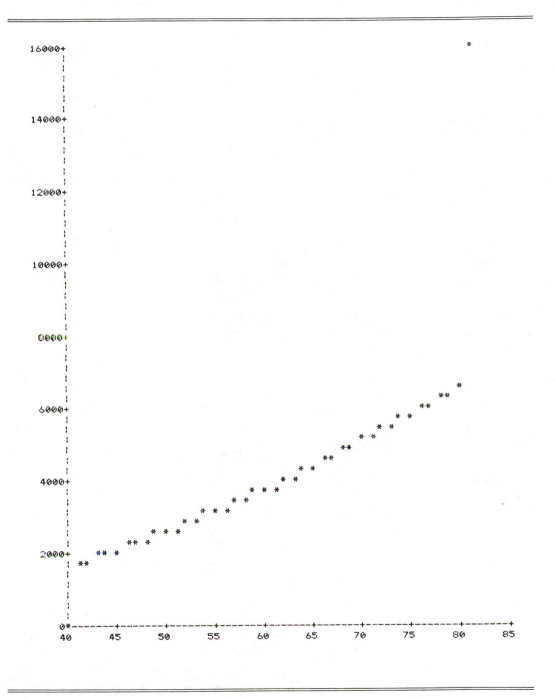

Figure 8–7.  (continued).

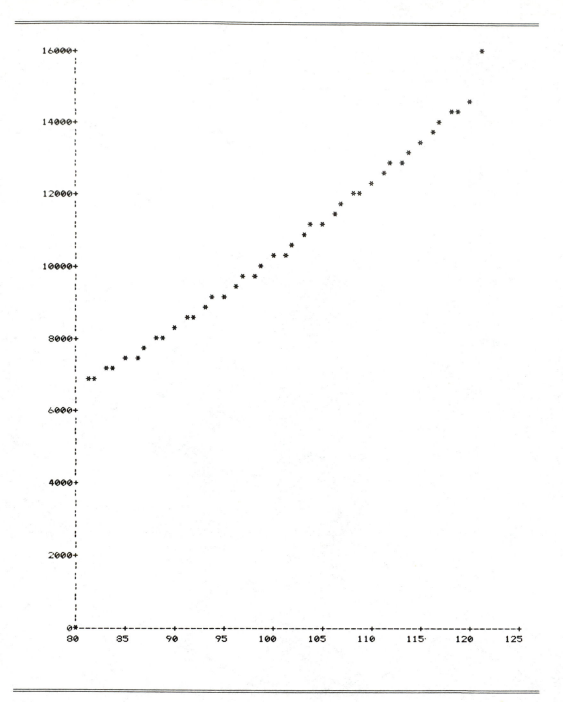

Figure 8–7.  (continued).

| St | Input | Name | Output | Unit | Comment |
|----|-------|------|--------|------|---------|
| L  |       | y1   | 6      |      |         |
|    | 1     | a    |        |      |         |
| L  | 1     | x1   |        |      |         |
|    | 2     | b    |        |      |         |
|    | 3     | c    |        |      |         |
| L  |       | y2   | 6      |      |         |
| L  | 1     | x2   |        |      |         |
| L  |       | y3   | 6      |      |         |
| L  | 1     | x3   |        |      |         |
| L  |       | x    |        |      |         |

| S | Rule |
|---|------|
| - | ---- |
|   | y1=a*x1^2+b*x1+c |
|   | y2=a*x2^2+b*x2+c |
|   | y3=a*x3^2+b*x3+c |

| Name | Elements | Unit | Comment |
|------|----------|------|---------|
| y1   | 42       |      |         |
| x1   | 42       |      |         |
| y2   | 42       |      |         |
| x2   | 42       |      |         |
| y3   | 42       |      |         |
| x3   | 42       |      |         |
| x    | 122      |      |         |

Figure 8–7. (continued).

To print all the sheets of a model, you must display each sheet or subsheet in succession, and enter the Print command from each one. The end result, assuming that you have included a few comments, is a complete description of the model.

# CHAPTER 9

# Functions

Up to this point, the discussion has concentrated on the operations you can perform with standard arithmetic operators—that is to say, you have only added, subtracted, multiplied, divided, and raised to a power. A serious engineering tool must offer more capabilities. Fortunately, TK!Solver provides exactly what you need.

TK!Solver enables you to use two types of functions: built-in functions and user-defined functions. Both types of functions allow you to build some powerful models. First, let's look at the built-in functions.

## BUILT-IN FUNCTIONS

The built-in functions can be subdivided into two groups: mathematical functions and TK!Solver functions. Since the mathematical functions are standard functions, we will merely list them and note any constraints imposed by the program. The functions are shown in capital letters, and the form of the function's argument is shown in small letters in parentheses. Then we will discuss the TK!Solver functions in detail.

All functions in the TK!Solver program follow the format

    function(argument, argument,...)

The program distinguishes functions from variable names by the absence of an operator between the function name and the parenthesis; the presence of a space between the function name and the

left parenthesis has no effect. Some functions accept a null argument—the function PI( ) returns the value of pi.

A function can be used as an element of an expression, or as
the argument of another function. However, you must still follow
the rules of the sheet at hand—the evaluated function call must
provide an acceptable value. For instance, if you wish to use the
value of pi as an input, you can place the cursor in the Input field of
the Variable sheet and type

3.14159..... &lt;ENTER&gt;

and enter the value to the accuracy of memory. Or, you can type

PI( )

and enter the value to an accuracy of ten decimal places.

Generally speaking, a function call can be used in a rule
wherever a variable name can be used. The program evaluates the
function just as it would a variable.

## Mathematical Functions

| | |
|---|---|
| ABS(x) | Returns the absolute value of the numeric value. |
| ACOS(x) | For $-1 \le x \le +1$, returns the inverse cosine of x in radians, between 0 and pi. |
| ACOSH(x) | For $x \ge 1$, returns the inverse hyperbolic cosine of x. |
| ASIN(x) | For $-1 \le x \le +1$, returns the inverse sine of x in radians, between $-pi/2$ and $pi/2$. |
| ASINH(x) | Returns the inverse hyperbolic sine of x. |
| ATAN(x) | Returns the inverse tangent of x in radians, bet-ween $-pi/2$ and $pi/2$. |
| ATANH(x) | For $-1 \le x \le +1$, returns the inverse hyperbolic tangent of x. |
| COS(x) | For x specified in radians, returns the cosine of x. |
| COSH(x) | Returns the hyperbolic cosine of x. |
| EXP(x) | Raises e(base of natural logarithms) to the x power. |
| LN(x) | For $x > 0$, returns the natural logarithm of x. |
| LOG(x) | For $x > 0$, returns the base 10 logarithm of x. |
| MOD(x1,x2) | For $x1,x2 > 0$, returns the remainder of x1/x2. |

SIN(x)          For x specified in radians, returns the sine of x.

SINH(x)         Returns the hyperbolic sine of x.

TAN(x)          Returns the tangent of x.

TANH(x)         Returns the hyperbolic tangent of x.

# TK!Solver Functions

Since the remaining built-in functions are called TK!Solver functions in the manual, they are referred to in the same way here. Most of them are familiar functions, but the names are not necessarily standard. Let's look at each in turn. Again, the name of each function is spelled out in capital letters, and the argument, if any, is indicated by small letters in parentheses. In addition to x, generally a number, arguments may comprise a series of numbers or expressions, as well as names of lists.

A series can be either a series of numeric values, or a series of variable names, and a list name can be the name of a list, or a variable that produces the name of a list as its value. The elements of a series must be separated by commas, and must follow other applicable rules—allowable characters for variable names, real numbers for numeric values, and so on. A simple use of a function might assign the largest value of a series of variables to another variable:

LARGEST = MAX(a,b,c,d,e)

or the smallest:

SMALLEST = MIN(a,b,c,−1)

APPLY('function, domain)

This function tells TK!Solver to apply a user-defined function, which we will examine later, to the value specified as domain. The term *domain* indicates that the second argument must fall within the domain of the function named in the first argument.

ATAN2(x1,x2)

This function returns the tangent of a number, as does ATAN(x). However, the number is specified as two arguments, and TK!Solver returns the correct value by quadrant. You can obtain values between 0 and 2*pi, rather than between −pi/2 and +pi/2, as with ATAN.

COUNT(series) or COUNT('listname)

This function is used to count the elements of a list or series.

DOT('listname1,'listname2) or DOT('listname1,series)

This function is used to find the dot product of two lists, or of one list and a series of values. The program multiplies each element of the first argument ('listname1) by the corresponding element of the second argument ('listname2 or a series of values), and then sums the individual products. It returns the sum.

E( )

This function relieves you of the task of looking up the value of e, the base of natural logarithms. The value returned is 2.7182818285, which is an approximation.

ELEMENT( ) or
ELEMENT('listname, element number) or
ELEMENT('listname, element number, x)

This function is used to obtain the value of an element of a list. If you use the first form, ELEMENT( ), the program returns the number of the element the List Solver is processing. If the List Solver is not processing a list, the function returns 0.

The second form, ELEMENT('listname, element number), is used to obtain the value of the nth element of the list named as 'listname.

The third form, ELEMENT('listname, element number, x), is used to ensure that a value is returned if the element specified by element number does not exist; x is a default function value.

GIVEN(variables, x1,x2)

You can use this function to determine whether a group of variables has been assigned input values. The variables can be symbolic constants or a series of expressions that will evaluate to a series of symbolic constants. Then you list two values, x1 and x2. The function will return x1 if all the variables named have been assigned input values, and x2 if one or more variables have not been assigned input values.

## INT(x)

This function produces integer values from real arguments. Since the function truncates everything to the fractional part of a real number, it rounds down for positive reals, and up for negative reals. Watch this function, as it will round 5.99999 to 5, not to 6, as you may wish at times.

## MAX(series) or MAX('listname)

This function is used to obtain the largest value in a list or series of values. Notice that it returns the actual value, not the position in the list of the largest value. If the series or list used as an argument contains symbolic values, they are ignored—only numeric values are considered.

## MIN(series) or MIN('listname)

You would use this function just as you would use the MAX function, but to find the smallest value rather than the largest.

## NPV(x, series) or NPV(x, 'listname)

This function is used to find the Net Present Value of a series of cashflow values, which will be specified as a series of values or as a list by name, at an interest rate of x. If you list a series of values rather than a list name, the program takes the first number to be the interest rate and the remaining values to be payments, beginning with the first payment.

## PI( )

Like E, this function is used to avoid memorizing important numbers. PI returns an approximate value of pi, which is accurate to ten decimal places.

### POLY(x, series) or POLY(x, 'listname)

This function enables you to evaluate a polynomial expression in a single variable. The value of the variable is specified as the first argument, $x$. You can specify the coefficients of the various terms by listing them in a series, as in the first form given, or by naming a list that contains the coefficients.

*Example:*

Evaluate $x^3 + 3x^2 + 4x + 2$ for $x = 3$, using POLY

VAL = POLY(3,1,3,4,2)     VAL = 68

### SGN(x)

This function is used to determine the sign of a value assigned to a variable. The function returns $-1$ if the value is negative, 0 if the value is 0, and 1 if the value is positive.

### SQRT(x)

You will use this function to find the square root of a number. You must specify a positive argument, not including 0.

### STEP(x1,x2)

This function is used to determine whether one variable has an assigned value that is larger than that assigned to another variable. If $x1 \geq x2$, then STEP(x1,x2) returns 1; if $x1 < x2$, STEP (x1,x2) returns 0.

### SUM(series) or SUM('listname)

You can use this function to find the sum of all the values listed as a series of values, or of all the elements of a list specified by name. If you specify an empty list, the function will return 0.

TK!Solver's built-in functions enable you to perform certain operations that would otherwise be difficult to accomplish. However, the systems engineers are interested in cannot always be modeled with the functions listed above. You will find that many relations must be determined experimentally and cannot be expressed in simple mathematical expressions. TK!Solver has a tool to deal with situations of this type—user functions.

# USER FUNCTIONS

User functions are functions that can be defined to express relationships that cannot be expressed as combinations of the built-in functions. You can specify both the relationship between two variables, and the domain over which the relationship is valid.

If you have used BASIC on your microprocessor, you have seen the term *user function* before. However, TK!Solver user functions are not the same as BASIC user functions. In BASIC, a user function is defined by a series of BASIC statements that perform a series of operations; the user function may or may not include some of BASIC's intrinsic, or built-in, functions.

A TK!Solver user function comprises two lists, which are known as the Range and Domain. In fact, there is a List subsheet for the Range, and one for the Domain. Before discussing them, however, let's look at functions in general, at least as they relate to TK!Solver. TK!Solver uses the following syntax to express functions:

$$f(x) = y$$

The letter $f$ is the name of the function, $x$ is the argument of the function, and $y$ is the value of the function. In TK!Solver's terminology, $x$ is a value from the Domain of the function, and $y$ is a value from the Range of the function. The Domain can be considered the independent variable, expressed as a list, and the Range the dependent variable, expressed as a list.

# THE USER FUNCTION SUBSHEET

When TK!Solver evaluates a built-in function, it performs operations on the argument and returns a value—lists are not involved. When it evaluates a user function, it looks up the specified argument in the Domain, and returns the corresponding value from the Range. The typical User Function subsheet shown in Figure 9–1 illustrates a user function.

You will first note a comment field. It is filled from the comment field of the User Function sheet, which we will examine later. As with all comment fields in TK!Solver models, this field is to record notes, explanations, instructions, and so forth.

The second field is the Domain List: field. Here you enter the name of the list you wish to use as the independent variable—the argument of the function. If you define a name in this field before you enter it on the List sheet, TK!Solver will make the entry on the List sheet for you; of course, the list will be empty until you fill it with values.

The third field is the Mapping: field. To understand this field, you must consider TK!Solver's user function in more detail. As mentioned above, user functions are defined by two related lists. At first glance you might think that these functions are not continuous, that they can only be evaluated for certain values in the Domain. Fortunately, this is not always the case.

There are three options for the Mapping: field:

T   Table Option
S   Step Option
L   Linear Option

The Table option describes the mode of operation you would expect from the definition of user functions. If you use a value that is not listed in the Range or Domain of a function, the program returns an

```
(1n) Name: sch40id                                                    192/!

==================== USER FUNCTION SHEET ====================================
Name        Domain      Mapping Range    Comment
----        ------      ------- -----    -------
sch40id     sch40       Step    sch40    inside dimensions of sch 40 steel pipe
noms40      sch40       Table   noms40   nominal schedule 40 pipe sizes
```

Figure 9–1. The User Function Subsheet.

error. Therefore, if you use the Table option, you define a discrete function; only discrete values are permitted.

If you enter either of the other two options, you can interpolate. If you enter the Step option, you can interpolate between values in the Domain. If you give the function a value that falls between two values in the Domain, the program will return the value from the range that corresponds to the previous value in the Domain—the Domain value that precedes the specified argument. If you were to plot the values returned against the arguments, specifying arguments at very small intervals, the function would appear to be a step function.

If you enter the Linear option, you can specify intermediate values in the domain, and TK!Solver will interpolate values from the Range values. The program uses linear interpolation, just as you would use it to interpolate values manually from the Steam Tables. If you were to plot such a function, it would appear to be a series of data points connected by straight line segments. Thus the Mapping: field lets you instruct TK!Solver how to interpolate when given an intermediate value as an argument.

Interpolation is important because it affects the way in which you can solve models. Earlier it was noted that TK!Solver can use the Direct Solver only under certain circumstances. The Direct Solver can solve for variables that appear as the arguments of functions, but only for functions that have unique inverses. If you have a model defined by the rule

$$y = ABS(x) + 2 * c$$

you can solve for y or c, but not for x; the reason is that either a positive or negative value of x would be a valid solution, and TK!Solver has no way to tell which is correct.

If you define a user function with the Linear option, you have a function that meets the Direct Solver's constraints. The program can interpolate to find a unique value whether it is solving for a value in the Range or the Domain. However, if you define the function with the Step option, you can only interpolate values in the Domain.

The fourth field is the Range List: field. This field is used to specify the list that will supply the values for the Range, that is, for the dependent variable.

After the four single fields, you will find three columns. First is the Element column, which gives the position of the values in the next two columns, relative to the beginning of the lists. You cannot move the cursor to the Element field; the program enters numbers in the column for you as you enter values in the next two columns.

Second is the Domain column. In this column you enter the values for the Domain elements. You can enter the values directly on the User Function subsheet or on the List subsheet of the same name. You can also fill this list with the Action command or the List Solver.

The third column is the Range column. You can fill this column in the same way that you fill the Domain column. We will examine a user function more closely after we examine the User Function sheet.

# THE USER FUNCTION SHEET

The User Function sheet summarizes the functions defined for the model in the same way that the List sheet summarizes the lists defined for the model. For each User Function subsheet, there is a group of entries on the User Function sheet (see Figure 9–2).

The User Function sheet comprises five columns. Each row on the sheet corresponds to a User Function subsheet.

The first column is the Name: column. In this field you enter the name of a User Function subsheet that defines the user function. You follow the same rules for user function names as for variable names, but you cannot use names of built-in functions, and each user function name must be unique. Note that while TK!Solver distinguishes between upper and lower case for variable and user function names, it does not for built-in functions. You cannot use sin, Sin, SIn, sIN, or siN for a user function because the program considers all those spellings equivalent to SIN, which is a built-in function.

The second column is the Domain: column. This is the same entry as the Domain field of the subsheet. If you enter it here first, it will appear automatically on the subsheet; if you enter it on the subsheet, it will appear automatically on this sheet.

In fact, the next three columns duplicate entries on the subsheets for the functions. You can make entries on the User Function sheet and automatically create entries on the subsheet, or you can make them on the subsheets and automatically create entries on the User Function sheet.

You can display the User Function sheet on the screen with the Select command

| = | | F |

```
(1d) Domain: 1.8855                                          192/!

==================== USER FUNCTION: noms40 ========================================
Comment:                        nominal schedule 40 pipe sizes
Domain List:                    sch40
Mapping:                        Table
Range List:                     noms40
Element Domain        Range
------- ------        -----
   1     1.8855       'twenty_fou
   2     1.567833333  'twenty_in
   3     1.406166667  'eighteen_i
   4     1.25         'sixteen_in
   5     1.09375      'fourteen_i
   6     .9948333333  'twelve_in
   7     .835         'ten_in
   8     .6650833333  'eight_in
   9     .5054166667  'six_in
  10     .4205833333  'five_in
  11     .3355        'four_in
  12     .2956666667  'three_$_ha
  13     .2556666667  'three_in
  14     .20575       'two_$_half
  15     .17225       'two_in
```

Figure 9-2. The User Function Sheet.

Then, to bring up a specific User Function subsheet, place the cursor in the function's row, and use the Dive command (©) just as you used it to reach the other subsheets, the Variable and List subsheets, from their respective sheets.

When do you need user functions? Not to construct simple models—simple Newtonian mechanics can be modeled with the ordinary operators and built-in functions, and so can many other processes. But real world engineering models beg us to include user functions.

The most obvious occasion for a user function is the application of relations determined experimentally. You take measurements, enter the measured values into TK!Solver lists, and associate pairs of lists as user functions.

You can also represent functions that are expressed by different equations over different intervals. For instance, the equations of state of some materials cannot be represented as a single equation. Instead, you must use a series of equations for the different temperature ranges. In some cases, there may be transitional regions where only experimental data are available.

You can also represent functions that are piecewise continuous with user functions. In one region the function may be a curve, in another it may be a straight line; and, there may be discontinuities between the various segments. Of course, you can construct such lists with the use of other TK!Solver facilities—the Block option of the List Solver, the built-in functions, and so on.

User functions are not necessarily limited to handling numbers. You can enter symbolic values and construct user functions that return names, formulas (when given a numeric argument), or numbers (when given a symbolic argument).

You can use the Apply built-in function in conjunction with a user function that returns names of other user functions. Figure 9–3 illustrates the use of the Apply function. You enter a shape and a dimension as inputs, and enter the Action command from the Rule sheet or the Variable sheet. The inner user function, area, returns a symbolic value, which happens to be the name of another user function. The Apply function then evaluates that function with the value of dim as its argument.

Notice that the first argument of the Apply function must evaluate to the name of a user function. If you try to construct a user function that returns the name of a built-in function for the argument of Apply, you will receive an error message.

TK!Solver's built-in functions do come in handy when you fill lists for user functions. You can type an expression for a list entry using the built-in functions, and TK!Solver will evaluate the expression and place its value in the list.

```
(2i) Input: 4                                                        202/

===================== VARIABLE  SHEET ===================================
St Input       Name      Output     Unit     Comment
-- -----       ----      ------     ----     -------
L              Asquare   16
L   4          d
L              Acircle   12.566371
               Area      16
      'square  shape


==================== RULE SHEET =========================================
S Rule
- ----
  Asquare = d^2
  Acircle = pi()*(d/2)^2

  Area = APPLY(Ashape(shape),d)
```

Figure 9–3.  User Functions and the Apply function.

# CHAPTER 10

# Communicating with the World

Now that you have examined all the sheets that can be part of a TK!Solver model, you have the tools to create some powerful models, indeed. You know how to create, use, and store models, and you know how to operate TK!Solver so that results can be presented as graphs and tables, both on the screen and the printer. However, those models have only been considered in relation to the TK!Solver program.

You will recall that tables and graphs can be sent to a disk file so that they can be printed later. Figures 10-1 and 10-2 have been prepared just that way. Figure 10-1 is a graph produced by TK!Solver and sent to a disk file called NPWRGRAF.PRF, rather than the printer, by entering NPWRGRAF in the Printer Device or Filename field of the Global sheet and then giving the Action command from the Plot sheet. Figure 10-2 was prepared in the same way, except that the Printer Device Name or Filename field was changed to NPWRTABL, and the Action command was given from the Table sheet. Remember that the Action command has different effects on different sheets.

You will find useful TK!Solver's capability to send a plot or a table to a disk file because the display can be incorporated into a report with the aid of a word processing program running on the same computer. Incidentally, when you send a plot or table to a disk file and read it later with the word processor, you will notice some strange characters at the end of the file. You can simply delete them, and everything will be fine—the plot or table will be left intact.

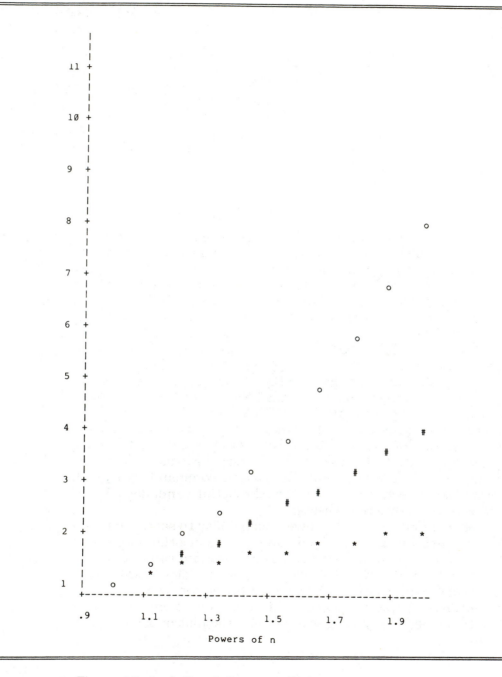

Figure 10–1. A Graph Store on Disk.

| List n | 1st power | 2nd power | 3rd power |
|---|---|---|---|
| 1 | 1 | 1 | 1 |
| 1.11111111 | 1.11111111 | 1.23456790 | 1.37174211 |
| 1.22222222 | 1.22222222 | 1.49382716 | 1.82578875 |
| 1.33333333 | 1.33333333 | 1.77777778 | 2.37037037 |
| 1.44444444 | 1.44444444 | 2.08641975 | 3.01371742 |
| 1.55555556 | 1.55555556 | 2.41975309 | 3.76406036 |
| 1.66666667 | 1.66666667 | 2.77777778 | 4.62962963 |
| 1.77777778 | 1.77777778 | 3.16049383 | 5.61865569 |
| 1.88888889 | 1.88888889 | 3.56790124 | 6.73936900 |
| 2 | 2 | 4 | 8 |

Figure 10-2. A Table Stored on Disk.

# THE DIF™ FILE

Occasionally you may want to interface your TK!Solver models with other programs, such as VisiCalc or one of its relatives. You can do so through the DIF format for data exchange between applications programs. Earlier you learned that one of the options of the Storage command is the # option. The # option enables you to save or load DIF files, which are data files that can be created or read by other programs. Notice that they are *data* files, not the files that make up the TK!Solver models. The files with the TK file extension are not portable.

Before delving into the DIF format, let's consider why such a convention is useful. Different programs and different languages read and write data differently. Some data base management systems use fixed-length fields and records, others use fields and records of varying lengths separated by special characters. High-level language compilers and interpreters have similar differences. Some languages and programs will recognize either type of data organization, but have to be told what to expect.

If you have two such dissimilar programs and want to use the output of one as the input for the other, you have a problem. You will have to retype the output of the first program as the input of the second, or perhaps you can write a program to convert the output of the first program so that it can be read by the second. If the two programs were your own, you could modify one or both so that they would use the same format, but if you are using a program such as TK!Solver, there is little chance of doing so successfully.

Software Arts, Inc. has tried to alleviate the situation by formulating and promoting DIF. Appendix C illustrates their success. Of course, VisiCalc and TK!Solver are products of Software Arts, Inc., but a considerable number of VisiCalc's competitors are on the list.

When TK!Solver records data in a DIF file, the data are organized as a table similar to the one produced by sending a table to disk with the Action command from the Table sheet. However, a DIF file includes additional information describing the data. The file is divided into two sections: the header section, which describes the table; and the data section, which actually contains the data.

A DIF file corresponds to a table with rows and columns, but rows are called "tuples," and the columns are called "vectors" (Figure 10–3). The Time values are a vector, as are the temperature, pressure, and density columns of values. A set of readings, composed of a time value, a temperature value, a pressure value, and a density value form a tuple.

Now let's consider how the table in Figure 10–3 would be described as a DIF file. The first section of the file will describe the structure of the table, including the titles of the vectors and of the table as a whole. Figure 10–4 shows the first section of the DIF file for the table, the Header Section. This Header Section contains eight items consisting of four fields each.

The first header item comprises the first three lines: TABLE, 0,1, "Temperature, Pressure, and Density History." The next three lines make up the second header item, and so on. The general structure of a header item is shown below:

Line 1:  Topic

Line 2:  Vector Number, Value

Line 3:  "String"

In Figure 10–4, the words TABLE, VECTORS, TUPLES, LABEL, and DATA are topics. TABLE refers to the table as a whole, VECTORS to the columns, TUPLES to the rows, LABEL to the column headings, and DATA to the actual data the table contains. The meaning of the second line of each item varies with the topic. Vector number refers to the vector, or column, to which the item refers. Of course, the topic TABLE does not refer to a specific vector, and in that case the DIF standard requires a 0 entry. The Value entry in the TABLE entry is the version number, which is 1. The "String" is the constant you want to use as the title of the table.

Temperature, Pressure, and Density History

| Time | Temperature | Pressure | Density |
|------|-------------|----------|---------|
| 1:00 | 100         | 14.7     | .075    |
| 1:15 | 110         | 14.8     | .075    |
| 1:30 | 115         | 14.9     | .075    |
| 1:45 | 118         | 14.9     | .075    |

Figure 10–3.  A DIF™ Table.

The two items, VECTORS and TUPLES, define the dimensions of the table. There is a 0 entry for the Vector Number field of each item, because neither refers to a specific vector. The Value field defines the number of vectors and tuples of data. Note that the column headings are not included in the tuple count, but are

```
TABLE
0,1
"Temperature, Pressure, and Density History"
VECTORS
0,4
""
TUPLES
0,3
""
LABEL
1,0
"Time"
LABEL
2,0
"Temperature"
LABEL
3,0
"Pressure"
LABEL
4,0
"Density"
DATA
0,0
""
```

Figure 10–4.  DIF™ File Header Section.

defined by the LABEL entries. Both the VECTORS and TUPLES items refer only to the data.

The LABEL items define the column headings. Since headings are associated with specific vectors, each LABEL item has a different number in the Vector Number field. The labels here occupy one line, and are given the value of 0; a 1 is equivalent to 0 in this case. If you wanted a two-line heading for each column, you could enter two LABEL items for each vector, each having the same entry in the Vector Number field, and the numbers 1 and 2 in their respective Value fields.

The last header item is always the DATA item. It indicates that the items that follow are the actual data. Since the item refers to the table as a whole and not to a specific vector, there is a 0 entry in the Vector Number field. The 0 entry in the value field, and the null string value are prescribed by the DIF standard.

Four header items are required for every DIF file: the TABLE, VECTORS, TUPLES, and DATA items. The LABEL items are optional, since not all programs use headings. As noted earlier, the tables constructed by TK!Solver are not required to have headings.

The second section of the DIF file is the Data Section (Figure 10–5). Each item in the Data Section consists of two lines, and is organized as follows:

Line 1:   Type Indicator, Number Value
Line 2:   String Value

All of the data are numeric, as indicated in each case by a type indicator of 0. Other possible type indicators are $-1$, for a special data value, and 1, for string data.

Some $-1$ Type indicators can be seen in Figure 10–5; they indicate the organization of the actual data values. If the type indicator is $-1$, the string value can have one of two possible values: BOT, for beginning of tuple; or EOD, for end of data. When the type indicator is $-1$, the number value is always 0.

Now that you understand the structure of DIF files, let's consider how you might put them to use.

One of the most powerful aspects of TK!Solver is its Iterative Solver. As noted earlier, many engineering problems cannot be solved directly and numerical methods involving iteration must be used. Given a list or lists of input, TK!Solver can produce a list or table of solutions to problems of this type using the Iterative Solver. However, if you want to produce a graph from the lists, TK!Solver can only achieve the resolution of the most dense

```
A>Ftype powersn.dif     V            0,64         -1,0
TABLE                   0,1          V            BOT
0,1                     V            -1,0         0,8
" "                     -1,0         BOT          V
VECTORS                 BOT          0,5          0,8
0,4                     0,2          V            V
" "                     V            0,5          0,64
TUPLES                  0,2          V            V
0,10                    V            0,25         0,512
" "                     0,4          V            V
LABEL                   V            0,125        -1,0
1,0                     0,8          V            BOT
"n"                     V            -1,0         0,9
LABEL                   -1,0         BOT          V
2,0                     BOT          0,6          0,9
"a"                     0,3          V            V
LABEL                   V            0,6          0,81
3,0                     0,3          V            V
"b"                     V            0,36         0,729
LABEL                   0,9          V            V
4,0                     V            0,216        -1,0
"c"                     0,27         V            BOT
DATA                    V            -1,0         0,10
0,0                     -1,0         BOT          V
" "                     BOT          0,7          0,10
-1,0                    0,4          V            V
BOT                     V            0,7          0,100
0,1                     0,4          V            V
V                       V            0,49         0,1000
0,1                     0,16         V            V
V                       V            0,343        -1,0
0,1                                  V            EOD
```

Figure 10–5. DIF™ File Data Section.

character pitch supported by the printer, usually about 16 characters per inch.

If you instruct TK!Solver to save its output in DIF format, you can produce a graph with much finer resolution using another program, such as VisiTrend/Plot® [1], to drive a graphics printer or

[1] VisiTrend is a registered trademark of VisiCorp.

plotter. By doing so, you may be able to produce a graph with a resolution of 600 to 1000 dots per inch.

On the other hand, you may use a FORTRAN program to gather data from a set of laboratory instruments, and record the data values on disk in the DIF format. Then you can read the data into a TK!Solver model using the /S#L command, and perform all the calculations you wish under TK!Solver.

You can run a simulation of a process using a TK!Solver model and store the input and output lists in DIF format. Then your process control program can read the data as inputs, and you will be able to compare actual behavior with the simulation. Appendix B demonstrates how a DIF file produced by a TK!Solver model can be read into an array in memory with another program.

You can also build an input data base in the DIF format using a data base management system. TK!Solver can read the DIF file and proceed with the solution of the model. Perhaps even more useful is its ability to generate a data base in the form of lists and tables, and then to sort, index, and retrieve the data.

In fact, you can use the DIF file to advantage even with programs that do not make explicit use of the DIF file format. Since the DIF file is a text file, you can load it with a word processor, just as you would any text file. You can create tables, lists, charts, by editing the DIF file.

## Saving the DIF™ File

You can construct a DIF file from a TK!Solver model with the DIF Save option of the Storage command. The command is entered as

At this point, TK!Solver asks a question it does not ask with other options:

DIF  Save:    Point to last list

Unlike the other options of the Save command, the DIF Save option is only available from the List sheet. In addition, when the command is given, the cursor must be positioned at the first list to be saved. When the prompt for the last list appears, you move the cursor to the last list using the arrow keys, or the Goto and Switch commands. When the cursor is positioned on the list, you indicate the last list by pressing the **ENTER** key.

After you indicate the last list, TK!Solver will prompt you for the name of the file you wish to use for the DIF file. As with the other options, you type the disk drive letter if you wish to save the file on a disk other than the default drive, and the file name; TK!Solver adds the file extension, DIF in this case. You can use Directory Scrolling with the option, as with the other Storage command options.

Saving a DIF file is somewhat slower than saving lists with the User Function Save option, and there is no real benefit in saving lists as DIF files, other than to transfer them to other programs. If you wish to exchange lists between TK!Solver models, the User Function Save option is the one to use.

## Loading the DIF File

If you have created a DIF file with another program such as VisiCalc, you can load it into a TK!Solver model with the DIF Load option of the Storage command. The manual states that the DIF Load option can only be used from the List sheet, as is true for the DIF Save option. However, at least under some conditions, the DIF Load option can be entered from other sheets.

The command for the DIF Load option is similar to the Save option:

| / | S | # | L |

TK!Solver immediately prompts you for the name of the file to load, with this message:

DIF Load:    Filename:

You can use Directory Scrolling with this option, as you can with other Storage command options.

If you load a DIF file into an empty List sheet, TK!Solver enters the names of the lists the DIF file contains, in the Name field of the List sheet. It also enters the number of elements in each list in the Element field, and the Display unit in the Unit field. If you load a DIF file into a model that already contains lists, the entries for the lists of the DIF file are appended at the end of the List sheet.

If you load a DIF file a second time, or if you load a DIF file that contains lists with names that duplicate those of lists already in the model, TK!Solver stores the lists, but leaves the Name field empty.

You can Dive to an unnamed list, add elements, change elements, and delete elements. You can return to the List sheet after Diving to an unnamed list, and you can Select other sheets, and return. There are apparently no ill effects from having lists without names. Of course, you can't reference those lists, or associate them with variables until you assign names to them.

This situation differs from that of loading a TK file over a model. If you load a TK file over a model, and that file contains variables whose names duplicate variables already in memory, the new values replace the existing values. It pays to be careful when loading files over existing models.

# VISICALC®

Transferring TK!Solver DIF files to and from VisiCalc worksheets is second nature to TK!Solver users—the Storage command syntax of VisiCalc is almost identical to that of TK!Solver. However, there are a few differences.

TK!Solver associates a name with each list, because you need to be able to associate lists and variables. VisiCalc does not work with lists in the same way. Instead, VisiCalc works with a table of "cells," which can contain numbers, formulas, or text in any order.

Whereas TK!Solver saves one or more lists from the List sheet, along with their names, VisiCalc saves an area of the worksheet. The position of the cursor when the command is entered determines the upper left corner of the area to be saved, and VisiCalc prompts for the lower right corner.

In a TK!Solver DIF file, a tuple is a list; in a VisiCalc DIF file, either a row or a column can be a tuple. Thus, VisiCalc asks whether a DIF file is to be saved or loaded in Row or Column order. The command to save a DIF file from VisiCalc goes like this:

Data save: File for Saving

filename <ENTER>

Data save: R C or ENTER

<ENTER>

**and the command to load a DIF file is**

Data load: File to Load

filename <ENTER>

Data load: R C or ENTER

<ENTER>

If you load a TK!Solver DIF file into a VisiCalc worksheet, VisiCalc will ignore the list names that TK!Solver stores with its lists. Similarly, if you load a VisiCalc DIF file into a TK!Solver model, there will be no list names, because VisiCalc does not save lists per se.

Consider, for example, the VisiCalc worksheet shown in Figure 10–6. Here you would save the worksheet in row order with the command

Then you would quit VisiCalc, load TK!Solver, and load the file, CHILLROW.DIF with the command

When TK!Solver has finished loading CHILLROW.DIF, the List sheet will show 15 unnamed lists of 7 entries each. The first and third lists are empty, because those rows were empty on the worksheet. The second list contains the labels from the second row of the worksheet, and the remaining lists contain the successive rows of numbers from the worksheet, as shown in Figure 10–7. A worksheet saved in this manner is not very useful in a TK!Solver model.

∧1

| | A | B | C | D | E | F | G | H |
|---|---|---|---|---|---|---|---|---|
| 1 | | | | | | | | |
| 2 | KW | KW/TON | EVAP PD | COND PD | MAX TON | KW@MAX | PRICE | |
| 3 | | | | | | | | |
| 4 | 261 | .696 | 19.5 | 16 | 436 | 322 | 80800 | |
| 5 | 259 | .691 | 19.5 | 16 | 416 | 319 | 80800 | |
| 6 | 258 | .688 | 19.5 | 16 | 437 | 338 | 80800 | |
| 7 | 256 | .683 | 19.5 | 16 | 435 | 322 | 80800 | |
| 8 | 253 | .675 | 19.5 | 13 | 416 | 312 | 80100 | |
| 9 | 253 | .675 | 19.5 | 13 | 437 | 330 | 80100 | |
| 10 | 252 | .672 | 19.5 | 13 | 436 | 315 | 80100 | |
| 11 | 251 | .669 | 19.5 | 10.6 | 420 | 313 | 82400 | |
| 12 | 247 | .659 | 19.5 | 19.8 | 417 | 304 | 81150 | |
| 13 | 247 | .659 | 19.5 | 19.8 | 435 | 302 | 81150 | |
| 14 | 245 | .653 | 19.5 | 19.7 | 436 | 315 | 81150 | |
| 15 | 243 | .648 | 19.5 | 19.7 | 491 | 304 | 81150 | |
| 16 | | | | | | | | |
| 17 | | | | | | | | |
| 18 | | | | | | | | |
| 19 | | | | | | | | |
| 20 | | | | | | | | |
| 21 | | | | | | | | |

Figure 10–6. A VisiCalc® Worksheet.

On the other hand, if you save the worksheet of Figure 10–6 in column order with the command

and then load the file into a TK!Solver model with the same command, you obtain a List sheet and List subsheets, as shown in Figure 10-8. This time you have 7 lists of 15 elements each, and with the exception of the two blank elements and the symbolic value, you have a group of useful lists. You can edit these lists to remove the blank elements and symbolic values, assign names, and use them in models. In fact, you can process the lists without editing them, with the Block option of the List Solver—after you assign a name to each list.

# OTHER DIF™–COMPATIBLE PROGRAMS

Several programs use the DIF file for data interchange, and each one has its own commands. Donald H. Beil's book, *The DIF™ File*,

```
type chillrow.dif      0,245              "   EVAP PD"
TABLE                  V                  1,0
0,1                    0,243              " "
" "                    V                  0,19.5
VECTORS                -1,0               V
0,15                   BOT                0,19.5
" "                    1,0                V
TUPLES                 " "                0,19.5
0,7                    1,0                V
" "                    "   KW/TON"        0,19.5
DATA                   1,0                V
0,0                    " "                0,19.5
" "                    0,.696             V
-1,0                   V                  0,19.5
BOT                    0,.691             V
1,0                    V                  0,19.5
" "                    0,.688             V
1,0                    V                  0,19.5
"         KW"          0,.683             V
1,0                    V                  0,19.5
" "                    0,.675             V
0,261                  V                  0,19.5
V                      0,.675             V
0,259                  V                  0,19.5
V                      0,.672             V
0,258                  V                  0,19.5
V                      0,.669             V
0,256                  V                  -1,0
V                      0,.659             BOT
0,253                  V                  1,0
V                      0,.659             " "
0,253                  V                  1,0
V                      0,.653             "   COND PD"
0,252                  V                  1,0
V                      0,.648             " "
0,251                  V                  0,16
V                      -1,0               V
0,247                  BOT                0,16
V                      1,0                V
0,247                  " "                0,16
V                      1,0                V
```

Figure 10–7.  Lists from a Worksheet Saved in Row Order.

```
0,16                    V                       -1,0
V                       0,417                   BOT
0,13                    V                       1,0
V                       0,435                   " "
0,13                    V                       1,0
V                       0,436                   "      PRICE"
0,13                    V                       1,0
V                       0,491                   " "
0,10.6                  V                       0,80800
V                       -1,0                    V
0,19.8                  BOT                     0,80800
V                       1,0                     V
0,19.8                  " "                     0,80800
V                       1,0                     V
0,19.7                  "      KW@MAX"          0,80800
V                       1,0                     V
0,19.7                  " "                     0,80100
V                       0,322                   V
-1,0                    V                       0,80100
BOT                     0,319                   V
1,0                     V                       0,80100
" "                     0,338                   V
1,0                     V                       0,82400
"   MAX TON"            0,322                   V
1,0                     V                       0,81150
" "                     0,312                   V
0,436                   V                       0,81150
V                       0,330                   V
0,416                   V                       0,81150
V                       0,315                   V
0,437                   V                       0,81150
V                       0,313                   V
0,435                   V                       -1,0
V                       0,304                   EOD
0,416                   V
V                       0,302
0,437                   V
V                       0,315
0,436                   V
V                       0,304
0,420                   V
```

Figure 10-7. (continued).

| Name | Elements | Unit | Comment' |
|------|----------|------|----------|
|      | 15       |      |          |
|      | 15       |      |          |
|      | 15       |      |          |
|      | 15       |      |          |
|      | 15       |      |          |
|      | 15       |      |          |
|      | 15       |      |          |

LIST:
| Element | Value |
|---------|-------|
| 2       | 'KW   |
| 4       | 261   |
| 5       | 259   |
| 6       | 258   |
| 7       | 256   |
| 8       | 253   |
| 9       | 253   |
| 10      | 252   |
| 11      | 251   |
| 12      | 247   |
| 13      | 247   |
| 14      | 245   |
| 15      | 243   |

LIST:
| Element | Value |
|---------|-------|
| 2       | 'KW   |
| 4       | .696  |
| 5       | .691  |
| 6       | .688  |
| 7       | .683  |
| 8       | .675  |
| 9       | .675  |
| 10      | .672  |
| 11      | .669  |
| 12      | .659  |
| 13      | .659  |
| 14      | .653  |
| 15      | .648  |

LIST:
| Element | Value  |
|---------|--------|
| 2       | 'EVAP  |
| 4       | 19.5   |
| 5       | 19.5   |
| 6       | 19.5   |
| 7       | 19.5   |
| 8       | 19.5   |
| 9       | 19.5   |
| 10      | 19.5   |
| 11      | 19.5   |
| 12      | 19.5   |
| 13      | 19.5   |
| 14      | 19.5   |
| 15      | 19.5   |

LIST:
| Element | Value  |
|---------|--------|
| 2       | 'COND  |
| 4       | 16     |
| 5       | 16     |
| 6       | 16     |
| 7       | 16     |
| 8       | 13     |
| 9       | 13     |
| 10      | 13     |
| 11      | 10.6   |
| 12      | 19.8   |
| 13      | 19.8   |
| 14      | 19.7   |
| 15      | 19.7   |

LIST:
| Element | Value |
|---------|-------|
| 2       | 'MAX  |
| 4       | 436   |
| 5       | 416   |
| 6       | 437   |
| 7       | 435   |
| 8       | 416   |
| 9       | 437   |
| 10      | 436   |
| 11      | 420   |
| 12      | 417   |
| 13      | 435   |
| 14      | 436   |
| 15      | 491   |

LIST:
| Element | Value    |
|---------|----------|
| 2       | 'KW@MAX  |
| 4       | 322      |
| 5       | 319      |
| 6       | 338      |
| 7       | 322      |
| 8       | 312      |
| 9       | 330      |
| 10      | 315      |
| 11      | 313      |
| 12      | 304      |
| 13      | 302      |
| 14      | 315      |
| 15      | 304      |

Figure 10–8.  Lists from a Worksheet Saved in Column Order.

```
LIST:
Element Value
------- -----

2       'PRICE

4       80800
5       80800
6       80800
7       80800
8       80100
9       80100
10      80100
11      82400
12      81150
13      81150
14      81150
15      81150
```

| KW | KWPERTON | EVAP_PD | COND_PD | MAX_TON | KW@MAXTON | PRICE |
|---|---|---|---|---|---|---|
| KW | KW | EVAP | COND | MAX | KW@MAX | PRICE |
| 261 | .696 | 19.5 | 16 | 436 | 322 | 80800 |
| 259 | .691 | 19.5 | 16 | 416 | 319 | 80800 |
| 258 | .688 | 19.5 | 16 | 437 | 338 | 80800 |
| 256 | .683 | 19.5 | 16 | 435 | 322 | 80800 |
| 253 | .675 | 19.5 | 13 | 416 | 312 | 80100 |
| 253 | .675 | 19.5 | 13 | 437 | 330 | 80100 |
| 252 | .672 | 19.5 | 13 | 436 | 315 | 80100 |
| 251 | .669 | 19.5 | 10.6 | 420 | 313 | 82400 |
| 247 | .659 | 19.5 | 19.8 | 417 | 304 | 81150 |
| 247 | .659 | 19.5 | 19.8 | 435 | 302 | 81150 |
| 245 | .653 | 19.5 | 19.7 | 436 | 315 | 81150 |
| 243 | .648 | 19.5 | 19.7 | 491 | 304 | 81150 |

Figure 10–8. (continued).

describes most of them. Still other programs have a data interchange facility different from that of the DIF file. Non-DIF files, although they may be readable, must be converted into DIF files before they can be loaded into a TK!Solver model.

# CHAPTER 11

# Controlling TK!Solver

One of the advertising slogans used to promote TK!Solver goes like this: "Now you don't have to think like a computer to use one!" It's true that once you have TK!Solver running on the machine, you don't have to know how to program a computer in order to solve complex problems in that you don't have to know any particular programming language. You don't have to know BASIC, or FORTRAN, or any other computer language. As you saw in the first few chapters, almost all you have to know is how to turn on the power, how to format spare disks, how to load TK!Solver into the machine, and how to give commands to TK!Solver.

However you do have to know how to "program" your models using TK!Solver's commands and the rules that describe the model mathematically. When you construct a TK!Solver model, you are actually programming a solution to a mathematical model. This concept should not cause any concern because it is no different from programming the solution if you had just designed a model to be solved on a spreadsheet by an engineering assistant who did not actually understand the model. You would write the equations of the model in such a form that they could be solved in steps, each intermediate step providing an output that is the input to another step. Each step would consist of simple calculations that could be checked easily and performed in a completely routine manner. That, of course, is programming.

The TK!Solver manual does not explain how the program actually performs its tasks, nor should this be of practical interest. You know that TK!Solver can solve equations that do not require symbolic integration and differentiation, you know that certain functions are available, and you know that the two solvers can handle equations that cannot be factored in all variables. You also know that, given models that are mathematically correct and consistent, TK!Solver will produce correct and consistent answers.

Most of the models presented in the manual and in the Software Arts SolverPacks are set up to be solved for a single instance. The Mechanical Engineering SolverPack, for example, includes a model that is based on a thorough treatment of fluid flow in real conduits and that calculates friction loss for fluids flowing in pipes. The model produces much more accurate results than those obtained from the friction charts and hydraulic tables used in most everyday engineering situations. However, if you were to use the model in its standard configuration to size a piping system, it would probably take longer than the old-fashioned method because the model is set up for a single solution.

Most real engineering problems involve a series of identical, or at least similar, calculations. To size a piping system, you need to calculate velocities and friction drop values for a series of pipe sections, each with a different volumetric flow rate. To design the steel structure of a building, you must perform hundreds of beam and column calculations. To check a building electrical system design, you must perform hundreds of voltage drop calculations.

# THE LIST SOLVER REVISITED

If you need to perform a series of identical calculations, TK!Solver's List Solver is the tool to use. You would first designate certain variables as list variables in order to associate a list with each variable. You would designate which lists are Input lists, and invoke the List Solver. Assuming that all the inputs will produce a solution, TK!Solver will run unattended, producing the output lists much faster than you could hope to do by repeatedly solving a single instance model.

You can also construct models incorporating lists in which each solution depends on the results of earlier solutions. Later, in the applications section, you will see a model that sizes air conditioning ducts using this type of procedure. You can look up the results of earlier solutions of a list-based model with the ELEMENT('listname, element number, default value) built-in function. For instance, you can calculate the factorial of a number with the model shown in Figure 11–1.

You need two lists for this model, an input list and an output list. The input list is a series of integers, beginning at 1 and ending at the last value for which you wish to calculate a factorial. Notice

```
(2i) Input: 1                                                       204/!

=================== VARIABLE SHEET =========================================
St Input      Name     Output    Unit    Comment
-- -----      ----     ------     ----    -------
L             fact
L   1         n




================== RULE SHEET =============================================
S Rule
- ----
  fact = n*element('fact,element()-1,1)


     FACTORIALS OF N

  n               fact

  1               1
  2               2
  3               6
  4               24
  5               120
  6               720
  7               5040
  8               40320
  9               362880
  10              3628800
```

Figure 11–1. Model to Calculate Factorials.

that the input list does not enter into the calculations. It's included for two reasons: first, TK!Solver needs an Input list to determine how many times to solve the model; and second, this list is used in creating a table to display the results. The actual inputs are the values stored in the output list, FACT. Where does TK!Solver get the input value for the first solution? It takes the value, 1, from the default value of the ELEMENT function call, because the element number term is evaluated as 0, a nonexistent list element, for the first solution. For each solution after the first one, TK!Solver can find a previous value in the output list, so that value is used rather than the default value.

The FACTORIAL model presents, somewhat disguised to the non-programmer, a method of making logical decisions with the computer. The ELEMENT function makes one decision, whether the list element exists or not, and either uses or ignores the default value accordingly. You have to use that particular form of the ELEMENT function as the second form—ELEMENT('listname, element number)—will choke on the FACTORIAL model and, produce an error message if the cursor is left in the Rule sheet status column, ELEMENT: Argument error.

## CONTROL STRUCTURES IN TK!SOLVER MODELS

In computer programmer jargon, following alternative courses based on the values of logical decisions is called "branching." The example just presented is an example of branching, but a restricted one. Since you can use branching to your advantage, we will explore the concept as it applies to TK!Solver.

If you are familiar with any type of high-level programming language, you have probably noticed that we have not run across any decision-making statements such as IF...THEN..., REPEAT... UNTIL..., WHILE...DO..., or even a lowly Goto statement. Those statements are found in most programming languages, but they do not appear in TK!Solver's repertoire of commands and functions. However, those constructs are handy for solving any problem, whether you are writing computer programs or even solving a problem manually, whether you call it programming or not. Fortunately, you can duplicate some of those features in a manner that TK!Solver accepts gracefully.

If you wanted to calculate the pressure drop of a fluid flowing in a pipe, you would first need to calculate a friction factor ($f$) to use in the equation

$$dPf = f*(L/D)*Pv.$$

Unfortunately, the friction factor is related to the Reynolds number (Re=rho$*D*V/(mu*gc)$—you'll see more of this in the applications section—by one of two different relations, depending on the value of the Reynolds number. If the Reynolds number is lower than 2000 (dimensionless), the flow is laminar, and the friction factor is given by this simple relation:

$$f=64/\text{Re}.$$

If the Reynolds number is greater than 2000, the flow is either transitional or turbulent, and the relation is given by the more complicated Colebrook equation

$$\frac{1}{\sqrt{f}} = 1.14 + 2\log{(D/E)} - 2\log{[1 + \frac{9.3}{\text{Re}\,(E/D)\sqrt{f}}]}$$

We'll consider this equation later, but for now we want to determine how to select the proper equation. If you have to use one model to determine whether the flow is laminar or turbulent, and then load the appropriate model, things will go slowly indeed.

Instead, you will put both equations in the model, but you have to give TK!Solver a way to decide which one to use. If you don't, the model will be overdefined.

You can use the STEP(x1,x2) function to help TK!Solver with its decision. Remember that the STEP function returns a 1 if x1 is greater than or equal to x2; if x2 is greater than x1, the function returns 0. With the rule

LAMINAR = STEP(2000,Re)

You can determine whether the Reynolds number indicates laminar flow or turbulent flow. A programmer would call LAMINAR a logical or Boolean variable; the value 1 corresponds to true, and 0 corresponds to false. Now you will use the value of LAMINAR to make a decision.

You want to use the laminar relation if LAMINAR is true, or 1, but not if it's false. The rule is written like this:

LAMINAR *        f = 64/Re        * LAMINAR

If LAMINAR is 1, the rule produces a nonzero value; if it's 0, both sides of the rule are 0, and no result is produced. Most high-level programming language compilers or interpreters would not accept such an expression, because a numerical value has been multiplied by a logical value—that is, data types have been mixed.

You use the same variable to select the Colebrook equation when the laminar relation does not apply. You write the rule like this:

```
(1−LAMINAR)    *    1/sqrt(f)   =
        −2*log(epsilon/(3.7*D) + 2.51/(Re*sqrt(f)))        *    (1−LAMINAR)
```

In this case, the equation disappears if LAMINAR is true, and produces a result if LAMINAR is false. For any given value of Re, one equation or the other, but only one, is used. This construct gives a fair approximation of the IF...THEN...ELSE statement that most high-level languages include.

# CASE–TYPE STRUCTURES

Often you may want to select an alternative from more than two possibilities. Some high-level programming languages do not have a single statement to accommodate this construct, other than to allow nested IF...THEN...ELSE statements, as follows:

```
IF A THEN Z
    ELSE IF B THEN Y
        ELSE IF C THEN X
            ELSE IF D THEN W
                ELSE IF E THEN V
                    ELSE U
```

Those languages that do offer a facility for selecting one of several alternative choices commonly call the statement the CASE statement, which looks like this:

```
CASE selector OF
        'A' THEN Z
        'B' THEN Y
        'C' THEN X
        'D' THEN W
        'E' THEN V
OTHERWISE U
```

(These two examples are not taken from any particular programming language, but instead are generalizations to illustrate the type of construct you want to duplicate.)

You can duplicate the CASE statement with the same general approach used to duplicate the IF...THEN...ELSE statement. Suppose that you have a model that includes three or more equations that describe different regimes of a real-world relation that cannot be described with a single equation. You need to determine which parameter governs the choice of which rule to apply, and then you need to construct a logical variable(s) to make the selection.

Assume that you have four equations, or rules, in your model that are mutually exclusive. For any solution of the model, only one of the four rules will be used, although many other rules may be used in all solutions. Assume, too, that the selection of the rule to use from the four depends only on temperature (T) in four regimes, 0 to 100'R, 101 to 400'R, 401 to 1000'R, and 1001'R and above.

The first and fourth regimes are straightforward: you use the STEP function just as you did to implement the IF... THEN...ELSE statement.

```
CASE#1 = STEP(100,T)
CASE#4 = STEP(T,1001)


CASE#1*    RULE#1__LEFT__SIDE = RULE#1__
           RIGHT__SIDE   *CASE#1
CASE#4*    RULE#4__LEFT__SIDE = RULE#4__
           RIGHT__SIDE   *CASE#4
```

The second and third regimes are not so simple, because you must test for both an upper and a lower bound. If you had an AND and an OR operator, as is available in many programming languages, things might be simpler, but you don't. However, you need only remember that Boolean operations have arithmetic counterparts. As long as you consider only logical values, 1 and 0, you can use the multiplication operator for the AND operator. The OR operator generally corresponds to the addition operator, but you must find a way to eliminate the possibility of a logical operation producing a result greater than 1. Fortunately, for the example at hand, you don't need the OR operator. You can construct the remainder of your control structure like this:

```
CASE#2 = STEP(T,101)*STEP(400,T)
CASE#3 = STEP(T,401)*STEP(1000,T)


CASE#2*    RULE#2__LEFT__SIDE = RULE#2__
           RIGHT__SIDE   *CASE#2
CASE#3*    RULE#3__LEFT__SIDE = RULE#3__
           RIGHT__SIDE   *CASE#3
```

Since the STEP function produces a value of 1 or 0, which you multiply, you need not worry about side effects on the rules. Regardless of the value of T, one and only one rule produces a nonzero result.

How could you duplicate the OR operator? It's a little trickier, but it can be done. Suppose you have a model with three regimes, two of which are described with the one rule, and the third of which is described with another. Furthermore, assume that the single exception is the second regime; you will use the same parameter for the governing variable, T (temperature in degrees Rankine). You want rule 2 to be applied if T is between 1000'R and 1100'R inclusive, and rule 1 to be applied otherwise.

You can develop the structure in steps to clarify the reasoning.

```
ABOVE1100 = STEP(T,1100)
BELOW1000 = STEP(1000,T)

CASE#1STEP(ABOVE1100 + BELOW1000,1)

CASE#1*    RULE#1__LEFT__SIDE = RULE#1__
           RIGHT__SIDE    *CASE#1
           (1–CASE#1)*    RULE#2__LEFT__SIDE = RULE#2__
           RIGHT__SIDE    *(1–CASE#1)
```

You could have nested the STEP function calls to eliminate calculating ABOVE1100 and BELOW1000, and made the Variable subsheet slightly less crowded. Sometimes it is better to show the extra steps while developing the model, and then eliminate them later, when the model has been proven.

The observant reader will also notice that the same selection could have been accomplished with two variables named ABOVE1000 and BELOW1100, as in the CASE statement above, but there will undoubtedly be instances where the OR operation is preferrable.

## USER MENUS

You can also duplicate another use of CASE-type statements—the menu, which is used to select program alternatives. For instance, if you have a model that performs several different calculations depending on the shape of a conductor, you can construct a model with a list of those shapes as the menu so that the user of the model can indicate which part of the model to execute by selecting the desired shape.

You can construct the menu using the GIVEN function, with the various shapes defined as variables (see Figure 11–2).

To use the menu shown in Figure 11–1, place an input in the Input field of one of the variables. If it would not make the model overdefined, you could select more than one shape, but the matter of whether the model would then be overdefined depends on the remainder of the model. This type of structure could also be used in a

```
(1r) Rule:
For Help, type ?
==================== VARIABLE SHEET ================
St Input      Name     Output    Unit      Comment
-- -----      ----     ------    ----      -------
              ROUND                         selects round formulas
              OVAL                          selects oval formulas
              SQUARE                        selects square formulas
              RECTAN                        selects rectangle formu



==================== Rule Sheet =======================
S Rule
- ----
   CASE_OF_ROUND=GIVEN('ROUND,1,0)
   CASE_OF_OVAL=GIVEN('OVAL,1,0)
   CASE_OF_SQUARE=GIVEN('SQUARE,1,0)
   CASE_OF_RECTAN=GIVEN('RECTAN,1,0)
```

Figure 11-2. A Menu in a TK!Solver Model.

complicated model for which it is desirable to make preliminary calculations for a number of cases, but to do only a few in detail. As for the fluid flow models, you might want to check the Reynolds number and then solve the model again, depending on whether the Reynolds number was in the range of interest. You can build a model with several levels of detail and solve to arbitrary levels.

On occasion you may wish to solve a model in steps, just as you would write a program in a programming language, without regard to the consistency of the model as a whole. You can do so with the help of the List Solver and a few logical constructs.

First, you must separate the rules of the model into groups corresponding to the steps or procedures of your "program." Then you apply your logic structure using the ELEMENT and the STEP functions, since TK!Solver does not include an EQUALS function.

Keep track of the steps you have executed with the ELE-MENT() function—remember that the ELEMENT function with no argument returns the element number the List Solver is processing for all the lists in the model. You can set up a variable that will always contain the number of the current element as follows:

$$CURRENT = ELEMENT( )$$

Then you can construct a series of variables that will control whether a rule or set of rules is executed, using CURRENT and the STEP function. The rules for these variables are shown below:

```
PASS#1 = STEP(1−CURRENT,0)*STEP(0,1−CURRENT)
PASS#2 = STEP(2−CURRENT,0)*STEP(0,2−CURRENT)
PASS#3 = STEP(3−CURRENT,0)*STEP(0,3−CURRENT)
PASS#4 = STEP(4−CURRENT,0)*STEP(0,4−CURRENT)
...
PASS#N = STEP(N−CURRENT,0)*STEP(0,N−CURRENT)
```

Here, you have taken advantage of the AND operation again. In each rule listed, the first STEP function returns a 1 up to and including the desired pass number, and the second STEP function returns a 0 until the pass before the desired pass number, returning a 1 on that pass number and after. Therefore, PASS#1 is 1 when CURRENT equals 1, PASS#2 is 1 when CURRENT equals 2, and so forth. Using the technique of multiplying each side of the rules by the appropriate PASS#n variables, you can control the order in which TK!Solver solves the various parts of the model.

If you try to control TK!Solver in the manner outlined above, you must make sure that all the Input lists contain a sufficient number of elements. Of course, you can make use of this technique in a model that does not actually use Input lists (that is, no variable of interest has an associated list). In that case, you need a single list associated with a dummy variable. The values of the list are of no interest, as long as a sufficient number of element numbers are defined. Here you are using the List Solver and the dummy list as a sort of a program counter to control the sequence in which the parts of the model are solved.

If you are using the List Solver to execute a "program," you do need Output lists, even though you may not be using Input lists. You need the Output lists for the storage of intermediate values. These intermediate values may be of interest per se, or they may have to be passed to subsequent steps as inputs, using the ELE-MENT function.

## REPEAT...UNTIL...

Earlier we mentioned the REPEAT...UNTIL... and WHILE... DO... constructs of some programming languages. These constructs are not available in all programming languages, but they are important nevertheless. Most modern languages include them, and with a little devious thinking, you can force TK!Solver to duplicate the constructs as well. First let's look at the REPEAT...UNTIL... statement.

The REPEAT...UNTIL... construct implies that a procedure will be executed, and that a logical statement will be evaluated at the conclusion of the procedure. If the logical statement is evaluated as false, the procedure will be executed again; if the statement is evaluated as true, the procedure will not be executed again, and the program will continue with the operations that follow the REPEAT...UNTIL... construct.

You can use the STEP function to make the logical test if you know from which direction the test variable will approach the test value, or you can use the method described above for testing for equality. You can either build the test value into the model by using a rule, or you can construct your test rule using a variable so that you can enter the UNTIL condition on the variable sheet.

With this version of TK!Solver, you must have at least one Input list long enough so that the program will reach the UNTIL condition before exhausting the control list (the longest INPUT list). You cannot force the List Solver to stop solving when a condition is reached; it solves the model for each element in the longest Input list. However, you can define the model so that all Output lists will be filled with blanks after the UNTIL condition has been reached. Then, when the List Solver is through processing, you can use the Table sheet and the Action command from the Table sheet to find the last nonblank entries, which is your solution. Notice that you do need Output lists even if you are only interested in the last solution; after the UNTIL condition has been satisfied, the List Solver will continue processing, producing values of no interest and displaying them in the Output fields of the Variable sheet.

Ideally, you should have a function that would allow you to fill lists an element at a time, as you can retrieve list elements with the ELEMENT function. Thus you could place an element in the control list, following the current element, until the UNTIL condition was reached. Then, the List Solver would stop processing after the element in which the UNTIL condition was satisifed. Perhaps that feature will be added in a future release.

## A LAST RESORT

Occasionally, you may have difficulty constructing a single model that will perform all the tasks you require. Error messages may abound, functions may produce argument error messages, and the model may be overdefined or underdefined. Usually, such a situation develops when a deadline is near.

The answer is to divide the model into segments, and store each as a separate model. You have already learned how to save segments of models using the options of the Storage command. To reiterate, you can save the data from the Variable sheet and the Variable subsheets with the Variable Save Option (/SV); the data from the Unit sheet with the Unit Save option (/SU); and the data from the User Function sheet, the List sheet, and the associated subsheets with the User Function Save option (/SF). As pointed out in Chapter 10, you can save lists in DIF form with the DIF Save option (/S#S). These options are provided by TK!Solver, and are described in the manual.

Other options are available but they are not described in the manual, and there are no special options of the Storage command dedicated to them. Instead, you have to do a little editing before you can save anything.

One option is to save the rules of a model separately. You can perform one set of calculations on a set of variables with one set of rules, clear the rule sheet, load another set of rules, and perform another set of calculations with the second set of rules. Unfortunately, there is no Save Rules option to the Storage command.

You will recall that if you load a model over an existing model, some entries will be duplicated, and some will be overwritten. You must be careful not destroy the results of earlier work. If you wish to load a new set of rules, you must make sure that the Variable sheet is not affected. If we are confident of the different rule sets, you can enter the different sets of rules in separate models without making entries in the Variable sheet by setting the Variable Insert ON: field of the Global sheet to NO.

Rule sets can be changed by resetting the Rule sheet with the Sheet option of the Reset command

and then loading the new rule set with the Load option of the Storage command

| / | S | L | **FILE NAME** |

To use the Sheet option of the Reset command to clear the Rule sheet, you must position the cursor in the initial cursor position of the Rule sheet, which is the first Rule field.

If you decide to save different rule sets as you develop a model, you can save the entire model at any point, reset all the sheets but the Rule sheet, and then save the Rule sheet with the Save Model option (/SS), as RULESET1.TK. Then you would reload the entire model, reset the Rule sheet, develop the second set of rules, and repeat the process for RULESET2. Finally, you would save the model without a Rule sheet. Then, you would have a model that you can load without rules, and a number of Rule sheets you can load over it.

You can use the same approach with other sheets, and thus can build, store, and use models in modules, just as you can remove and insert sheets in a loose-leaf binder. Remember, TK!Solver is a tool that is as productive as you make it. You can construct the same model over and over, or you can develop standard modules—standard Unit sheets by discipline, standard Variable sheets by field, standard User Function sheets—and build models quickly and efficiently.

# CHAPTER 12

# Advanced Mathematics

This chapter is about applying TK!Solver to engineering mathematics. As you noticed earlier, TK!Solver includes a number of mathematical functions, including the sine, cosine, tangent, arc sine, arc cosine, arc tangent, and their hyperbolic counterparts. Having these functions available greatly simplifies the solution of a number of engineering problems. Any self-respecting BASIC interpreter will include the ordinary trig functions, but few programming language packages will add the hyperbolic functions, and even fewer will offer a DOT Product function and a POLY Polynomial Evaluation function.

## VECTORS AND MATRICES

You will recall that dot products are found in matrix and vector algebra. The dot product is also referred to as the scalar product or the inner product.

The term *vector* can have two slightly different meanings, depending on the field in which it appears. In physics and engineering, a vector is normally considered to be a directed quantity, that is, a quantity that has both magnitude and direction. Velocity is a vector quantity—60 miles per hour in a northerly direction, for example, clearly does not have the same meaning as 60 miles per hour in a southerly direction. Force and acceleration are two other vector qualities because they also possess direction and magnitude.

Quantities such as volume and mass are not vector quantities because direction is not associated with them. Magnitude is sufficient to describe such quantities or properties. Quantities that can be completely specified by magnitude are scalar quantities.

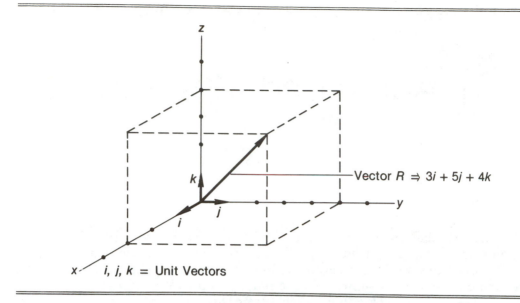

Figure 12–1. Vectors and Unit Vectors.

A vector can be specified by its magnitude and direction (for example, 60 knots north, or 30 knots north by northwest). More commonly, a vector is specified by its components. That is, a list of vectors is specified in standard directions, which, when added according to the rules of vector algebra, will produce the single vector. In mechanics problems, a vector is specified by its components in the *x, y,* and *z* directions, as shown in Figure 12–1.

Everyday quantities such as lengths in three dimensions, velocities, forces, and so on, can be specified by three elements as long as they are constant over time. You can easily specify a vector quantity in a TK!Solver model with a list of three elements, each one of which corresponds to a component of the vector.

Although you can perform many operations on vector quantities by performing the operations on their elements, you may wish to find the length of a vector. First, however, you'll have to recall the formula for finding the length of a line in three dimensions:

$$L = \sqrt{(x\char94 2 + y\char94 2 + z\char94 2)}$$

*or*

$$L\char94 2 = x\char94 2 + y\char94 2 + z\char94 2.$$

The dot product of two vectors is defined as the sum of the products of the corresponding elements of the two vectors. Thus, the length of a vector is the square root of the dot product of the vector and itself.

TK!Solver provides you with a built-in function for the dot product, DOT('listname1,'listname2) or DOT('listname1,series). The DOT function multiplies each element of listname1 by the corresponding element in listname2, or series, and then sums the individual products. You can calculate the length of a vector represented as a TK!Solver list with this rule:

LENGTH = SQRT(DOT(VECTOR,VECTOR))

By using the List Solver and a few control structures, as described in Chapter 11, you can manipulate a series of vectors in the form of lists, and then find the magnitude, or length, of the vectors of interest in the last step.

Not all vector quantities of interest are three-dimensional quantities from physics and engineering. A matrix is a generalized form of the vector, or the vector may be a specialized form of the matrix. You can multiply two matrices together if they are conformable in the order in which they appear in the multiplication. Two matrices are conformable in the order in which they appear if the number of columns in the first matrix is equal to the number of rows in the second matrix (Figure 12–2).

If you consider two matrices represented as TK!Solver lists, you can form the dot product of the two matrices in one step with the DOT function:

DOTPRODUCT = DOT(COLUMN__MATRIX,ROW__MATRIX)

You can form the product of two conformable matrices in general by using the List Solver in conjunction with the DOT function—but, remember, the matrices must be conformable. The model appears in Figure 12–3.

$$\begin{Vmatrix} 2 & 1 \\ 3 & 2 \\ 4 & -1 \end{Vmatrix} \cdot \begin{Vmatrix} 1 & 2 & 4 & 1 \\ 3 & -1 & 2 & 0 \end{Vmatrix} = \begin{Vmatrix} 5 & 3 & 10 & 2 \\ 9 & 4 & 16 & 3 \\ 1 & 9 & 14 & 4 \end{Vmatrix}$$

Figure 12–2. Two Conformable Matrices.

```
(5i) Input:                                                        200/!

==================== VARIABLE SHEET ===================================
St Input       Name     Output    Unit    Comment
-- -----       ----     ------    ----    -------
                                          MATRIX MULTIPLICATION
                                          *********************
L   'X         B
L              C_ROW1
L              C_ROW2
L              C_ROW3
L   1          A_ROW1
L   1          A_ROW2
L   1          A_ROW3
L   1          B_COL1
L   1          B_COL2
L   1          B_COL3
L              CURRENT
L              CONTROL




(1r) Rule: CURRENT=CONTROL                                         200/!

==================== RULE SHEET =======================================
S Rule
- ----
* CURRENT=CONTROL

  C_ROW1=DOT('A_ROW1,B)
  C_ROW2=DOT('A_ROW2,B)
  C_ROW3=DOT('A_ROW3,B)
```

Figure 12–3.  Matrix Multiplication Model.

Each row of the first matrix, A, is a list, and each column of the
second matrix, B, is a list. Therefore, you can form any element of
the result matrix, C, by taking the dot product of the corresponding row of A and column of B. You need a separate rule for each row
of C, and then you need a method to determine which column list of
matrix B to use. The simplest way is to make another list of the
names of the column lists of matrix B, as follows:

(1v) Value:

```
============================== LIST: B       ==============================
Comment:          names of the column lists of matrix B
Display Unit:
Storage Unit:
Element Value

------- ----
    1  'BCOLUMN1
    2  'BCOLUMN2
    3  'BCOLUMN3
```

•
•
•

n    'BCOLUMNn

Each successive pass of the List Solver will make use of the corresponding column list of matrix B and will generate an element in each of the row lists of matrix C, the result matrix. You can display the C matrix in familiar form by creating a table in the horizontal format (see Figure 12–4).

With some similar TK! trickery you can also take the transpose of a matrix. The transpose of a matrix, of course, is formed by interchanging the rows and columns, as shown in Figure 12–5.

```
C_ROW1  18          24          30
C_ROW2  27          36          45
C_ROW3  36          48          60
```

Figure 12–4. The Table Display of the Matrix Multiplication Model.

You can consider a matrix to be a list, each member of which is another list. This notion is entirely consistent with matrices in general—a matrix may have elements that are in turn matrices. You can construct a TK!Solver matrix by constructing a list, each element of which is the name of another list, a row of the matrix

$$\left\| \begin{matrix} a_{11} & a_{12} & a_{13} & a_{14} \\ a_{21} & a_{22} & a_{23} & a_{24} \\ a_{31} & a_{32} & a_{33} & a_{34} \\ a_{41} & a_{42} & a_{43} & a_{44} \end{matrix} \right\| \quad \text{Matrix}$$

$$\Downarrow$$

$$\left\| \begin{matrix} a_{11} & a_{21} & a_{31} & a_{41} \\ a_{12} & a_{22} & a_{32} & a_{42} \\ a_{13} & a_{23} & a_{33} & a_{43} \\ a_{14} & a_{24} & a_{34} & a_{44} \end{matrix} \right\| \quad \text{Transpose}$$

Figure 12–5. A Matrix and Its Transpose.

(you could just as easily use the names of the columns of the matrix). Since each element of the first line is the name of another list, each is a symbolic constant, as in the previous example.

To take the transpose of the matrix, use the model shown in Figure 12-6 and the List Solver. On each pass of the List Solver, the innermost ELEMENT function forms an index that corresponds to the number of the pass. The next ELEMENT function uses that index as a pointer into list B, selecting the first element on the first pass, the second element on the second pass, and so on. From the current row list, the outermost ELEMENT function extracts the proper element to be inserted into the row lists of the transpose matrix.

Note that the row lists of the transpose are output lists, but the row lists of the original matrix are not really input lists. The only input list you need is the control list, which must have the correct number of elements (equal to the number of elements in the row lists), but it need not enter into the actual operation of transposing the matrix.

It should be obvious that you could transpose a matrix faster manually than with this model, but the matrix operations to follow will be a different story. You should consider this operation an example of TK!Solver techniques that are not immediately apparent from the first reading of the manual.

The next logical step in the study of matrices is the formation of the inverse of a matrix. (We have skipped the product of a matrix and scalar, since it should be a trivial model.) Since the inverse of a matrix is formed by dividing the adjoint of the matrix by the determinant of the matrix, we must first look at determinants.

```
BTRANS_ROW_1=ELEMENT(ELEMENT('B,ELEMENT()),1)
BTRANS_ROW_2=ELEMENT(ELEMENT('B,ELEMENT()),2)
BTRANS_ROW_3=ELEMENT(ELEMENT('B,ELEMENT()),3)
BTRNAS_ROW_4=ELEMENT(ELEMENT('B,ELEMENT()),4)
 .
 .
 .
BTRANS_ROW_N=ELEMENT(ELEMENT('B,ELEMENT()),N)
```

Figure 12-6. Model To Take the Transpose of a Matrix.

$$\begin{vmatrix} a_{11} & a_{12} & a_{13} \\ a_{21} & a_{22} & a_{23} \\ a_{31} & a_{32} & a_{33} \end{vmatrix} = \begin{array}{l} a_{11}a_{22}a_{33} + a_{12}a_{23}a_{31} + a_{13}a_{21}a_{32} \\ - a_{13}a_{22}a_{31} - a_{11}a_{23}a_{32} - a_{12}a_{21}a_{33} \end{array}$$

Or

$$\begin{array}{ccc} (+) & (+) & (+) \end{array}$$

$$\begin{vmatrix} a_{11} & a_{12} & a_{13} \\ a_{21} & a_{22} & a_{23} \\ a_{31} & a_{32} & a_{33} \end{vmatrix} \begin{array}{cc} a_{11} & a_{12} \\ a_{21} & a_{22} \\ a_{31} & a_{32} \end{array}$$

$$\begin{array}{ccc} (-) & (-) & (-) \end{array}$$

Figure 12–7. Expansion of a Third Order Determinant.

## DETERMINANTS

The term *determinant* refers to an array of elements that expresses a function. Second and third order determinants are used to solve systems of two and three simultaneous equations in elementary algebra. The array must be square. A determinant may be either the array of the elements or the expansion of the array. The expansion of a third order determinant is shown in Figure 12–7.

Once you proceed to fourth and higher order determinants, the diagonal expansion process from high school algebra is no longer valid, and you have to expand the determinant in terms of the elements of a row or column, and the cofactors of those elements. If the cofactors are of fourth order or higher, they must in turn be expanded by the same process. Only when you obtain cofactors of the second or third order can you reduce the expansion to a single number by diagonal expansion.

Let's look at a rule that expands a second order determinant, and one that expands a third order determinant. You will recall that the value of a second order determinant is the difference between the product of the elements on the principal diagonal and the product of the elements on the other diagonal. This algorithm is easily expressed as a TK!Solver rule:

```
DET = ELEMENT('ROW1,1)*ELEMENT('ROW2,2)
        −ELEMENT('ROW2,1)*ELEMENT('ROW1,2)
```

If you refer to the determinant with a list that names the rows, as you did for matrices, you can also write the rule this way:

```
DET2 = ELEMENT(ELEMENT('A,1),1)*ELEMENT(ELEMENT('A,2),2)
      −ELEMENT(ELEMENT('A,2),1)*ELEMENT(ELEMENT('A,1),2)
```

Furthermore, you can write a single rule to expand a third order determinant. Here you would use the names of the rows—as in the first example of the expansion of a second order determinant—rather than extract them from a list with the Element function:

```
DET3 = ELEMENT('R1,1)*ELEMENT('R2,2)*ELEMENT('R3,3)
      − ELEMENT('R3,1)*ELEMENT('R2,2)*ELEMENT('R1,3)
      + ELEMENT('R1,2)*ELEMENT('R2,3)*ELEMENT('R3,1)
      − ELEMENT('R3,2)*ELEMENT('R2,3)*ELEMENT('R1,1)
      + ELEMENT('R1,3)*ELEMENT('R2,1)*ELEMENT('R3,2)
      − ELEMENT('R3,3)*ELEMENT('R2,1)*ELEMENT('R1,2)
```

Of course, this rule would be written on a single line and would be only partly visible. You need not worry about a rule that is too long to be displayed on the screen—TK!Solver rules can be 200 characters long, even though the screen only displays 80 characters including the status field.

You cannot expand a fourth or higher order determinant with a single rule. You must duplicate the manual procedure of multiplying each element of a row or column by its cofactor, and then applying the same procedure to each cofactor until you finally obtain second or third order cofactors, which can be expanded by one of the rules given above. The general approach is given below:

```
DET4 = ELEMENT('R1,1)*COFACTOR11−ELEMENT('R1,2)*COFACTOR12
      + ELEMENT('R1,3)*COFACTOR13−ELEMENT('R1,4)*COFACTOR14
```

```
COFACTOR11 = ELEMENT('R2,2)*ELEMENT('R3,3)*ELEMENT('R4,4)
            − ELEMENT('R4,2)*ELEMENT('R3,3)*ELEMENT('R2,4)
            + ELEMENT('R2,3)*ELEMENT('R3,4)*ELEMENT('R4,2)
            − ELEMENT('R4,3)*ELEMENT('R3,4)*ELEMENT('R2,2)
            + ELEMENT('R2,4)*ELEMENT('R3,2)*ELEMENT('R4,3)
            − ELEMENT('R4,4)*ELEMENT('R3,2)*ELEMENT('R2,3)
```

```
COFACTOR12 = ELEMENT('R2,1)*ELEMENT('R3,3)*ELEMENT('R4,4)
            − ELEMENT('R4,1)*ELEMENT('R3,3)*ELEMENT('R2,4)
            + ELEMENT('R2,3)*ELEMENT('R3,4)*ELEMENT('R4,3)
            − ELEMENT('R4,3)*ELEMENT('R3,4)*ELEMENT('R2,3)
```

+ ELEMENT('R2,4)*ELEMENT('R3,1)*ELEMENT('R4,3)
− ELEMENT('R4,4)*ELEMENT('R3,1)*ELEMENT('R2,3)

... and so on.

Obviously, entering the rules for expanding a large determinant is a forbidding task, but manually expanding the same determinant is even more forbidding. You should note at this point that you cannot formulate a general model for expanding a determinant of any size, and simply enter zeroes in the unneeded elements, because a determinant that has all zeroes in any row or any column will expand to zero. You must construct a separate model for each size of determinant you wish to expand, but you only need to do so once, as you can store the models on disk and thus use them again and again.

By applying all the techniques described above, you can take the inverse of a matrix. You cannot expand its determinant, however, unless the matrix is square. You must form the adjoint by constructing a matrix in which each element of the original matrix is replaced by its expanded cofactor, and then taking the transpose of the resulting matrix. To form the inverse, you would then divide each element of the adjoint by the determinant of the original matrix.

You can test the inverse by multiplying it by the original matrix—if the inverse was formed correctly. The result will be the unit matrix:

```
1 0 0 0 0 0
0 1 0 0 0 0
0 0 1 0 0 0
0 0 0 1 0 0
0 0 0 0 1 0
0 0 0 0 0 1
```

Matrices and determinants are used primarily in the solution of simultaneous equations. Fortunately, you do not have to go through an exercise like the one above when using TK!Solver to solve systems of equations. If a model consisting of a number of simultaneous equations is neither overdefined nor underdefined, TK!Solver will solve the model without your manipulation. But, if you had to solve the same model manually, you would be faced with the task of expressing the system of equations as a coefficient matrix and manipulating that matrix. It's comforting to know that TK!Solver can not only solve these systems of equations, but can help you study the systems as well.

# DIFFERENTIATION AND INTEGRATION

Let's leave matrices, determinants, and vectors for a moment in order to consider two other topics—differentiation and integration of functions. Even though these operations may not be performed on a daily basis in all branches of engineering, both are basic to almost every field.

Although TK!Solver cannot yet perform symbolic integration or differentiation, it can perform approximations of both operations with the aid of the List Solver.

You will recall that the derivative of a function, $f(x)$, is another function, $g(x)$, which describes the instantaneous rate of change of $f(x)$. Calculus texts typically introduce the concept of derivatives with the example of calculating the slope of a curve at various points along the curve. First, an average slope is calculated using an arbitrary interval, as shown in Figure 12–8; although we normally think of the slope of a curve as its tangent, the slope calculated by this method only approximates the tangent.

As the interval of the independent variable is reduced, the slope that is calculated approaches the true tangent. The same concept is used in developing the derivative of a function. You define the average rate of change of a function as shown below:

$$g(x) = \frac{f(x + h) - f(x)}{h}$$

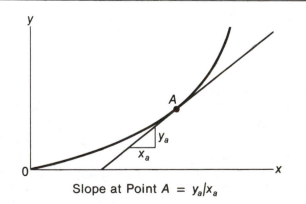

Slope at Point $A = y_a/x_a$

Figure 12–8. The Slope of a Curve.

Then, you allow the difference between the two values of the independent variable to approach 0. The limiting function that results is the derivative.

When you follow the reasoning that leads to the derivative of a function, you work with symbols in a way that TK!Solver cannot duplicate. You begin with a function—a rule—and form another rule without using numerical values. TK!Solver cannot perform that type of symbolic manipulation.

However, TK!Solver can construct the derivative of a function, expressed as a rule, and form a numerical approximation of its derivative in list form. Using Coulomb's law as an example (the force of attraction between two unlike charges is inversely proportional to the difference between the charges), you would enter the following rules:

```
F = k/r^2
dF = (F−ELEMENT('F,ELEMENT( )−1,0))/(r−ELEMENT('r,ELEMENT( )−1,1))
```

The variables, F, dF, and r, must be list variables; r is an input list, and dF and F are output variables. You set up the r list by diving (>) to it, and using the Fill List feature. Remember that you can fill a list by indicating the first and last elements, entering values for those two elements, and then entering the Action (!) command from the list subsheet. In the example here, you would determine the interval of interest, and the number of discrete points you wish to calculate. Although TK!Solver can only plot the resulting list, dF, with the accuracy of the printer, it can calculate the list to 12 significant digits. The output lists of the Coulomb's law model are shown in Figure 12-9.

You can form the function, Fprime, as a TK!Solver User function by using the two lists, dF and r (dF/dr would be a logical choice, but TK!Solver will not allow you to use the / sign in any variable, list, or function name). Then, you can use the function just as you would any other function:

$$Fdot = Fprime(r)$$

Of course, if you wish to be able to interpolate between values of r, you must use Linear mapping for the User Function Fprime.

You can approach integration in the same manner. You will recall that the definite integral of a function is often introduced as the area under a curve. The function of interest is plotted as a two-dimensional curve (see Figure 12-10), and the area under the curve is calculated as a series of rectangles.

```
(6i) Input: 60000000000                                               204/

=================== VARIABLE SHEET ======================================
St Input        Name   Output       Unit    Comment
-- -----        ----   ------       ----    -------
                                            COULOMB'S LAW
                                            *************
    .000088     F                   newton  Electric force between 2 charges
    2.306E-28   k                   newton-m^ proportionality factor
    6E10        q1                  elem_chg electric charge on particle 1
    6E10        q2                  elem_chg electric charge on particle 2
                r      .09712691    m       separation between charges

=================== RULE SHEET ==========================================
S Rule
- ----
  F=k*q1*q2/r^2
```

```
(5i) Input: 1                                                         202/!

=================== VARIABLE SHEET ======================================
St Input        Name   Output       Unit    Comment
-- -----        ----   ------       ----    -------
                                            COULOMB'S LAW
                                            *************
L               F                   NEWTON
    2.306E-28   k                   NEWTON-M^
L   1           r                   M
L               dF                  NEWTON

=================== RULE SHEET ==========================================
S Rule
- ----
  F=k/r^2
  dF=(F-ELEMENT('F,ELEMENT()-1,0))/(r-ELEMENT('r,ELEMENT()-1,1))
```

```
(1n) Name: F                                                          202/!

=================== LIST SHEET ==========================================
Name       Elements  Unit     Comment
----       --------  ----     -------
F          20        NEWTON
r          20        M
dF         20        NEWTON
```

Figure 12-9.  Output Lists from the Coulomb's Law Model.

| LIST: F | | LIST: dF | | LIST: r | |
| Element | Value | Element | Value | Element | Value |
| ------- | ----- | ------- | ----- | ------- | ----- |
| 1 | 2.306E-22 | 1 | -2.30830831E-22 | 1 | .001 |
| 2 | 5.765E-23 | 2 | -1.7295E-19 | 2 | .002 |
| 3 | 2.562222222E-23 | 3 | -3.20277778E-20 | 3 | .003 |
| 4 | 1.44125E-23 | 4 | -1.12097222E-20 | 4 | .004 |
| 5 | 9.224E-24 | 5 | -5.1885E-21 | 5 | .005 |
| 6 | 6.405555556E-24 | 6 | -2.81844444E-21 | 6 | .006 |
| 7 | 4.706122449E-24 | 7 | -1.69943311E-21 | 7 | .007 |
| 8 | 3.603125E-24 | 8 | -1.10299745E-21 | 8 | .008 |
| 9 | 2.84691358E-24 | 9 | -7.56211420E-22 | 9 | .009 |
| 10 | 2.306E-24 | 10 | -5.40913580E-22 | 10 | .01 |
| 11 | 1.905785124E-24 | 11 | -4.00214876E-22 | 11 | .011 |
| 12 | 1.601388889E-24 | 12 | -3.04396235E-22 | 12 | .012 |
| 13 | 1.364497041E-24 | 13 | -2.36891848E-22 | 13 | .013 |
| 14 | 1.176530612E-24 | 14 | -1.87966429E-22 | 14 | .014 |
| 15 | 1.024888889E-24 | 15 | -1.51641723E-22 | 15 | .015 |
| 16 | 9.0078125E-25 | 16 | -1.24107639E-22 | 16 | .016 |
| 17 | 7.979238754E-25 | 17 | -1.02857375E-22 | 17 | .017 |
| 18 | 7.117283951E-25 | 18 | -8.61954804E-23 | 18 | .018 |
| 19 | 6.387811634E-25 | 19 | -7.29472316E-23 | 19 | .019 |
| 20 | 5.765E-25 | 20 | -6.22811634E-23 | 20 | .02 |

Figure 12–9. (continued).

In the study of calculus you are not necessarily interested in finding the area under a curve described by a function, as much as in finding the integral as a function. Again, TK!Solver will not allow you to manipulate symbols and produce another function expressed as a rule. You can begin with a function expressed as a rule, but you must be satisfied with a numerical approximation of the integral.

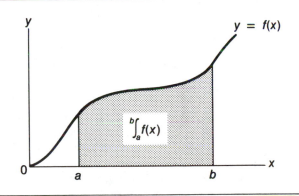

Figure 12–10. A Definite Integral—Area under a Curve.

Using Coulomb's law again, you can calculate the work done in moving a charge in the presence of another, unlike, charge. Work is the product of force and distance. The model includes the following rules:

```
F = k/r^2
delW = F*(r—ELEMENT('r,ELEMENT( )—1,0)
W = SUM('delW)
```

To solve this model with the List Solver, use r as the input list and F and delW as the output lists. With r and delW as the two lists, you can also define a User function, Wdot, as follows:

```
POWER = Wdot(r)
```

Again, you should set the mapping of Wdot to Linear, so that you can interpolate in either direction.

## MAXIMA AND MINIMA

You can locate relative maximum or minimum points of a curve after you have constructed the list describing its derivative. Again, remember that if a curve displays a maximum or minimum point, as in Figure 12–11, the first derivative will have a value of 0 at that point. The converse is not necessarily true: it is possible for the first derivative to have a value of 0 at a point of inflection that is not a maximum of minimum point (see Figure 12–12).

You can construct a model for the first derivative of the curve for a parabola with these rules:

```
X^2 = 2*P*Y
dYdx = (Y—ELEMENT('Y,ELEMENT( )—1,0))/
       (X—ELEMENT('X,ELEMENT( )—1,1))
```

To find a maximum or minimum point, you must find a point where the slope of the first derivative is 0, or in other words, where the tangent to the curve is horizontal. Finding such a point does not guarantee that a maximum or a minimum exists—additional tests are needed to make that determination.

It is tempting to simply test for a value of 0 in the dYdx list, but there are two problems with that approach. First, your model may not produce a value of 0 in the list because it produces a series

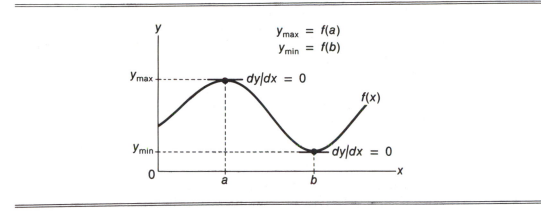

$$y_{max} = f(a)$$
$$y_{min} = f(b)$$

Figure 12–11. Maximum and Minimum Points on a Curve.

of discrete values, even though a value of 0 would appear if the list were graphed as a continuous function of X. This problem could possibly be eliminated by choosing a fine enough increment for the values of X, but even that is not a guarantee.

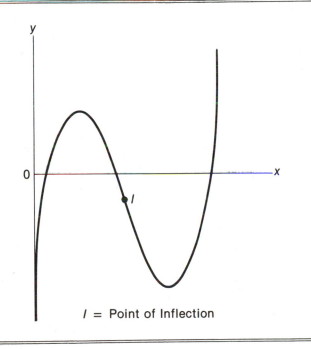

I = Point of Inflection

Figure 12–12. Points of Inflection.

The second problem is that TK!Solver does not have a built-in function that returns the element number when given a value. Even the functions MAX and MIN return the largest or smallest value in a list, not the position of the value. As a result, you have to construct your own function that will return the position of a list element that meets your conditions, using a User function.

The first step in constructing the function needed is a rule that will create an output list of the possible positions:

$$POS = ELEMENT( )$$

The second step is a User function that relates the variable you will test to the position list, which is shown below as an entry in the User Function sheet:

(1n) Name:

=============================== USER FUNCTION SHEET ===============================

| Name | Domain | Mapping | Range | Comment |
| --- | --- | --- | --- | --- |
| POSITION | dYdx | Step | POS | returns the element number |

You must include a rule to invoke the Position function, and you must make provisions for function arguments that do not fall within the bounds of the domain list. The rule to invoke the function is

$$MAXORMIN = POSITION(0)$$

If you leave the lists of the function empty, TK!Solver will fill them as the List Solver processes the model. However, TK!Solver expects the argument to any user function to fall within the bounds of the lists that comprise the function. If you begin with empty lists, you may have a model that spends more time beeping than calculating. The answer is to fill the lists with default values, as shown in Figure 12–13; the number of elements in the two lists should be one more than the number of elements in the input list, X.

As the List Solver processes the model, it will fill the output list dYdx with new values. No difficulty arises from filling an output list with default values. Each time the List Solver processes a set of elements—that is, on each pass—the rule, MAXORMIN=POSITION(0), will look for 0 in the dYdx list, the domain list, and will not find it. However, because the mapping has been

```
    dYdX          POS

    -10            1          80950.4762  18
    4752.38095     2          85712.8571  19
    9514.76191     3          90475.2381  20
    14277.1429     4          95237.6191  21
    19039.5238     5          100000
    23801.9048     6
    28564.2857     7
    33326.6667     8
    38089.0476     9
    42851.4286    10
    47613.8095    11
    52376.1905    12
    57138.5714    13
    61900.9524    14
    66663.3333    15
    71425.7143    16
    76188.0952    17
```

(2c) Comment:                                                    200/!

```
======================== VARIABLE SHEET =================================
St Input        Name      Output     Unit     Comment
-- -----        ----      ------     ----     -------
                                              MAX OR MIN POINTS ON CURVE
                                              ****************************

L   1           X
    4           P
L               Y
L               dYdX
L               POS
L               MAXORMI
================== RULE SHEET ============================================
S Rule
- ----
  X^2=2*P*Y

  dYdX=(Y-ELEMENT('Y,ELEMENT()-1,0))/(X-ELEMENT('X,ELEMENT()-1,1))
  POS=ELEMENT()
  MAXORMIN=POSITION(0)
```

```
    dYdX          POS

    -1.1363636     1          1.375      17
    -2.375         2          1.625      18
    -2.125         3          1.875      19
    -1.875         4          2.125      20
    -1.625         5          2.375      21
    -1.375         6
    -1.125         7
    -.875          8
    -.625          9
    -.375         10
    -.125         11
    .125          12
    .375          13
    .625          14
    .875          15
    1.125         16
```

Figure 12-13. The Default Values of the Position Function.

set to Step, it will interpolate between the next to the last and the last element in the domain list, and return the value of the next to the last element of the range list, 0. That is, it will do so until the List Solver places a value in the domain list that is 0 or just less than 0. Then, the function will return a value that gives the position of the element that produced the approximately 0 value of dYdx.

You could extend this concept further and actually identify maxima and minima, but that task can be a "homework" exercise.

# CHAPTER 13

# More Mathematics

Now you know that with a little coaxing TK!Solver will perform some mathematical chores not mentioned in the manual. Normally, you don't have to worry about such matters, but it's helpful to know that you can investigate mathematical phenomena if you want to.

Since engineering formulas can usually be found for the construction of your models, you do not often have to rely on theoretical derivation. TK!Solver is well-suited to solving models that fall in that class. Another class of models that are not so easily handled, however, are those based on experimental data that are undoubtedly related by some function, but for which you do not have the actual equations. The textbooks contain formulas and equations that represent only ideal cases. You know from these ideal cases that the function represented by your experimental data could be differentiated or integrated to provide other useful relations, but you cannot produce these derivatives or integrals without the equation(s) that your experimental data represent. Or can you?

## FINITE DIFFERENCES

Many engineering problems can be represented as differential equations, or systems of differential equations. For many important problems, these differential equations are linear, and their solutions can be obtained by methods that resemble those used to solve simultaneous algebraic equations. You have seen that TK!Solver can easily handle the solution of large systems of simultaneous equations.

When you try to solve complicated differential equations or models represented by experimental data, you must restore to numerical methods. One class of numerical techniques comprises those known as finite-difference methods. These methods are employed in the numerical solution of differential equations, numerical differentiation and integration, curve fitting, smoothing of data, and interpolation. Since the derivation of these methods could easily fill a volume, this discussion will consider only how to implement several well-known methods using TK!Solver. For additional information, the interested reader may wish to examine some of the texts listed in the bibliography.

Having performed an experiment and recorded the data, you usually have at your disposal one or more functions presented in tabular form as shown below:

| $X$ | $F(X)$ |
|---|---|
| $x0$ | $f(x0)$ |
| $x1$ | $f(x1)$ |
| $x2$ | $f(x2)$ |
| $x3$ | $f(x3)$ |
| $x4$ | $f(x4)$ |
| • | • |
| • | • |

You should immediately be able to see that TK!Solver will be useful—the equation has exactly the same form as a TK!Solver User function! You can enter the function into a TK!Solver model just as you would any other user function, either from the List subsheets or the User Function subsheet. If you stopped after entering the data and defining the function on the User Function sheet, you would be able to use table lookup or to interpolate values, depending on the mapping selected. But you can go further with various finite-difference methods.

The first operation to consider in connection with the function is the first divided difference, which is defined by the following formula:

$$f(xi, xj) = \frac{f(xi) - f(j)}{xi - xj}$$

The second divided difference is defined by a similar formula:

$$f(xi,xj,xk) = \frac{f(xi,xj) - f(xj,xk)}{xi - xk}$$

A series of divided differences of a function is usually presented in the form of a difference table, in which each succeeding divided difference is a column and its elements are positioned vertically midway between the elements of the preceding divided difference. Following is the difference table for the above function:

| X  F(X) | | | | | |
|---|---|---|---|---|---|
| x0  f(x0) | | | | | |
| | f(x0,x1) | | | | |
| x1  f(x1) | | f(x0,x1,x2) | | | |
| | f(x1,x2) | | f(x0,x1,x2,x3) | | |
| x2  f(x2) | | f(x1,x2,x3) | | f(x0,... | |
| | f(x2,x3) | | f(x1,x2,x3,x4) | | |
| x3  f(x3) | | f(x2,x3,x4) | | f(x1,... | |
| | f(x3,x4) | | f(x2,x3,x4,x5) | | |
| x4  f(x4) | | f(x3,x4,x5) | | | |
| | f(x4,x5) | | | | |
| x5  f(x5) | | | | | |

Here is a specific numerical example:

| X | $X^3$ | | | |
|---|---|---|---|---|
| 0 | 0 | | | |
| | | 1 | | |
| 1 | 1 | | 4 | |
| | | 13 | | 1 |
| 3 | 27 | | 8 | 0 |
| | | 37 | | 1 |
| 4 | 64 | | 14 | 0 |

|   |   | 93 |   | 1 |
|---|---|----|----|---|
| 7 | 343 |   | 20 |   |
|   |   | 193 |   |   |
| 9 | 729 |   |   |   |

If you have a TK!Solver user function, you can easily construct a model of the first divided difference with a few rules, and use the List Solver to perform the calculations. The most difficult part of the task will be to achieve the triangular presentation. The first rule is fairly straightforward; for each element in the output list, TK!Solver must use the current element in each of the input lists, as well as the current+1 element in each.

$$DD1 = (FX - ELEMENT('FX,ELEMENT( ) + 1,0))/$$
$$(X-ELEMENT('X,ELEMENT( ) + 1,1))$$

Since the number of elements in the first divided difference is one less than that in the function, you should stop calculating at the n−1 element. As noted earlier, you have no way to force TK!Solver to stop processing before the input lists have been exhausted, but you can use logical conditions to prevent it from entering elements in the output lists. You can formulate a general rule that will work with input lists of any length, as follows:

$$OK = STEP(COUNT('X)-1,ELEMENT( ))$$

If the function (both lists of which are the input lists) is n elements long, the variable OK will be 1 for the first n−1 elements of the input lists, and 0 for the last element. If you multiply both sides of the first rule by OK, no result will be placed in the nth element of the output list, DD1.

There is no need to form the first divided difference before forming the second divided difference, nor the second before the third, although it may be clearer to form them in steps. You can form all the possible divided differences directly by entering a rule for each one, following the example of the rule for the first divided difference. The rule for the second divided difference is shown below, along with the logical condition rule:

$$OK2* DD2 = ((FX-ELEMENT('FX,ELEMENT( ) + 1,0)$$
$$-(ELEMENT('FX,ELEMENT( ) + 1,0)$$
$$-ELEMENT('FX,ELEMENT( ) + 2)))$$

$$/(X-ELEMENT('X,ELEMENT(\ )+2,1))*OK2$$

$$OK2 = STEP(COUNT('X)-2,ELEMENT(\ ))$$

Usually, the values of the independent variable, the domain variable in the TK!Solver user function, will be equally spaced, and you will be able to do without the divided differences. Instead, you will form the differences of the function:

$$\text{delta}\_f_k = t_{k+1} - f_k$$

The function will follow the form:

$$x_k = x_0 + kh \qquad Where: \quad k = \ldots -2, -1, 0, 1, 2, \ldots$$
$$Y_k = f(x_0 + kh)$$

You can construct these differences with a model similar to that use for divided differences, and you can also construct the successive differences simultaneously:

$$OK1 = STEP(COUNT('X)-1,ELEMENT(\ ))$$

$$OK2 = STEP(COUNT('X)-2,ELEMENT(\ ))$$

$$OK1*D1 = (ELEMENT('FX,ELEMENT(\ )+1,0)-FX)*OK1$$

$$OK2*\ D2 = ((ELEMENT('FX,ELEMENT(\ )+2,0)-ELEMENT('FX,ELEMENT(\ )\\ +1,0))-(ELEMENT('FX,ELEMENT(\ )+1,0)-FX))*OK2$$

Note the differences and antidifferences are analogous to derivatives and integrals, and proceed to some practical uses of finite-difference methods. In some of these methods you may refer to the difference operator delta, which refers to the differences presented above.

## INTERPOLATION

Earlier it was noted that, when given an intermediate domain value as the argument of a user function, TK!Solver can interpolate between values in the range if the mapping is set to Linear. The program can also interpolate between values in the Domain list if the argument of the function is unknown and an intermediate range value is known. Specifically, TK!Solver uses linear interpolation, which means that it interpolates as if the function were a series of straight line segments connecting the discrete points of the user function, as shown in Figure 13-1.

Linear interpolation is acceptable in most practical situations; the loss of accuracy is usually several orders of magnitude less than the safety factor that is applied after the calculations are complete. If a function presented in tabular form is calculated at small enough intervals, little would be gained by a more accurate method of interpolation.

On the other hand, if the data change rapidly, or if the intervals between successive values are large, linear interpolation may produce unacceptable errors. For such a function you need a more accurate method of interpolation.

One method of increasing the accuracy of interpolation is to plot the function, using the values that are available, and then to "fit" a curve between the points of the plot. We could use arcs of circles, sections of parabolas, or other curves for which equations are known to fit curves through successive groups of three points (see Figure 13-2), or you could simply "eyeball" a curve that appears to be correct. Then, you would use the fitted curve to read the values from the scales of the graph.

The other method is to use an interpolation formula that makes use of the differences of the function. Several important formulas of this type that can be used in a TK!Solver model are discussed next.

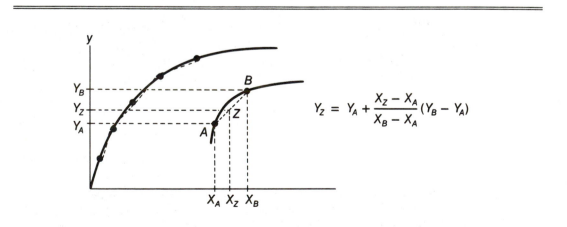

$$Y_Z = Y_A + \frac{X_Z - X_A}{X_B - X_A}(Y_B - Y_A)$$

Figure 13-1. Linear Interpolation.

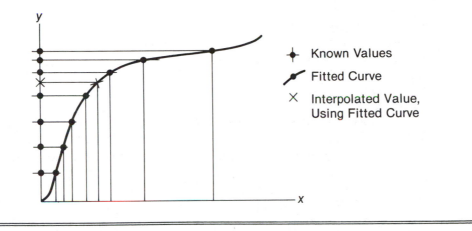

Figure 13–2. Graphical Interpolation.

One of the most fundamental interpolation formulas is Newton's divided difference formula:

$$f(x) = f(x0) + (x-x0)f(x0,x1) + (x-x0)(x-x1)f(x0,x1,x2) + \ldots$$
$$+ (x-x0)(x-x1)\ldots(x-x_{n-1})f(x0,x1,x2,\ldots,xn)$$
$$+ (x-x0)(x-x1)\ldots(x-x_n)f(x,x0,x1,x2,\ldots,xn)$$

This formula can be expressed as a TK!Solver rule as follows, given a user function comprising a Domain list, X, and a range list, F, and assuming that the Domain list is of the form $-2, -1, 0, 1, 2, 3, \ldots, n$:

```
F01 = (F(0)−F(1))/(0−1)
F12 = (F(1)−F(2))/(1−2)
F23 = (F(2)−F(3))/(2−3)
F34 = (F(3)−F(4))/(3−4)
F45 = (F(4)−F(5))/(4−5)

    . . .
F012 = (F01−F12)/(0−2)
F123 = (F12−F23)/(1−3)
F234 = (F23−F34)/(2−4)
F345 = (F34−F45)/(3−5)

    . . .
F0123 = (F012−F123)/(0−3)
F1234 = (F123−F234)/(1−4)
F2345 = (F234−F345)/(2−5)

    . . .
F01234 = (F0123−F1234)/(0−4)
```

$$F12345 = (F1234-F2345)/(1-5)$$

. . .

$$F012345 = (F01234-F12345)/(0-5)$$

. . .

$$FX = F(0) + (x-0)*F01 + (x-0)*(x-1)*F012 + (x-0)*(x-1)*(x-2)*F0123$$
$$+ (x-0)*(x-1)*(x-3)*F01234 + (x-0)*(x-1)*(x-2)*(x-3)*(x-4)*$$
$$F012345$$

This model can be expanded to greater accuracy, depending on the number of values in the user function. Note that the model will not include the last term of Newton's divided difference formula, as it includes the value x, and so does not appear in the difference table created in the first section of the model. This term is usually referred to as the remainder after $n+1$ terms, or just the error term. Also note that you could generalize the model by referring to the domain elements by using the ELEMENT function rather than the actual values.

You can also use Lagrange's interpolation formula, which is closely related to Newton's divided difference formula, and takes the following form:

$$f(x) = \frac{(x-x1)(x-x2)...(x-xn)}{(x0-x1)(x0-x2)...(x0-xn)}\, f(x0)$$

$$+ \frac{(x-x0)(x-x2)...(x-xn)}{(x1-x0)(x1-x2)...(x1-xn)}\, f(x1)$$

$$+ \frac{(x-x0)(x-x1)...(x-xn)}{(x2-x0)(x2-x1)...(x2-xn)}\, f(x2)$$

$$...............................................$$

$$+ \frac{(x-x0)(x-x1)...(x-x_{n-1})}{(xn-x0)(xn-x1)...(xn-x_{n-1})}\, f(xn)$$

Since you do not have to form the divided differences, you can express this formula as a single rule by using only four points for the function:

$$FX = (((x-1)*(x-2)*(x-3)*(x-4))/((0-1)*(0-2)*(0-3)*(0-4)))*F(0)$$
$$+ (((x-0)*(x-2)*(x-3)*(x-4))/((1-0)*(1-2)*(1-3)*(1-4)))*F(1)$$
$$+ (((x-0)*(x-1)*(x-3)*(x-4))/((2-0)*(2-1)*(2-3)*(2-4)))*F(2)$$

$$+ \ (((x-0)*(x-1)*(x-2)*(x-4))/((3-0)*(3-1)*(3-2)*(3-4)))*F(3)$$
$$+ \ (((x-0)*(x-1)*(x-2)*(x-3))/((4-0)*(4-1)*(4-2)*(4-3)))*F(4)$$

Again, you can generalize the model by using the element function to extract the values from the domain list.

## NUMERIAL DIFFERENTIATION AND INTEGRATION

If a function is displayed in tabular form, you can use various interpolation formulas to find an arbitrary derivative of the function, given a sufficient number of values to form the required differences. One formula that was not considered earlier for interpolation, but that can be used for differentiation, is the forward Gregory-Newton formula. It can be derived from the Newton divided difference formula for functions based on regular intervals, that is, uniform intervals in the domain list. This formula is particularly well-suited to the lower values of the domain list as it makes use of succeeding values of the domain, but not of preceding values. The backward Gregory-Newton formula would be appropriate for the higher values of the domain list. Its derivation is similar to that of the forward version.

Using the forward Gregory-Newton formula, you can derive the following formulas for successive derivatives at the value x0 in the domain

$$f'(x0)=(1/h)(\Delta f0 - (1/2)\Delta^2 f0 + (1/3)\Delta^3 f0 - (1/4)\Delta^4 f0 + ...)$$
$$f''(x0)=(1/h^2)(\Delta^2 f0 - \Delta^3 f0 + (11/12)\Delta^4 f0 - ...)$$
$$f'''(x0)=(1/h^3)(\Delta^3 f0 - (3/2)\Delta^4 f0 + ...)$$

You can combine these or similar formulas derived from other interpolation formulas with the difference models and perform differentiation on your user functions, whether or not you have an actual equation for the function. Assuming that you have constructed a list for each difference DELTA1, DELTA2, DELTA3, DELTA4,..., and have defined user functions DIF1, DIF2, DIF3, DIF4,... for each of the lists, you can write the TK!Solver rule to find the derivatives at an arbitrary domain value in the lower part of the list (lower values).

```
h = ELEMENT('X,1)−ELEMENT('X,2)
Fprime = (1/h)*(DIF1(N)−(1/2)*DIF2(N) + (1/3)*DIF3(N)−(1/4)*DIF4(N))
F2prime = (1/h^2)*(DIF2(N)−DIF3(N) + (11/12)*DIF4(N))
```

Of course, you must input a value for N in the input column of the Variable sheet. If you wished to find the derivative at the higher domain values, you could construct a similar model using formulas derived from the backward Gregory-Newton formula.

To perform numerical integration, you would use Gregory's formula of numerical integration, which is derived from the Euler-McLaurin summation formula—again, the derivation is left to the mathematical textbooks.

$$\int_{x_o}^{x_n} f(x)dx = h(\frac{f_o}{2} + f_1 + \ldots + f_{n-1} + \frac{fn}{2})$$

$$- \frac{h^2}{12}[\frac{1}{h}(\Delta f_{n-1} + \frac{\Delta^2 fn-2}{2} + \frac{\Delta^3 fn-3}{3} + \frac{\Delta^4 fn-4}{4} + \ldots)$$

$$- \frac{1}{h}(\Delta f_o - \frac{\Delta^2 f_o}{2} + \frac{\Delta^3 f_o}{3} - \frac{\Delta^4 f_o}{4} + \ldots)]$$

$$+ \frac{h^4}{720}[\frac{1}{h^3}(\Delta^3 f_{n-3} + \frac{3}{2}\Delta^4 f_{n-4} + \ldots)$$

$$- \frac{1}{h^3}(\Delta^3 f_o - \frac{3}{2}\Delta^4 f_o + \ldots)]$$

$$+ \ldots\ldots$$

$$= h(\frac{f_o}{2} + f_1 + \ldots + f_{n-1} + \frac{fn}{2}) - \frac{h}{12}(\Delta f_{n-1} - \Delta f_o)$$

$$- \frac{h}{24}(\Delta^2 f_{n-2} + \Delta^2 f_o) - \frac{10h}{720}(\Delta^3 f_{n-3} - \Delta^3 f_o)$$

$$- \frac{3h}{160}(\Delta^4 f_{n-4} + \Delta^4 f_o) - \ldots$$

First you must form the differences of the function you wish to integrate, just as you did for differentiation. With the same list names as above, the Gregory numerical integration model would appear thus on an interval from 1 to 4:

IFX = h*(FX(1)/2+FX)2)+FX(3)+FX(4)/2)−(h/12)*(DIF1(3)−DIF1(1))
−(h/24)*(DIF2(2)+DIF2(1))−(19*h/720)*(DIF3(1)−DIF3(1))

If you wished to integrate over a longer interval, you would use a greater number of terms with higher order differences; in fact, the last term in the model above will vanish because of the small interval. Notice that, if you ignore the correction terms (That is, the terms that include the differences), you are left with the trapezoidal rule of integration encountered in your first calculus course.

Returning to the List Solver, you can compute the running integral of a tabular function using the trapezoidal rule

$$\int_a^b f(x)dx = h(\frac{f_o}{2} + f_1 + f_2 + \ldots + f_{n-2} + f_{n-1} + \frac{f_n}{2})$$

The running integral is simply another output list that can be computed along with several of the other lists covered earlier. The rule for computation of the running integral is

```
OK = STEP(ELEMENT( ),2)
OK*   RX = ELEMENT('RX,ELEMENT( )−1,0) + h*((FX + ELEMENT('FX,
      ELEMENT( )−1,0))/2)   *OK
(1−OK)*   RX = 0   *(1−OK)
```

The logical variable is added so that you will not attempt to calculate an area for the first table entry, which is the lower boundary of the integral.

## LEAST SQUARES CURVE FITTING

In working with real-world models, you often have to determine a formula that describes data gathered experimentally, or data calculated at intervals and presented in tabular form. If you are confident that the data contain no significant errors, as could be the case with calculated values, you can relate the various points with interpolation formulas. If the data have been gathered experimentally, however, they may contain errors of measurement—in other words, the data are uncertain. In this case, interpolation formulas could complicate matters unnecessarily by defining a function that passes through all the points. For instance, you might measure the pressures and flow rates of a pump and obtain a series of points as shown in Figure 13-3. If you constructed the equation of a curve that passes through all the points, you would not produce an accurate model—the pump curve should be a smooth curve. The bumps in the curve are caused by errors of observation.

The reason for producing a pump curve is to predict performance at points other than those at which the pump is tested. The pump's performance is actually described by a smooth curve that

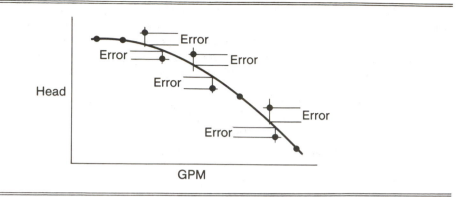

Figure 13–3. A Pump Curve with Errors of Observation.

passes through some of the measured points, and between others. You need to construct a curve that fits the experimental data "as close as possible."

What is meant by "as close as possible?" The answer to that question depends on a number of factors, such as the accuracy of measurements, the method of interpolation used in reading instruments, the method of rounding off values, and so on. If all these factors conspire to produce experimental values that are consistently high, the correct curve might pass through the lowest points. In some cases, the data may appear to be following a curve when the correct curve is actually a straight line.

In the remainder of the discussion it will be assumed that experimental data are obtained in a manner that produces a set of values containing realistic, random errors of observation. In this case, the generally accepted criterion of the best fit is the least squares criterion. The method of applying the criterion to data is called the method of least squares.

The objective of the method of least squares is to define a function, an equation, so that, for each value of the independent variable, the aggregate of the differences between the experimental values of the dependent variable and the calculated values of the dependent variable is as small as possible. You can't use the sum of the errors for each value—the distance between the experimental value and the value calculated with the equation—because large positive and negative errors could cancel out, and give the false impression of a good fit. Instead, you should use the sum of the squares of the errors, and try to make that sum as small as possible.

To apply the method of least squares, select an arbitrary equation, such as a straight line or a parabola, using variables for the coefficients of the various terms of the equation. Then find the coefficients, and finally, find the sum of the squares of the errors. You can repeat the process for several different forms, and select the equation with the lowest total error if you do not know the form the equation should take. Of course, theoretical considerations may dictate the general form of the equation, in which you would only fit an equation of that form.

Again, let's skip the derivation of the method of least squares and proceed directly to a description of TK!Solver's implementation of the method. Consider the following data stored in a TK!Solver model as two lists, X and Y, and related by FX, a user function:

$$
\begin{array}{c|ccccc}
x & -3 & -2 & 0 & 3 & 4 \\
\hline
y & 18 & 10 & 2 & 2 & 5
\end{array}
$$

You wish to fit a parabolic equation of the form

$$y = a + bx + cx^2$$

to the data for x and y. If you substitute the values from the two lists, you will obtain a set of equations in which $a$, $b$, and $c$ are the variables:

$$
\begin{aligned}
a - 3b + \phantom{0}9c &= 18 \\
a - 2b + \phantom{0}4c &= 10 \\
a \phantom{- 2b + 00c} &= \phantom{0}2 \\
a + 3b + \phantom{0}9c &= \phantom{0}2 \\
a + 4b + 16c &= \phantom{0}5
\end{aligned}
$$

Unfortunately, you cannot solve this system of equations: TK!Solver will give the error, Overdefined. However, you can derive three normal equations from the five equations, and solve for $a$, $b$, and $c$ using the system of the three equations. To form the normal equations, you would construct an equation corresponding to each of the variables. To form the first normal equation, you would multiply each of the equations by the coefficient of the first variable in that equation, and add the equations thus formed. The result is the first normal equation. Now continue forming a normal equation for each variable.

In the system above, the first normal equation is simply the sum of the equations, since the coefficient of $a$ is 1 in each of the equations. To form the second normal equation, you would multiply each equation by the coefficient of $b$ in that equation; that is, you would multiply the first equation by $-3$, the second by $-2$, the third by 0, the fourth by 3, and the last by 4. The sum of the equations is

$$2a + 38b + 56c = -48$$

which is the second normal equation. The third normal equation, formed by using the coefficients of $c$, is

$$38a + 56b + 434c = 300$$

You can easily solve this systems of equations, finding

$$a = 1.82, \qquad b = -2.56, \qquad c = 0.87$$

and consequently, the equation for the curve is

$$y = 1.82 - 2.56x + 0.87x^2$$

You can let TK!Solver handle these manipulations for you with a few simple rules, so that you have only to enter the lists of data. In fact, several curves can be fitted at once, or, the form of the curve you wish to fit can be selected from a series of alternatives by means of the menu techniques discussed earlier.

Let's work backwards from the rules for the normal equations, which you can enter immediately.

```
Y1 = C1a*a + C1b*b + C1c*c
Y2 = C2a*a + C2b*b + C2c*c
Y3 = C3a*a + C3b*b + C3c*c
```

Once you have supplied values for the coefficients of a, b, and c, and for Y1, Y2, and Y3, solving the system of equations is a snap for TK!Solver. Entering the rules the first time requires a little typing, but a pattern soon emerges, and you have a model that can be solved without the aid of the List Solver. Enter these rules above the three rules for the normal equations:

```
Y1 = SUM(Y)

Y2 = ELEMENT('Y,1)*ELEMENT('X,1) + ELEMENT('Y,2)*ELEMENT('X,2)
   + ELEMENT('Y,3)*ELEMENT('X,3) + ELEMENT('Y,4)*ELEMENT('X,4)
   + ELEMENT('Y,5)*ELEMENT('X,5)

Y3 = ELEMENT('Y,1)*ELEMENT('X,1)^2 + ELEMENT('Y,2)*ELEMENT('X,2)^2
```

```
  + ELEMENT('Y,3)*ELEMENT('X,3)^2 + ELEMENT('Y,4)*ELEMENT('X,4)^2
  + ELEMENT('Y,5)*ELEMENT('X,5)^2

C1a = 5

C1b = SUM('X)

C1c = ELEMENT('X,1)^2 + ELEMENT('X,2)^2 + ELEMENT('X,3)^2
    + ELEMENT('X,4)^2 + ELEMENT('X,5)^2

C2a = SUM('X)

C2b = C1c

C2c = ELEMENT('X,1)^3 + ELEMENT('X,2)^3 + ELEMENT('X,3)^3
    + ELEMENT('X,4)^3 + ELEMENT('X,5)^3

C3a = C2b

C3b = C2c

C3c = ELEMENT('X,1)^4 + ELEMENT('X,2)^4 + ELEMENT('X,3)^4
    + ELEMENT('X,4)^4 + ELEMENT('X,5)^4
```

TK!Solver will solve this model in one pass and provide the coefficients to the equation of the fitted curve. However, you can also add an input list for x (*not* X, which is your list of data) and an output list for y, and generate the fitted curve point by point. You can then plot the fitted curve and the original data points on the same graph. However, you need to add the following rules:

$$y = a + b*x + c*x^2$$
$$YI = FX(x)$$

If you have set the mapping of FX to Linear, TK!Solver can plot the data points connected by straight line segments, along with the curve of the equation $y = a + bx + cx^2$. The plot sheet for this model appears in Figure 13-4, and the plot produced appears in Figure 13-5.

At this point we will leave the subject of curve fitting, but the reader is encouraged to investigate other techniques, such as orthogonal polynomials, and to investigate the fitting of other types of curves.

```
Name        Elements  Unit     Comment       Screen or Printer:        Printer
----        --------  ----     -------        Title:                    CURVE FITTING MODEL
Y           5                                 Display Scale ON:         Yes
X           5                                 X-Axis:                   x
x           20                                Y-Axis     Character
YI          20                                ------     ---------
y           20                                 y          *
                                               YI         #
```

```
     St Input       Name     Output   Unit    Comment
     -- -----       ----     ------   ----     -------
                                               CURVFIT2: CURVE FITTING MODEL
                                               ****************************
        1.8246628  A
        -2.645472  B
        .87283237  C

                    Y1       37
                    C1A      5
                    C1B      2
                    C1C      38
                    Y2       -48
                    C2A      2
                    C2B      38
                    C2C      56
                    Y3       300
                    C3A      38
                    C3B      56
                    C3C      434
                    X
                    Y
     L              y
     L   1          x
     L              YI
```

```
S Rule
- ----
* Y1 = SUM('Y)
* Y2 = ELEMENT('Y,1)*ELEMENT('X,1)+ELEMENT('Y,2)*ELEMENT('X,2)+ELEMENT('Y,3)*EL
* Y3 = ELEMENT('Y,1)*ELEMENT('X,1)^2+ELEMENT('Y,2)*ELEMENT('X,2)^2+ELEMENT('Y,3

* C1A = 5
* C1B = SUM('X)
* C2A = SUM('X)

* C1C = ELEMENT('X,1)^2+ELEMENT('X,2)^2+ELEMENT('X,3)^2+ELEMENT('X,4)^2+ELEMENT
* C2C = ELEMENT('X,1)^3+ELEMENT('X,2)^3+ELEMENT('X,3)^3+ELEMENT('X,4)^3+ELEMENT
* C3C = ELEMENT('X,1)^4+ELEMENT('X,2)^4+ELEMENT('X,3)^4+ELEMENT('X,4)^4+ELEMENT

* C2B = C1C
* C3A = C2B
* C3B = C2C

* Y1 = C1A*A+C1B*B+C1C*C
* Y2 = C2A*A+C2B*B+C2C*C
* Y3 = C3A*A+C3B*B+C3C*C

* y = A + B*x +C*x^2
* YI = FX(x)
```

Figure 13-4.  The Plot Sheet for the Curve-fitting Model.

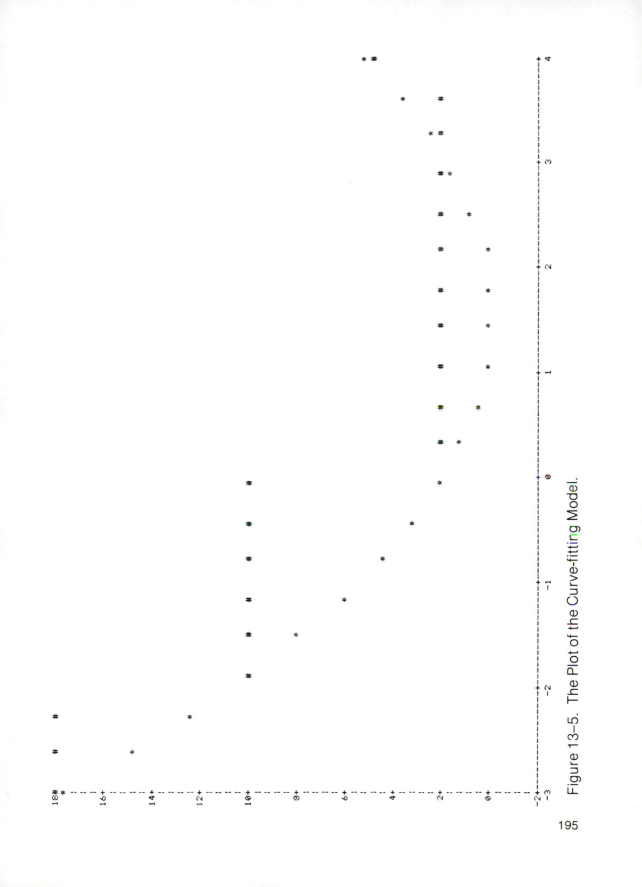

Figure 13–5.  The Plot of the Curve-fitting Model.

# CHAPTER 14

# Periodic Functions, Fourier Series, Laplace Transforms

Many engineering problems, regardless of the discipline, involve periodic functions—vibration problems in mechanical engineering and alternating current problems in electrical engineering are just two of the most obvious examples. In fact, the various problems involving periodic functions are so similar that a single model can represent a variety of systems, given the proper units in each case. Furthermore, vibrating mechanical systems can be modeled physically with electrical circuits, with the electrical measurements providing exact values of forces and displacements in the corresponding mechanical system when scaled with appropriate factors. Although there would be little value in doing so, electrical circuits could also be modeled with mechanical devices because of their similarity in mathematical terms.

## LINEAR DIFFERENTIAL EQUATIONS

Engineering problems usually present themselves in the form of differential equations, and their solutions are usually integrals of the functions that appear in the equations. For instance, you may wish to know the distance a projectile will travel, given its initial velocity and a function describing its drag in terms of velocity. The velocity function integrated over time gives the distance traveled, and the velocity is a function of the forces that act on the projectile—gravity and drag.

As mentioned earlier, TK!Solver cannot perform symbolic differentiation and integration; you cannot input a function and obtain the integral or derivative of the function. However, TK!Solver can help you with either task.

About the simplest differential equation in engineering problems is the linear differential equation with constant coefficients. The general form of the second order differential equation is

$$y'' + P(x)y' + Q(x)y = R(x)$$

You may recall that this equation is nonhomogeneous because of the term $R(x)$. If the term $R(x)$ is 0, the equation is homogeneous.

The solution of the general equation, in either the homogeneous or the nonhomogeneous form, is possible only in special cases, and they are beyond the scope of this book. But, when the coefficients $P(x)$ and $Q(x)$ are constant, the equations can be solved in either form by a number of methods.

A complete solution of a linear differential equation consists of two parts. The first part is the solution to the homogeneous equation, and it is called the complementary function. The second part is the particular integral, which is any solution to the nonhomogeneous equation. The complete solution takes the form

$$y = c_1 y_1 + c_2 y_2 + y$$

where the first two terms on the right are the complementary function and the last term is the particular integral.

If a linear differential equation is of the standard form

$$ay'' + by' + cy = f(x)$$

you might be tempted to find a solution of the form $y = e^{mx}$, because all derivatives of that function are identical, except for numerical constants. Thus a solution would take the form

$$am^2 e^{mx} + bm e^{mx} + c e^{mx} = 0$$

When you factor $e^{mx}$ from each term, you are left with the characteristic equation

$$am^2 + bm + c = 0$$

which can be solved for m by inspection, or the quadratic formula.

The characteristic equation of a differential equation is formed by replacing $y''$ with $m^2$, $y'$ with $m$, $y$ with 1, and $f(x)$ with 0, thereby obtaining the equation

$$am^2 + bm + c = 0$$

immediately.

Now, it would be useful to be able to type the characteristic equation into a TK!Solver model and let the program do the work. Unfortunately, neither the Direct Solver nor the Iterative Solver knows what to do with a quadratic equation. However, you can enter the quadratic formula in both alternative forms and obtain the two roots of the characteristic equation:

$$m1 = (-b + \text{SQRT}(b\hat{} 2 - 4*a*c))/(2*a)$$
$$m2 = (-b - \text{SQRT}(b\hat{} 2 - 4*a*c))/(2*a)$$

The solution to the homogeneous equation is

$$y = c_1 e^{m1x} + c_2 e^{m2x}$$

At this point, you should recall that TK!Solver cannot handle imaginary numbers or complex numbers. Therefore, you need to protect your model with a logical variable.

If the quantity $b^2 - 4*a*c$ is negative, the roots of the characteristic equation will be conjugate complex quantities. You can use the STEP function to form a logical variable, and can multiply each side of the two rules for m1 and m2 so that they will only compute roots:

$$\text{real} = \text{STEP}(b\hat{} 2 - 4*a*c, 0)$$

$$\text{real}* \quad m1 = (-b + \text{SQRT}(b\hat{}2 - 4*a*c))/(2*a) \quad *\text{real}$$
$$\text{real}* \quad m2 = (-b - \text{SQRT}(b\hat{}2 - 4*a*c))/(s*a) \quad *\text{real}$$

All is not lost if the roots to the characteristic equation are complex numbers—the solution merely takes a different form, that of

$$y = e^{px}(A \cos qx + B \sin qx).$$

The TK!Solver rules for finding $p$ and $q$ are

$$p = -b/(2*a)$$
$$q = \text{SQRT}(4*a*c - b\hat{}2)$$

The use of linear differential equations can be illustrated by a mechnical system that can be described with an equation of the form considered above. An electrical system would also suffice.

# THE TRANSLATIONAL–MECHANICAL SYSTEM

Consider now the mechanical system shown in Figure 14–1, which consists of a weight, a spring, and a damper. The displacement of the weight is described by the differential equation

$$\frac{w}{g}\frac{d^2y}{dt^2} + c\,\frac{dy}{dt} + ky = F_0 \cos \omega t$$

If the right side of the equation is set to 0, it describes the free motion of the weight, and corresponds to the homogeneous equation.

The term on the right side of the equation is known as the forcing function, and the equation as shown describes the forced motion of the weight. This is the nonhomogeneous equation.

You can determine the damping characteristics of the system by solving the characteristic equation

$$\frac{w}{g}\,m^2 + cm + k = 0$$

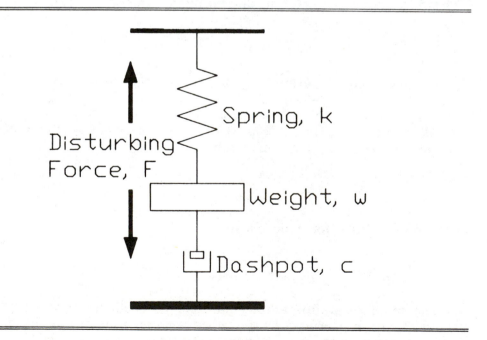

Figure 14–1. A Translational-Mechanical System.

using the TK!Solver rules

$$m1 = -(c*g)/(2*w) + (g/(2*w))*SQRT(c^2 - 4*k*w/g)$$
$$m2 = -(c*g)/(2*w) - (g/(2*w))*SQRT(c^2 - 4*k*w/g)$$

and

$$y = A*e(\hat{})^{m1*t} + B*e(\hat{})^{m2*t}$$

If m1 and m2 are real and unequal, the system is overdamped, and the weight, when displaced, will return to a neutral position without oscillation.

If m1 and m2 and real and equal in this model, the system is said to be critically damped. The exact value of damping that produces a critically damped system, the critical damping, is given by the rule

$$cc = 2*SQRT(k*w/b)$$

The difference between overdamped and critically damped systems is one of degree; in either case, the displaced weight eventually returns to a neutral position without oscillation. A plot of the motion of such a system is shown in Figure 14-2, which uses arbitrary constants.

If the quantity $c^2 - 4kw/g$ is less than 0, the system is said to be underdamped. As noted above, the solution to the differential equation takes a different form:

$$y = e^{-pt}(A \cos qt + B \sin qt)$$

*Where:*

$$p = \frac{cg}{2w} \quad \text{and} \quad q = \frac{g}{2w} V \frac{\text{/}4kw}{g} - c^2$$

Since you are solving the homogeneous equation (without a forcing function), you obtain the free motion of an underdamped system, also known as damped oscillation. The motion of such a system, simulated by a TK!Solver model, using arbitrary constants for A and B is shown in Figure 14-3.

Although the actual motion of the weight is of interest, the envelope of the function—that is, the curve defined by successive extreme displacements on the same side of the equilibrium position—is often more interesting. If you were to generate a list of

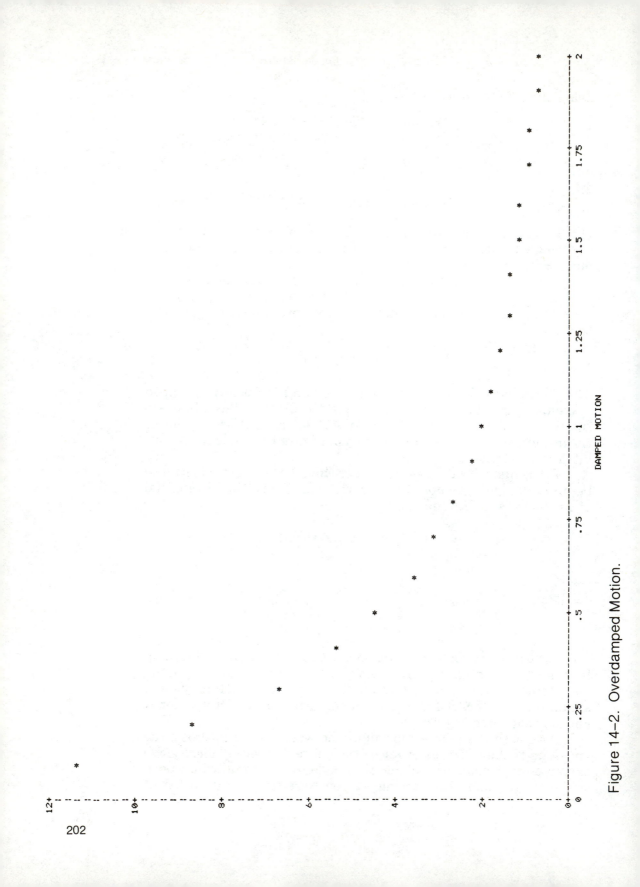

Figure 14–2. Overdamped Motion.

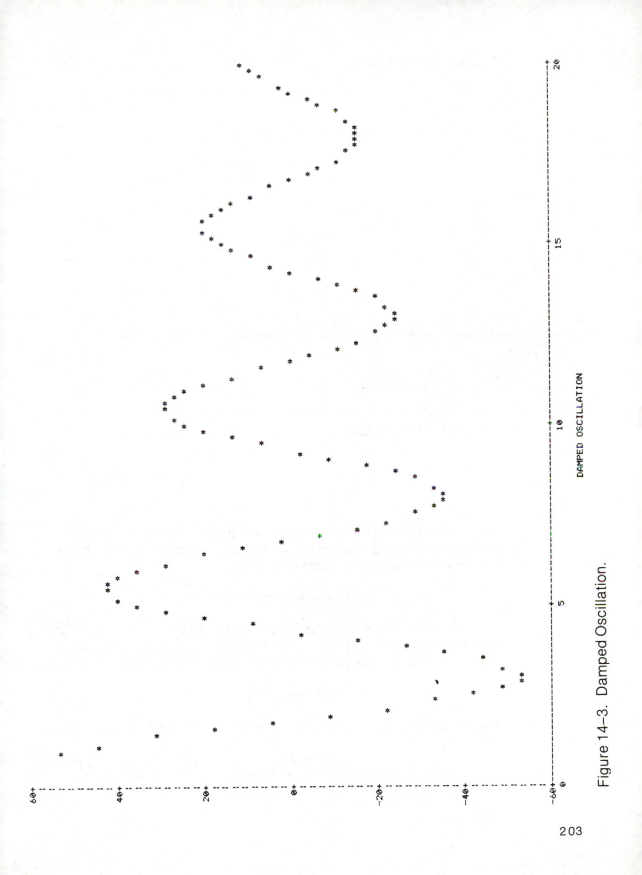

Figure 14-3. Damped Oscillation.

```
(1c) Comment: DAMPED OSCILLATION                                    198/!

===================== VARIABLE SHEET =================================
St Input      Name    Output      Unit    Comment
-- -----      ----    ------      ----    -------
                                          DAMPED OSCILLATION

              p       .08
    100       w
    .5        c
    32        g
              q       1.2623787
    5         k
L             y
L   .1        t
    50        A
    45        B



(1r) Rule: p=c*g/(2*w)                                              198/!

===================== RULE SHEET ====================================
S Rule
- ----
* p=c*g/(2*w)
* q=(g/(2*w))*sqrt((4*k*w/g)-c^2)
* y=e()^(-p*t)*(A*cos(q*t)+B*sin(q*t))
```

Figure 14–3. (continued).

these extreme displacements, you would find that the ratio of extreme positive or negative displacements is constant, and that it can be computed from values at hand:

$$\text{yextreme:yextreme}+2 = e^{2\,pi\,p/q}$$

The free motion of a system is only part of the story—you need to consider the forced motion, which means finding a particular integral of the general equation. If you assume a solution of the form

$$Y = A\,\cos\omega t + B\,\sin\omega t$$

then with a little manipulation you will arrive at the rules for A and B:

A = (k−omega^2*(w/g))*F0/((k−omega^2*(w/g))^2 + (omega*c)^2
B = (omega*c)*F0/((k−omega^2*(w/g))^2 + (omega*c)^2

# FOURIER SERIES

Periodic phenomena are common in the engineering world, but the solutions are usually not simple sine and cosine functions. For example, the computer you use to run TK!Solver makes use of many periodic waveforms, but most of them are not sinusoidal. Instead, they are pulses or square waves (Figure 14–4).

    Although these functions can be described verbally or graphically, it is only with difficulty that they can be manipulated mathematically because they are not continuous. Life would be much simpler if such functions could be described in terms of sines and cosines. For instance, you might wish to describe a square wave function as the series

$$f(t) = \frac{1}{2}a0 + a1\cos t + a2\cos 2t + a3\cos 3t + \ldots an\cos nt$$

$$+ b1\sin t + b2\sin 2t + b3\sin 3t + \ldots bn\sin nt$$

The investigation of the existence of such series and the determination of their exact forms are known as Fourier analysis. One method of formulating a series of this type is called the Fourier series. As usual, the derivation of the formulas used will not be presented here—instead, their use is illustrated in the TK!Solver environment.

    If you wish to describe a function as a Fourier series using the form given above, you must determine the coefficients $a0$, $a1$, $a2, \ldots, an$, $b1$, $b2, \ldots, bn$. You can accomplish this objective with the Euler (or Euler-Fourier) formulas:

$$a0 = \ldots \quad \frac{1}{\pi} \int_{d}^{d+2\pi} f(t)dt$$

$$an = \ldots \quad \frac{1}{\pi} \int_{d}^{d+2\pi} f(t)\cos nt\, dt$$

$$bn = \ldots \quad \frac{1}{\pi} \int_{d}^{d+2\pi} f(t)\sin nt\, dt$$

Each of these formulas contains an integral of f(t), which must be evaluated. You can perform this integration manually, using tables of integrals, or, at worst, you can do it by some numerical method.

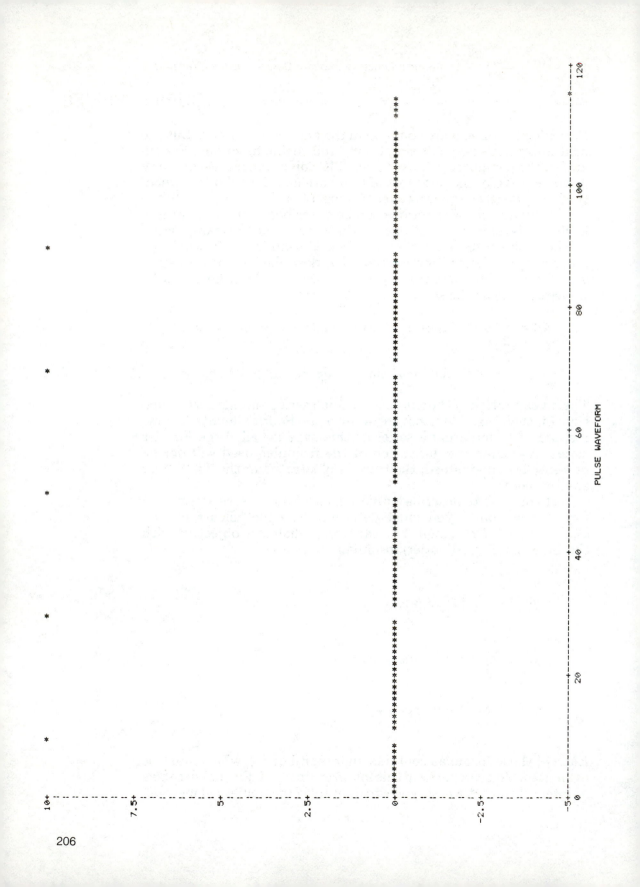

PULSE WAVEFORM

SQUARE WAVE

Figure 14–4. Pulses and Square Waves.

Once you have determined the coefficients, it is a simple matter to formulate the TK!Solver rule for the series. Figure 14–5 illustrates the model of the function

$$f(t) = \begin{cases} 0, & -pi < t < 0 \\ \sin t, & 0 < t < pi \end{cases}$$

and the plot produced by the model.

Unlike most of the models you have seen thus far, the Fourier series model is best solved in two steps. First, you use an input list, n, and the List Solver to compute the coefficients for the number of terms you wish to use. Of course, the number of terms affects the accuracy with which the series represents the function. The more terms used, the closer the approximation will be. In this respect, TK!Solver offers a tremendous advantage over manual methods—it is no more trouble to construct a series with hundreds of terms than it is to construct one having only a few terms.

The second step is to evaluate the series using t as an input list, and the List Solver to expand the series, producing the function as an output list that can be displayed as a table or a plot.

# THE LAPLACE TRANSFORM

If you remember, certain differential equations can be solved by algebraic methods used with their characteristic equations. Although the characteristic equation method works only with that type of equation, the Laplace transformation offers the same advantages in the solution of a more general class of equations. The Laplace transformation is defined as follows:

$$L[f(t)] = \int_0^\infty f(t)e^{-st}dt$$

When you apply the Laplace transformation to a function of t, you transform it into a function of s:

$$F(s) = L[f(t)]$$

Of course, the advantage of using the Laplace transformation is that the transformed equation in s can be solved by strictly algebraic means. You begin with a differential equation, or possibly an integrodifferential equation, in $t$, for which you seek the

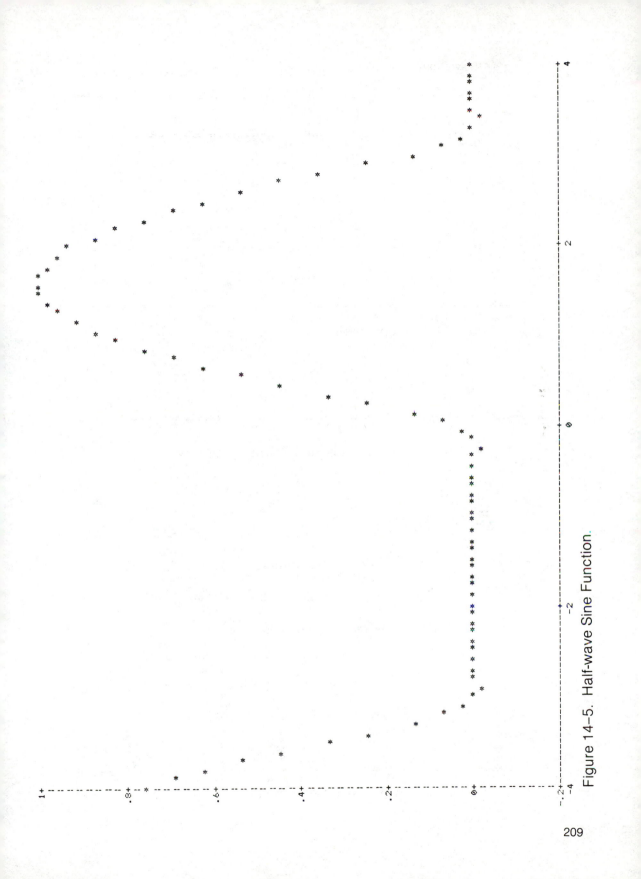

Figure 14–5. Half-wave Sine Function.

```
(2s) Status: L                                                          199/!

=================== VARIABLE SHEET ============================================
St Input        Name    Output      Unit    Comment
-- -----        ----    ------      ----    -------
L               f       -.0047163
L    0          t




================== RULE SHEET ================================================
S Rule
- ----
* f=1/pi()+sin(t)/2-(2/pi())*(cos(2*t)/3+cos(4*t)/15+cos(6*t)/35+cos(8*t)/63)
```

Figure 14–5. (continued).

**Differential Equation and Initial Conditions:**

$$\frac{dx}{dt} + 3x = 0$$

$$x(o) = 2$$

1. **Transform differential equation:**

$$sx(s) - 2 + 3x(s) = 0$$

2. **Solve transformed equation:**

$$x(s) = \frac{2}{s + 3} = 2 \ \frac{1}{s + 3}$$

3. **Find inversion of transform:**

$$x(t) = 2\epsilon^{-3t}$$

Figure 14–6.  Solution of a Differential Equation—Laplace.

complete solution. You transform this differential equation in $t$ to one in $s$ by the various theorems of the Laplace transformation, and thus obtain an equation with no derivatives. When you have reduced the transformed equation in s to one of a number of standard forms, you can take the inverse Laplace transform of the equation obtaining the solution a function in t, which, when substituted into the original differential equation, will reduce to an identity (see Figure 14–6).

Obviously a considerable amount of algebraic manipulation is necessary when the Laplace transformation is applied. Furthermore, TK!Solver appears to be of limited use in the actual manipulation of symbols—or is it?

TK!Solver was designed to be a generic tool, one that will eventually become indispensable. It allows the user to employ techniques that are impractical manually, or even with the use of traditional programming languages. Sooner or later, the user becomes addicted to the more sophisticated methods, and can no longer work without TK!Solver. (Why do you think they call us users?)

The traditional method of using the Laplace transformation begins in a mathematical handbook such as *Standard Mathematical Tables,* published by CRC Press. This handbook includes a comprehensive list of functions and their Laplace transforms. You begin implementation of the Laplace transformation method by storing this table, or as much of it as is commonly used, in two TK!Solver lists. Then you relate the two lists with a user function. The List subsheets, the User Function sheet, and the User Function subsheet are shown in Figure 14–7. You need only one rule to use this function:

$$Fs = L(ft)$$

You enter the function in $t$ into the Input field of $ft$ using the single quotation mark to signify that it is a symbolic value, and give the Action command. The desired transform appears in the Output field of $Fs$.

Once you have simplified the transformed equation by whatever means, you can enter the terms in s into the Input field of Fs one at a time, and obtain terms in t as the output. Notice that the mapping of L has been left in its default, Table, so that each entry has a unique inverse.

TK!Solver can help you manipulate the algebraic equation in s, even if it cannot do so unattended. Typically, you will be presented with an equation containing a fraction of two

| Function | Graph | Transform |
|:---:|:---:|:---:|
| $\mu(t)$ | | $\dfrac{1}{s}$ |
| $t\mu(t)$ | | $\dfrac{1}{s^2}$ |
| $t^n\mu(t)$ | | $\dfrac{n!}{s^{n+1}}$ |
| $e^{-at}\mu(t)$ | | $\dfrac{1}{s+a}$ |
| $t^n e^{-at}\mu(t)$ | | $\dfrac{n!}{(s+a)^{n+1}}$ |
| $\sin kt\,\mu(t)$ | | $\dfrac{k}{s^2+k^2}$ |

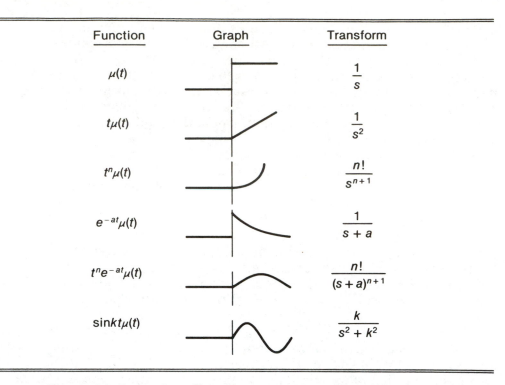

Figure 14–7. Laplace Transforms.

polynomials in s—at best, only the denominator will be a polynomial, and usually both the numerator and denominator will be polynomials. Although you could eventually compile a comprehensive user function L, you must generally be able to find the roots of the polynomials, and to separate them into sums of simpler forms.

If you have obtained an equation in $s$, you should first try to find the inverse with your user function L; if there is no entry for the function, TK!Solver will place a $>$ in the Status field of the output variable. Then you must resort to factoring the polynomials and applying the method of partial fractions to separate the fraction into a sum of simpler fractions. For instance, consider this equation:

$$L[y(t)] = (s + 1)/(s^2 + s - 6)$$

Although you might place this function in a general form into your user function, after you solve it the first time, you will notice that it is not in the table listed above. The first step, after arriving at this equation, is to find the roots of the denominator of the right side of the equation, so that you can factor it. Of course, you can easily find the roots of this expression by inspection, but in general, you can use the quadratic formula to find the roots. You can apply a similar method to cubic equations (Figure 14–8).

Cubic Equation:

$$y^3 + py^2 + qy + r = 0;$$

substitute x $-$ $\dfrac{p}{3}$ for $y$,

$$x^3 + ax + b = 0$$

*Where:*

$$a = \frac{1}{3}(3q - p^2)$$

$$b = \frac{1}{27}(2p^3 - 9pq + 27r);$$

*Let:*

$$A = \sqrt[3]{+\frac{b}{2} + \sqrt{\frac{b^2}{4} + \frac{a^3}{27}}}$$

$$B = -\sqrt[3]{+\frac{b}{2} + \sqrt{\frac{b^2}{4} + \frac{a^3}{27}}}$$

*Then:*

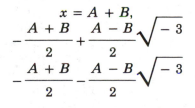

$$x = A + B,$$
$$-\frac{A + B}{2} + \frac{A - B}{2}\sqrt{-3}$$
$$-\frac{A + B}{2} - \frac{A - B}{2}\sqrt{-3}$$

Figure 14–8. Solution of the Cubic Equation.

TK!Solver lets you implement another method in solving the cubic equation. You will recall that every cubic equation has at least one real root. If you find that root, the resulting factor can be factored out, leaving a quadratic equation that can be solved with the quadratic formulas. The facility that TK!Solver offers is the POLY( ) function. Note that when one of the real roots is substituted for s, the expression in s will be equal to 0. To use the POLY( ) function, enter this rule:

$$ZERO = POLY(s,a,b,c,d)$$

Next, enter the coefficients a, b, c, and d, and enter an input list for s. Then fill the s list with the values that range from a high guess to a low guess, and invoke the List Solver, with ZERO set up as the output list. Examine the ZERO list when the List Solver has completed processing. If you find a 0 in the list, that element of the s list is the root. If you find that the ZERO list has changed from positive to negative, but has not produced a value of 0, the root lies between the two elements of the s list that correspond to the elements of the ZERO list on either side of the changed sign; then generate another s list using those two values as the first and last elements, and repeat the process.

When you have obtained the factors of the polynomial in the denominator, you are ready to apply the method of partial fractions in order to separate the function. You can then write the equation in the example above as follows:

$$\frac{s + 1}{(s - 2)(s + 3)} = \frac{A}{(s - 2)} + \frac{B}{(s + 3)}$$

$$= \frac{A(s + 3) + B(s - 2)}{(s - 2)(s + 3)}$$

Now you can multiply both sides of the equation by $(s - 2)(2 + 3)$, collect terms, and solve for $A$ and $B$ using one of the number of techniques. Although you may have to resort to pencil and paper to simplify the algebraic equation, you are still way ahead of the game. Then you set the separated fractions equal to the Laplace transform and proceed with the inversion of each simple term. In this case, you will obtain

$$L[y(t)] = \frac{1}{5} \left( \frac{3}{s - 2} + \frac{2}{s + 3} \right)$$

Inversion of this function is straightforward, with or without the user function:

$$y(t) = \frac{1}{5}\,(3e^{2t} + 2e^{-3t})$$

Now you can enter the inverted function as a TK!Solver rule and use it to model the system that gave rise to the differential equation.

This example highlights an aspect of TK!Solver that we have not emphasized so far. TK!Solver is an interactive tool in every sense of the word. Although it will not perform some tasks, such as symbolic differentiation or integration or the factoring of polynomials, it is nonetheless useful in problems that require those tasks to be performed.

Furthermore, each time that you solve a differential equation using these methods, you can generally enter the results into your user function for future use. Eventually, you may have almost every Laplace transform required for a given type of problem recorded on disk, ready for a speedy retrieval.

This chapter has merely touched on topics that can easily fill volumes, but, of course, the aim here is to survey several applications that do not seem to fall within TK!Solver's power. As you have seen, TK!Solver can indeed solve many more problems than the obvious ones.

# CHAPTER 15

# More on Input and Output

By now it should be obvious that TK!Solver is a powerful tool for creating and solving mathematical models of physical processes. As long as a model can be expressed in numerical terms (no imginary numbers or symbols), TK!Solver can solve it. However, the creation and solution of a model is only part of an engineer's task. Data must be entered, and results must be presented.

You should now be able to enter data directly on the Variable sheet, the Variable subsheets, the List sheet, and the List subsheets. You can speed the process in some cases by using the Action command from the List sheet, which invokes the Fill List option.

You also know that two primary options are available for the presentation of the output data—tables and plots. Tables and plots can be displayed on the screen or printed on the printer. Since TK!Solver will not take advantage of the graphics capabilities that many printers have, the resolution of the plots is limited to the pitch of the printer. With a line printer set for a horizontal pitch of 15 to 16.5 characters per inch, and a vertical pitch of 8 characters per inch, you can obtain reasonably smooth plots, but they will not be line graphs, just isolated points.

## VISITREND/PLOT®

Engineers love graphs, or plots, as they are called in the TK!Solver manual. Graphs are an effective way to present data in a report, and they are useful in many calculations. TK!Solver can generate

VisiTrend/Plot is a registered trademark of VisiCorp.

the information required to construct a smooth, continuous line graph—it just can't plot the graph. To achieve the graph, you need another program, such as VisiCorp's VisiTrend/Plot.

VisiTrend/Plot is a business graphics package that incorporates a number of statistical analysis and forecasting functions. It is definitely a business-oriented product, as its graphing format is generally tailored to produce graphs as functions of time expressed in years, months, or weeks.

The hardware requirements for running VisiTrend/Plot are almost identical to the requirements for running TK!Solver. If you use VisiTrend/Plot on an IBM PC to PC-XT, you will need 128K RAM, compared with 96K for TK!Solver. You will also need a graphics display board and a graphics printer. With those qualifications, an IBM PC that will run TK!Solver will also run VisiTrend/Plot.

VisiTrend/Plot is organized as three subprograms: the Main subprogram, the Plot subprogram, and the Trend subprogram. Any subprogram can be reached from either of the other two. The entire program is operated through menus. The choices available at any stage are displayed at the top or bottom of the screen.

## The Main Subprogram

The Main subprogram is used to perform file management and data editing functions. The available menu choices are Load, Plot, Display, Clear, Files, Edit, Trend, Save, Quit, and Device. Plot and Trend are used to select the other two subprograms.

Load selection is used to load a series (list) from disk into memory. When the program is first loaded, it contains no data, just as TK!Solver is initially loaded without a model. The operation of VisiTrend/Plot's Load command differs from TK!Solver's mainly in that the command is issued by placing the cursor over the word Load in the menu and pressing **ENTER**. After the program lists all the files on the data disk, you move the cursor to the desired file and press **ENTER** again.

You use the Save command to save data on disk, just as you use TK!Solver's Save command to save a model for future use. This command operates in the same manner as the Load command. You place the cursor over the word Save and press **ENTER**. The program first asks whether the data should be saved in Normal, or DIF format. It then lists all the series in memory so that you can select the lists to be saved in the data file. You can save all the lists in memory, one of the lists, or selected lists.

The display command is used to display the series in memory. You simply place the cursor over the word Display and press **ENTER**. The program displays the name of each series, the period of the data, the start date, the end date, and the number of data points in the series. While the list of series is on the screen, you have two additional choices: Exit, and Print. Exit returns you to the Main subprogram menu, and Print sends the contents of the screen to the printer.

The Clear command clears all data from memory; it corresponds to the Reset All command of TK!Solver. As with TK!Solver, the command only clears memory—the data remain on disk if they were previously saved.

The Files command allows you to format disks for data storage, and to delete files on the data disk without leaving the VisiTrend/Plot program. The format function of the Files command has no counterpart in TK!Solver, although you can delete a TK!Solver file with the Delete File option of the Storage command.

The Device command enables you to change the name of the data disk drive. The VisiTrend/Plot program disk must be in Drive A:, but the data disk can be in drive A:, B:, C:, or D:.

The Quit command terminates the operation of VisiTrend/Plot program, and returns control of the computer to the operating system.

## The Plot Subprogram

You can invoke the Plot subprogram from either the Main subprogram or from the Trend subprogram by placing the cursor on the word Plot and pressing **ENTER**. The Plot menu includes the choices Plot, Select, Window, Print, Screen, Options, Overlay, New, Text, and Exit.

The Plot command of the Plot subprogram menu produces another menu that allows you to select the type of graph you wish to display. You can select Line, Bar, Area, Pie, Hi-Lo-Close, or XY from the submenu. The remaining commands on the Plot subprogram menu allow you to format the graph; you can enter titles, select scales, select portions of the series to plot, and add or delete grid lines from the graph.

A VisiTrend/Plot list is a series of data points indexed by a Date, rather than by an Element number, as in TK!Solver, and it's called a series. You construct a series by entering the Edit function, and instructing VisiTrend/Plot that you will be constructing a new series. The program prompts you for the name of the series so that it can store the series on disk as a file. Then it prompts you for the period, which is the number of subdivisions in the basic time unit (see Figure 15-1). The assumed basic time unit is the year, and you can use values from 1 to 99 for the period.

After you enter the period, you must enter the start date. Start dates can range from 0 to 2499. If the period is 1, the start date consists of only the year, and if the period is greater than 1, the start date consists of a year and a period value.

Once you have entered the period and the start date, the start date appears in the Date column, which corresponds to the Element field of the TK!Solver List subsheet, and the cursor appears to the right of the Date, in a column with the name of the series as its title. From here on, entering the values is just like entering values in a TK!Solver list. You enter values in the series column, terminating each entry with **ENTER** or the down arrow key. As with TK!Solver lists, you can insert elements, change elements, and delete elements; VisiTrend/Plot keeps track of the entries in the Date column.

## PLOTTING WITH VISITREND/PLOT®

Plotting a graph with VisiTrend/Plot is similar to plotting with TK!Solver—instead of using a Plot sheet, you use a Plot menu (actually a series of menus). You have the option of plotting any VisiTrend/Plot series as a function of time, as bar graphs, area graphs, line graphs, pie charts, high-low-close charts, and combinations of all but the pie chart, which can only be plotted alone (see Figure 15-2). Of course, you have the XY plot mentioned above, which is analogous to the TK!Solver plot. Each point on an XY plot is plotted using the value of one series element as its X-coordinate, and the value of the corresponding element of another series as its Y-coordinate.

Of course, the value in the Date column need not be interpreted as the date of a data point; it could just as easily be interpreted as time in seconds, or microseconds, or it could be some linear measurement. It's all a matter of interpretation.

The most convenient way to transfer data from a TK!Solver model to VisiTrend/Plot for graphing is to use the DIF file that both programs support. However, VisiTrend/Plot series files are text files, which could be constructed on a word processor or a simple text editor from a TK!Solver .PRN file produced by the Action command given from the Table sheet.

The principal advantage of using VisiTrend/Plot to produce graphs from data generated by a TK!Solver model is that the pro-

LIST:
| Element | Value |
| --- | --- |
| 2 | 'KW |
| 4 | 261 |
| 5 | 259 |
| 6 | 258 |
| 7 | 256 |
| 8 | 253 |
| 9 | 253 |
| 10 | 252 |
| 11 | 251 |
| 12 | 247 |
| 13 | 247 |
| 14 | 245 |
| 15 | 243 |

LIST:
| Element | Value |
| --- | --- |
| 2 | 'KW |
| 4 | .696 |
| 5 | .691 |
| 6 | .688 |
| 7 | .683 |
| 8 | .675 |
| 9 | .675 |
| 10 | .672 |
| 11 | .669 |
| 12 | .659 |
| 13 | .659 |
| 14 | .653 |
| 15 | .648 |

LIST:
| Element | Value |
| --- | --- |
| 2 | 'EVAP |
| 4 | 19.5 |
| 5 | 19.5 |
| 6 | 19.5 |
| 7 | 19.5 |
| 8 | 19.5 |
| 9 | 19.5 |
| 10 | 19.5 |
| 11 | 19.5 |
| 12 | 19.5 |
| 13 | 19.5 |
| 14 | 19.5 |
| 15 | 19.5 |

Use <Esc> for commands

| Date | Quantity | Date | Quantity |
| --- | --- | --- | --- |
| 0 | 1103 | 0 | 1103 |
| 1 | 1303 | 1 | 1303 |
| 2 | 1506 | 2 | 1506 |
| 3 | 1810 | 3 | 1810 |
| 4 | 2068 | 4 | 2068 |
| 5 | 2389 | 5 | 2389 |
| 6 | 2670 | 6 | 2670 |
| 7 | 2980 | 7 | 2980 |
| 8 | 3201 | 8 | 3201 |
| 9 | 3506 | 9 | 3506 |
| 10 | 3885 | 10 | 3885 |

| Date | floppy sales |
| --- | --- |
| 1973 | 1 |
| 1974 | 10 |
| 1975 | 60 |
| 1976 | 120 |
| 1977 | 220 |
| 1978 | 290 |
| 1979 | 330 |
| 1980 | 400 |
| 1981 | 500 |
| 1982 | 325 |
| 1983 | 700 |
| 1984 | 840 |

Figure 15-1. VisiTrend Periods.

LINE GRAPH

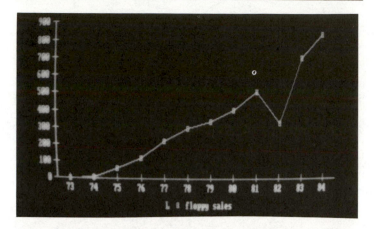

AREA GRAPH

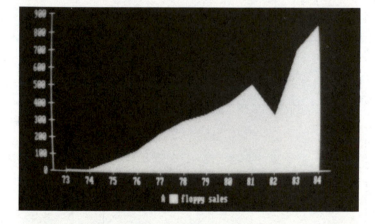

BAR CHART

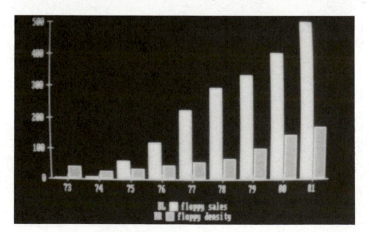

**BAR CHART**

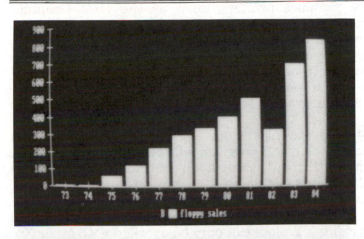

**PIE CHART**

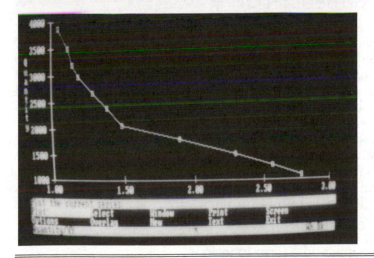

**X-Y GRAPH**

Figure 15–2.  VisiTrend/Plot®  Graph Formats.

gram supports graphics printers, such as the IBM Graphics Printer. You can plot graphs in which the data points are indicated as symbols and are connected with line segments, rather than isolated points; you can also plot continuous lines without data point symbols. VisiTrend/Plot also allows you considerably more latitude in adding titles to your graphs. You can add multiple legends, position the graph's title on the graph, and add notes to the graph.

If you attempt to plot more than 40 data points on a VisiTrend/Plot XY plot, the program will not automatically plot data point symbols. Instead, it will plot a continuous line, up to the maximum of 250 points per series.

You can overlay a number of charts, or graphs, if they are compatible. Area, line, bar, and high-low-close charts are compatible, and can be overlaid. XY plots can only be overlaid with other XY plots, and pie charts cannot be overlaid.

VisiTrend/Plot provides you with an excellent tool for the presentation of your TK!Solver-produced data with another function, the Text function, which will enable you to construct and save bit-mapped screen images of graphs enhanced with shading, large font lettering, reverse video, and so on. You can either photograph the screen image to produce slides, or print the screen image on the graphics printer. (Sorry, VisiTrend/Plot does not support plotters.)

## ANALYSIS WITH VISITREND/PLOT®

VisiTrend/Plot also provides you with a powerful set of statistical analysis tools. The Trend subprogram includes several regression and forecasting functions, statistical calculation functions, and formatting functions. It also allows series to be transformed, and new series to be generated automatically through mathematical and logical operations on existing series.

You can peform all of these functions with TK!Solver, of course. You have only to enter the correct rules. But, as long as the tools are available, there is no reason to ignore them.

The Trend subprogram, like the Plot subprogram, is used with a menu. The Trend subprogram menu provides eight functions: Edit, Analyze, Main, Display, Function, Xform, Clear, and Plot. The functions that are unique to the Trend subprogram are the Analyze, Function, and Xform functions.

The Analyze function has a submenu of five choices: Table, Statistics, Regression, Format, and Exit. The Table command allows you to present several lists, or series, as a table, just as you would present TK!Solver tables using the Action command from the Table sheet. As with TK!Solver, VisiTrend/Plot will display the table on the screen or the printer.

The Statistics function displays several statistical measurements on the series in memory:

- Minimum and maximum values in each series.

- Arithmetic mean of each series.

- Variance for each series.

- Standard deviation for each series.

- Coefficients of correlation for selected combinations of series.

The Regression function performs linear multiple regression, using the method of least squares. The function can relate the dependent variable (range) to five independent variables (domain). The function generates two new series: a fitted series, and a series of the differences between the original dependent variable series and the fitted series. In addition, the Regression function calculates several statistical measures of the regression it performs.

The Function command is used to generate new series from actual data series. The choices on the Function submenu are Moving-Average, Smoothing, Percent-Change, Difference, Lag, Lead, Total, and Exit. These functions are built-in and allow you to specify certain factors or parameters.

The Xform function allows you to generate a new series using an existing series as the source, and a transformation formula consisting of arithmetical and logical operators and functions.

VisiTrend/Plot has a number of capabilities, but you can duplicate almost all of them with TK!Solver models. However, VisiTrend/Plot can add considerable visual impact to the presentation of a TK!Solver model.

## VISICALC®

You simply cannot ignore the first of the electronic tool programs, VisiCalc, which also bears a striking resemblance to TK!Solver. You can interchange data between the two through the DIF format.

You can use VisiCalc as a "front-end" or "back-end" program for TK!Solver; in other words, you can use it to enhance input and output much as you can use VisiTrend/Plot to enhance graphic displays of TK!Solver lists.

TK!Solver's input mechanism is somewhat restricted, particularly with regard to lists. At times, you can use the Fill List option to fill a list quickly, but the option will only generate a linear list. If you want to enter experimental test data, you must enter the data points one at a time in each list. Furthermore, you can only have two List subsheets on the screen at a time, one in each of the two windows.

You can use VisiCalc to enter several lines at a time, and have them on the screen simultaneously. The default display is eight columns, each nine characters wide (see Figure 15-3). VisiCalc is similar to TK!Solver in that the screen display is a window into a much larger virtual space—VisiCalc's total work space extends from cell A1 (the upper left corner) to BK254 (the lower right corner).

You can make three types of entries in VisiCalc cells: labels, which correspond to TK!Solver's symbolic values; numeric values, which are the same as TK!Solver's numeric values; and formulas, which have no permanent counterpart in a TK!Solver list. (Remember that you can enter a value in a TK!Solver list by typing in a formula consisting of numbers, operators, and functions, but TK!Solver evaluates the value immediately, and does not store the formula.)

Entering a single value in a VisiCalc cell is no quicker than marking a single entry in an element of a TK!Solver list; you type the value, and press return or the arrow key, in either case. The advantage VisiCalc offers is that you can see several lists at a time, and you can move from one list to another by using the arrow keys.

You must remember that TK!Solver works with lists per se: they are named objects. VisiCalc does not work with lists, but with cells. You assign the list attribute mentally when you use VisiCalc to store lists. Because this association exists only in your mind, you can either work with 63 lists (each holding a maximum of 254 elements) by treating each column as a list; or you can work with 254 lists (each holding a maximum of 63 elements) by treating each row as a list.

The example of transferring lists between VisiCalc and TK!Solver presented in Chapter 10 indicated that the VisiCalc could be stored in a DIF file in either row or column order. If you save the DIF file from a VisiCalc worksheet with the command /S#S...R, each row will be a tuple; when you load that file into a

Figure showing a VisiCalc® spreadsheet printout (main worksheet) with an overlapping window inset.

**Main worksheet**

| | | 10 | 25 | 10 | 16 | 10 | 13 | 10 | 50 | 10 | 16 | 10 | 23 | 10 | 12 | 10 | 200 | 10 | 100 |
|---|---|---|---|---|---|---|---|---|---|---|---|---|---|---|---|---|---|---|---|
| EQUIV LG | SIZE | PIPE | STD 90 | QTY | LR 90 | QTY | STD 45 | QTY | T-SIDE | QTY | T-OUT | QTY | REDR | QTY | GATE | QTY | GLOBE | QTY | CHECK | QTY |
| 91.2 | .5 | 60 | 1.6 | 4 | 1 | 2 | .8 | 4 | 3 | 3 | 1 | 5 | 1.4 | 5 | .7 | 2 | 18 | 2 | 6 | |
| 133.4 | .75 | 120 | 2 | 4 | 1.4 | | .9 | 6 | 4 | | 1.4 | | 1.9 | | .9 | | 22 | | 8 | 3 |
| 148.6 | 1 | 30 | 2.6 | 6 | 1.7 | | 1.3 | 5 | 5 | 3 | 1.7 | 3 | 2.3 | 3 | 1 | 2 | 29 | | 10 | |
| 67.8 | 1.25 | 45 | 3.3 | 2 | 2.3 | | 1.7 | 7 | 7 | | 2.3 | | 3.1 | | 1.5 | | 38 | 3 | 14 | |
| 104.7 | 2 | 90 | 5 | | 3.3 | 3 | 2.6 | | 10 | 6 | 3.3 | 6 | 4.7 | | 2.3 | | 55 | | 20 | |
| 150 | 2.5 | 35 | 6 | 3 | 4.1 | | 3.2 | | 12 | | 4.1 | | 5.6 | | 2.8 | | 69 | 3 | 25 | |
| 229 | 3 | 80 | 7.5 | 4 | 5 | | 4 | 3 | 15 | | 5 | | 7 | 5 | 3.2 | | 84 | | 30 | |
| 188.4 | 4 | 100 | 10 | 6 | 6.7 | 2 | 5.2 | 2 | 21 | | 6.7 | 6 | 9 | | 4.5 | 4 | 120 | 4 | 40 | 1 |
| | 6 | | 16 | | 10 | | 7.9 | | 30 | | 10 | | 14 | | 7 | | 170 | | 60 | |
| | 2 | | 5 | | 3.3 | | 2.6 | | 10 | | 3.3 | | 4.7 | | 2.3 | | 55 | | 20 | |
| | 2.5 | | 6 | | 4.1 | | 3.2 | | 12 | | 4.1 | | 5.6 | | 2.8 | | 69 | | 25 | |
| | 3 | | 7.5 | | 5 | | | | 15 | | 6.7 | | 7 | | 3.2 | | 84 | | 40 | |
| | | | 10 | | 6.7 | | 5.2 | | 21 | | 6.7 | | 9 | | 4.5 | | 120 | | 120 | |
| | | | | | | | | | | | 3.3 | | 4.7 | | 2.3 | | 55 | | 20 | |

**Window inset**

A19

C
382

| | A | B | C | D | E | F | G | H |
|---|---|---|---|---|---|---|---|---|
| | | | 10 | 25 | 10 | 16 | 10 | 13 |
| 19 | | SIZE | PIPE | STD 90 | QTY | LR 90 | QTY | STD 45 |
| 20 | | | | | | | 2 | | |
| 21 | | | | | | | | | |
| 22 | ZEQUIV LG | SIZE | PIPE | STD 90 | QTY | LR 90 | QTY | STD 45 |
| 23 | | .5 | 60 | 1.6 | 4 | 1.4 | 2 | .8 |
| 24 | 91.2 | .75 | 120 | 1.6 | 4 | 1.4 | 2 | .9 |
| 25 | 133.4 | 1 | 30 | 2.6 | 6 | 1.7 | | 1.3 |
| 26 | 148.6 | 1.25 | 45 | 3.3 | 2 | 2.3 | | 1.7 |
| 27 | 67.8 | 2 | 90 | 5 | 3 | 3.3 | | 2.6 |
| 28 | 104.7 | 2.5 | 35 | 6 | 3 | 4.1 | | 3.2 |
| 29 | 150 | 3 | 80 | 7.5 | 4 | 5 | | 4 |
| 30 | 229 | 4 | 100 | 10 | 6 | 6.7 | | 5.2 |
| 31 | 188.4 | 6 | | 16 | | 10 | 10 | 7.9 |
| 32 | | 2 | | 5 | | 3.3 | 3.3 | 2.6 |
| 33 | | 2.5 | | 6 | | 4.1 | 4.1 | 3.2 |
| 34 | | 3 | | 7.5 | | 5 | 5 | 4 |
| 35 | | 4 | | 10 | | 6.7 | 6.7 | 5.2 |
| 36 | | 1.5 | | 4 | | 2.6 | 2.6 | 2.1 |
| 37 | | 2 | | 5 | | 3.3 | 3.3 | 2.6 |
| 38 | | 2.5 | | 6 | | 4.1 | 4.1 | 3.2 |
| 111.3. | | | | | | | | |

Figure 15–3. The VisiCalc® Worksheet and Window.

TK!Solver model, each tuple becomes a list. If you use the command /S#S...C, each column becomes a tuple, and, in turn, a TK!Solver list. In either case, once the worksheet is saved, the data are formatted into lists.

You must also remember that a VisiCalc worksheet contains formulas, formatting data, and so on. This information is stored with the worksheet when it is stored as a VisiCalc file. When the worksheet is stored as a DIF file, only values are stored. Specifically, if a VisiCalc cell contains a formula, which it displays as a numeric value on the screen, the displayed value will be stored in the DIF file, and the formula will not be stored. This means that you can expand lists for input to TK!Solver with a VisiCalc worksheet beginning with one or more lists, or perhaps with only a few elements, and you can produce more lists, or lists with more elements.

As with VisiTrend/Plot's analysis functions, VisiCalc will not perform any tasks that you could not duplicate with a TK!Solver model, but it can streamline the user interface.

You can also use a VisiCalc worksheet to advantage in transforming lists for input to a TK!Solver model. VisiCalc allows you to enter formulas that make explicit use of logic functions and operators:

| | |
|---|---|
| < | Less then |
| > | Greater than |
| = | Equal |
| < = | Less then, or equal |
| > = | Greater than, or equal |
| < > | Not equal |
| @NOT | |
| @AND | |
| @OR | |
| @IF | |
| @ISNA | |
| @ISERROR | |

Although TK!Solver has no explicit logic operators or logic functions, you can emulate logical operations, but this can become a complicated task.

VisiCalc also allows you to perform table lookup by means of the @CHOOSE and @LOOKUP functions. This task can be performed with TK!Solver by means of the User function, but at times it is handier to perform the Lookup function and transform one type of value to another, using the VisiCalc worksheet rather than the TK!Solver model (see Figure 15–4). As an example, you might use a VisiCalc worksheet to record a list of piping system components in preparation for input to a TK!Solver friction loss calculation model. You could enter the components—tees, elbows, valves—as labels (symbolic values in a TK!Solver model), and let VisiCalc produce another list of equivalent footages using the @LOOKUP function. The resulting list would be a composite of actual pipe footages for straight lengths, and equivalent footages for valves and fittings, which is precisely the form of input the TK!Solver model would require.

You can also use VisiCalc to enhance the display of the output of a TK!Solver model. Again, you would transfer the lists from the TK!Solver model to the VisiCalc worksheet through the DIF disk file. When you create the DIF file using TK!Solver's storage command, each list selected for storage is saved as a tuple in the DIF file; you have no options, as you have with VisiCalc's storage command. However, you do have an option when you load the DIF file into the VisiCalc worksheet.

VisiCalc will normally load each tuple of a DIF file as a row. You can specify that it load the tuples as columns by ending the storage command with C, instead of R or just **ENTER**. You can also specify the upper left corner of the area where the lists are loaded. This feature allows you to leave room for titles, borders, notes, and so on.

TK!Solver gives very little flexibility in preparing tables with the Table sheet. By using VisiCalc to prepare the final version of the table, you can enhance the table in a number of ways, for example, by including some crude graphic displays. The current version of VisiCalc for the IBM PC includes an enhancement called StretchCalc™ [1], which adds more sophisticated graphic capabilities.

VisiCalc enhanced with StretchCalc enables you to plot TK!Solver lists as area graphs, bar graphs, component graphs, dot graphs, high-low-close graphs, line graphs, pie charts, and scatter graphs (Figure 15–5).

---

[1] StretchCalc is a trademark of Multisoft Corporation.

**EQUIVALENT LENGTHS OF FITTINGS AND VALVES**

Look Up Tables

Input Columns

Output Columns

Saves Output Columns as DIF File

Quit VisiCalc

Load TK!Solver

/ S/ # S B : EQUIVLG ⟨ENTER⟩ C

/ S Q Y

TK ⟨ENTER⟩

1113.1

Load DIF File into TK!Solver Model

Select List and Table Sheets

Name the Lists

Table Display
of Transformed
Lists

```
/  S  #  L  B  :  EQUIVLG  ⟨ENTER⟩

=  L  ;  =  T
```

```
(1n) Name:                                    204/

Screen or Printer:   Screen
Title:
Vertical or Horizontal:   Vertical
List    Width First Header
======= LIST SHEET =======
Name      Elements  Unit   Comment
----      --------  ----   -------
            29
            27
```

```
(21) List: SIZE                               204/1

Screen or Printer:   Screen
Title:   LIST TRANSFORMATION
Vertical or Horizontal:   Vertical
List    Width First Header
======= TABLE SHEET =======

EQUIV_LG  10   1
SIZE      10   1

======= LIST SHEET =======
Name      Elements  Unit   Comment
----      --------  ----   -------
EQUIV_LG    29
SIZE        27
```

```
        LIST TRANSFORMATION

EQUIV_LG    SIZE

EQUIV       SIZE

91.2        .5
133.4       .75
148.6       1
67.8        1.25
104.7       2
150         2.5
229         3
188.4       4
0           6
0           2
0           2.5
0           3
0           4
0           1.5
0           2
0           2.5
0           3
0           2
0           2.5
0           4
0           2.5
0           3
0           4
0           2

1113.1
```

Figure 15–4.  List Transformation with VisiCalc®.

```
NF(N)=5*N+3

 1  ********
 2  ************
 3  ******************
 4  ************************
 5  ******************************
 6  **********************************
 7  ******************************************
 8  ************************************************
 9  ******************************************************
10  ************************************************************
```

```
A12    (V)  10                                                          C
                                                                       400

 1                              A
 2                                                          N
 3
 4                                                          1
 5                                                          2
 6                                                          3
 7                                                          4
 8                                                          5
 9                                                          6
10                                                          7
11                                                          8
12                                                          9
13                                                         10
14
15
16
17
18
19
20
21
```

```
B12  /F*   (V)  5*A12+3                                                 C
                                                                       400

                                 B
 1F(N)=5*N+3
 2
 3  ********
 4  ************
 5  ******************
 6  ************************
 7  ******************************
 8  **********************************
 9  ******************************************
10  ************************************************
11  ******************************************************
12  ************************************************************
13
14
15
16
17
18
19
20
21
```

Figure 15–5.  Graphs Displayed with VisiCalc®  and StretchCalc™.

VisiCalc and StretchCalc provide another capability for polishing up a worksheet before printing it. The VisiCalc command is the Move command, and the StretchCalc version is the Advanced Move command. VisiCalc's Move command allows you to rearrange rows and columns on the worksheet. StretchCalc's Advance Move command allows you to move several columns at a time.

StretchCalc also allows you to sort the rows of a worksheet according to the values in one column. This capability is akin to the features of database management systems. Needless to say, building a sorting model in TK!Solver is possible, but would take some time. StretchCalc enables you to accomplish the sorting task with just a few keystrokes.

VisiCalc or VisiTrend/Plot cannot be treated adequately in a single chapter, but it should be stressed that both are valuable companions to TK!Solver. With all three in your arsenal, you can tackle some formidable engineering problems and produce some impressive models. Fortunately, the operation of the program is similar. VisiCalc uses a command structure that is much like that of TK!Solver, and VisiTrend/Plot's use of lists should be easy to master after you have worked with TK!Solver's list functions.

# CHAPTER 16

# The Model-Building Approach

It is now time to reconsider TK!Solver's effect on your problem-solving methods. In particular, it should be pointed out that not all textbook methods are suitable for TK!Solver models.

You have seen that some equations must be rearranged from their normal presentation for use as rules in a TK!Solver model. Specifically, you saw that an equation in which a variable name appears twice cannot be solved for that variable with the Direct Solver, although the Iterative Solver may be able to accomplish the task. Remember, too, that TK!Solver cannot perform integration, differentiation, or some transformations directly.

However, TK!Solver will attempt to satisfy all the rules the current model contains. You can construct useful models that contain rules that cannot be satisfied simultaneously, but in general you should construct models that can be completely satisfied. You saw that you can add pseudological variables that will allow two conflicting rules to be satisfied simultaneously.

## SETTING UP A SOLUTION— DIAGRAMS AND SKETCHES

When you use the paper-and-pencil approach to solving engineering problems, you usually take several standard steps. First, you draw a sketch to illustrate the system in question. In a mechanics problem, you would draw a free-body diagram showing the body in question and the forces that act on the body (Figure 16–1). In a thermodynamics problem, you would draw a diagram of the control

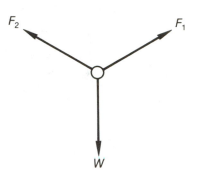

Figure 16-1. Free Body Diagram.

volume; the control volume illustrates the space of interest, and the forms of energy and mass crossing the boundaries of that space (Figure 16-2). Other problems also have appropriate diagrams.

You draw sketches and diagrams, of course, to ensure that you understand the problem. You need to determine what is known, and what is unknown. Furthermore, you need to know what is relevant to the problem, and what is not; engineering problems are typically presented with both relevant and irrelevant information.

As you draw the sketch, you introduce suitable notation for the knowns and unknowns, and indicate the known values. Each field of engineering has its own standard notation, and you should use that notation.

Having TK!Solver at hand does not eliminate the need for these preliminary steps of the problem-solving approach. You still need the free body diagram, the control volume diagram, the energy balance system diagram, the cash flow diagram. And you still need to introduce the notation appropriate to the problem, and to the field—so that others can follow the solution.

In fact, it is more important to draw sketches and diagrams for a TK!Solver model than for old-fashioned methods. Why? Because it is too tempting to sit down at the keyboard and begin entering rules.

Although the word "programming" is not to be used in connection with the TK!Solver program—remember, you don't have to be a programmer to use it—practically every programming text recommends that a program be diagrammed before the code is written. Before the modern era of "structured programming," people were encouraged to draw flow charts of the proposed program.

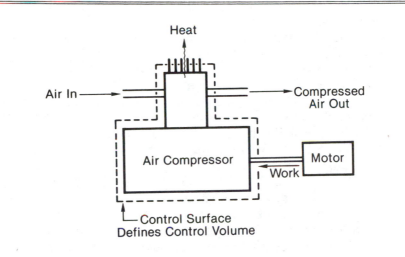

Figure 16–2. Control Volume Sketch.

Modern structured languages have made it possible to outline a program in "pseudo-code," an abbreviated outline of the program in close to standard English. By following this approach, you will tend to write more coherent programs.

A TK!Solver model is no different, in that it deserves some preliminary thought. If you don't understand the problem or the theory or phenomena involved, or even if you aren't sure that you do, turn the computer off. Draw a sketch of the problem, consult the handbooks and textbooks, write down the equations in their customary form. If possible, or applicable, use free-hand methods to prepare a graph of any functions that may be involved.

## PLANNING THE SOLUTION

Once you have a complete understanding of the problem, you have to develop a plan of solution. Again, this phase of the problem-solving process is necessary, whether you use paper, pencil, and slide rule, or TK!Solver. Before you begin to type rules into TK!Solver's rule sheet, you should determine the method of solution. The equations found in a textbook only describe the relations between variables—they don't prescribe a method of solution.

For instance, most problems of interest to engineers involve equations that contain integrals and derivatives at some level of abstraction. In practice, equations are seldom used in the integrodifferential form; specialized forms of those equations, which can be solved by purely algebraic methods, are used instead. In some cases, exact solutions are impossible, and almost exact solutions are costly in terms of time. In these cases, approximate or assumed values are typically used for some variables in order to obtain "workable" solutions at reasonable cost.

With TK!Solver at hand, you have some additional latitude in planning a solution because the program will carry out some tasks unattended in far less time than you could accomplish them otherwise. You can apply numerical methods that would require special programming to be run on a computer, or that could only be solved for a limited number of points manually. You have to analyze the problem to determine how to invest your time, now that you have a different tool.

You also have to recognize TK!Solver's limitations and idiosyncrasies. Other applications programs, such as spreadsheet programs, word processors, data base management programs, also have limitations—things they can't do, or ways things must be done. However, those limitations are usually only mildly irritating; they don't affect the actual problem solution.

TK!Solver performs a substantially more complicated task than the tasks performed by the types of programs mentioned above; although the manual doesn't give a clue as to how it performs its task, it does provide a brief explanation of the operation of the Direct Solver's method of solving simultaneous equations. However, it does not explain the way in which the Iterative Solver works. Thus, we often have no way of determining why an interative solution won't converge, or why an error message appears in the Status field of the Variable sheet.

You must carefully plan how the model will accommodate such things as imaginary numbers: they are almost inevitable in engineering problems, and TK!Solver cannot handle them. Whenever an imaginary number appears in a TK!Solver expression, the program will return an error, or it will fail to converge under the Iterative Solver.

You must be certain that a model is complete. When you work with models manually, you can be satisfied with finding the value of a ratio, or an expression. TK!Solver cannot stop at some arbitrary point and present the value of an expression. It must find a value for each variable, and it must have enough conditions provided—input values, or rules—to do so.

When you use the Iterative Solver, you usually have to provide guess values for one or more variables. You must not only determine which variables require guesses, but whether or not the "goodness" of the guess is important. In some models, the solution will converge regardless of the guess, and in others, the guess will affect convergence.

You can control the way TK!Solver works to some extent. You can arrange the sequence of rules, both the order in which they appear on the Rule sheet, and with logical variables. You should note, however, that imposing a structure on a TK!Solver model is not a cut-and-dried, foolproof process. On occasions, the additional rules and variables will produce a model that is overdefined and cannot be solved. A thorough understanding of the model, as well as TK!Solver, is the only defense.

You also need to plan for the system of units your model will use. Remember that TK!Solver will keep track of units, but again, there can be subtle problems. You must remember that the program does not actually understand units in the same way that you do. TK!Solver only does the bookkeeping for you, so that it is up to you to ensure that the units you use are consistent. For instance, TK!Solver does not know that if you assign the unit FT to the variables, Length and Width, the product, Area, will have the unit FT^2, unless you tell it so. When you do define the unit FT^2 for Area, the presence of the operator in the unit name has no significance to the program.

Not only do you need to plan for a consistent set of units, but you also need to plan for conversion factors, and constants. Not all textbooks present equations with commonly used units. Nor do they always present them with the constants required. It is always advisable to perform a dimension analysis of the model to ensure that the answer will have the correct units.

Remember that lists and variables use units in slightly different ways. Specifically, you must remember that User functions are based on the display units of the lists associated by the User Function subsheet.

## ENTERING THE MODEL

Once the problem has been defined and the solution planned, you can enter the rules, variables, units, lists, and functions that make up the model. If you have done the preliminary work thoroughly, entering the model is routine, but there are still things to watch for.

The order in which the rules are entered determines the order in which the variables appear on the Variable sheet. This order may not be ideal for solving problems efficiently. You can wait until you have entered all the rules, and then rearrange the variables. If you develop a model in several steps, this method is almost unavoidable.

If you can enter the entire model at one time, you can avoid rearranging the variables by entering the variables first. TK!Solver will not duplicate a variable, regardless of when it was entered.

Variable names should be meaningful. Long, complicated models with single-letter variable names can be very difficult to follow when problems arise (they always do). Greek symbols are common in engineering formulas and equations, and usually have commonly accepted meanings within a field. Of course, the IBM PC doesn't have Greek symbols on the keytops, and they can't be entered from the keyboard, even though both the computer and the printer are capable of displaying them. You can, however, spell out the Greek symbol name, and obtain a meaningful variable name.

TK!Solver allows you to enter comments, and it is wise to make liberal use of them. You can explain the meaning of each variable, and of each rule; those comments come in handy when you use a model only once in a great while.

There is ample room for comments, with room for 32,000 entries in all the sheets that contain columns of entry fields—the Rule sheet, Variable sheet, List sheet, List subsheet, Unit sheet, User Function sheet, User Function subsheet, Table sheet, and Plot sheet. You can enter the equivalent of several pages of text as comments, eliminating the need for notebooks of instructions on how to use the models. In fact, you can set up one model whose only function is to explain other models.

## PROVING THE MODEL

Even though you may use proven relationships for the rules of a model, the operation of the model itself should be proven. Most models of interest in engineering will involve lists, and many of the lists will be long. Although TK!Solver is faster than other methods, substantial amounts of time may be required to process a large model that uses long lists. Thus, you need to prove the model with a few isolated points.

A model that uses lists for input and output can still be solved for individual points by using the Action command or the Solve command, instead of the List command. Of course, if the model accesses previous list elements, as you might do to calculate factorials or integrals, you may have to partly fill some lists that would normally be output lists. Even so, you need to know a model will produce valid results before turning it loose on an input list of several thousand elements.

You can ease the process of proving the model by incorporating rules that produce intermediate values that may not be of interest in the final solution. These values and rules should be such that they can easily be checked manually. That's right—don't throw the paper and pencil away for good.

Once you know that the model produces valid results on isolated points, you may wish to solve a block of elements using the Block option of the List Solver. With the Intermediate Redisplay ON field of the Global sheet set to Yes, you can monitor the List Solver as it processes a block of elements. Then, you can allow the model to run unattended while you do something else.

## MODELS THAT GROW

There is no need to begin with complicated models just because you know that TK!Solver can handle them. If simple models were acceptable when you used a notebook, pencil, and calculator, they should still be acceptable. It's best to begin with models, methods, and problems that you understand. If an assumed value for the friction factor is acceptable for piping head loss calculations when you use hydraulic tables, you should begin with a model that uses a friction factor that can be entered as an input value.

First, construct a model that will be productive, even though it may leave something to be desired in accuracy. When the model can size a piping system, producing standard pipe sizes, and can calculate the losses using the approximate friction factor, you can proceed to refine the rules for increased accuracy.

When you develop a model in steps, be careful to save the model at various stages in its development process. Of course, you should save a model at operational points, but sometimes you will want to save a model that obviously does not work, just for reference (or perhaps to get some sleep).

There is no need to refine a model to the nth degree before beginning to use it. You can add refinements gradually, as they become necessary to solve a particular problem, or as they occur to you. As you become familiar with the TK!Solver program, adding refinements becomes easier in a way that old-fashioned program maintenance never can.

# TK!SOLVERPACKS

Software Arts provides packages of models on disk called TK!SolverPacks. These packages are libraries of models for various fields, and are furnished with manuals similar in style to the TK!Solver manual. Although TK!SolverPacks may not suit your needs exactly, they contain useful models that can be further refined and expanded.

There is no need to wait for a TK!SolverPack that happens to fit your needs. You can develop your own libraries of models, and store them for repeated use. The TK!Solver program, and a library of models can replace stacks of paper, and save some wear and tear on those handbooks.

# PART II

# TK!SOLVER SESSIONS

Now that we have surveyed the TK!Solver program's organization, facilities, and capabilities, we can proceed to more comprehesive examples of its use. We will present examples as sessions, describing for each session the problem to be solved, the theory of the solution, and the TK!Solver model used to solve the problem. We present sample inputs and outputs exactly as used in the development of the models.

As you read through these examples, bear in mind that there may well be other ways to solve the problems presented. We are simply presenting examples to demonstrate the power of the TK!Solver program. In many cases, the examples are chosen and structured to demonstrate particular aspects of the program, and not necessarily to produce an elegant or rigorous solution.

Of course, the practice of engineering does not always permit elegant or rigorous solutions to problems, but instead demands expedient solutions. The TK!Solver program is a blessing in such situations, as it will permit the use of solutions that could not be achieved manually in the allotted time.

You can develop a TK!Solver model in steps, beginning with the equations, formulas and methods you would use manually. Then, if time permits, you can refine the model to provide more detail and more precision than you could normally obtain. There is no question that the TK!Solver program can bring the power of the computer to any engineer's desk.

Be prepared to spend some time with the program. It may be "user-friendly" compared to learning a progamming language, but it's still rather terse. The TK!Solver program is designed to be used on a regular basis, just like a slide rule or a calculator. With regular use, operation of the program becomes second nature; occasional use can be frustrating. The program's terse nature provides a benefit—operation of the program is generally fast enough that it will accept commands as fast as you can type them. Solution of models and plotting of multiple curves are not so fast, but are still much faster than the old way.

# CHAPTER 17

# Piping System Analysis

The first session is the analysis of a piping system through which a fluid is pumped, as shown in Figure 17–1. We want to size the piping, calculate the total flow and pressure loss through the piping system, and select a pump that will produce the flow with a reasonable residual pressure at the last outlet. To simplify matters, we will assume that the fluid is water, and we will only consider one path through the system. A complete analysis would require a similar analysis of each path through the system, but we would use the same procedure for each path, so nothing is lost by the simplification.

A typical quick and dirty solution to this type of problem begins with tabulating the flows in all sections of the piping system, which is a necessity in any event. Label each point in the system where the flow is diverted or mixed; these labels allow you to refer to a node, where branches leave or join mains or other branches, and sections between nodes. The flow in a section is equal to the sum of the flows entering the previous node. Begin at the end of each branch, and work backward, adding the flows at each node until you arrive at the total system flow in the section into which the pump discharges.

Next, select the path to the most remote outlet or appliance by inspection, with the assumption that it will have the greatest pressure loss. Then size the piping in that path, using hydraulic tables, as shown in Figure 17–2, or a "system sizer" slide rule, as shown in Figure 17–3. You should attempt to size each section of pipe so that its velocity and friction loss (per 100 ft) falls within a certain range. The two constraints are: (a) too small a pipe size will produce noise, wear, and excessive energy use, and (b) too large a pipe size increases the cost of the system. Since pipe is supplied in

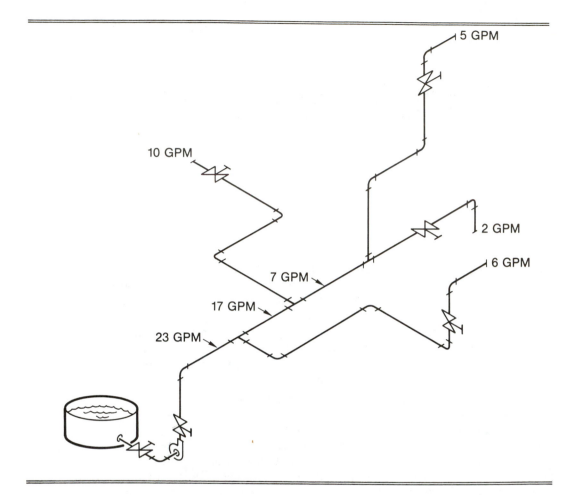

Figure 17-1. A Piping System.

standard sizes, flow velocity and pressure loss will vary considerably from one section to another.

Once all the pipe in the longest path is sized, you can make a preliminary pump selection, if you calculate the losses in that path. However, the prudent designer will size all paths and calculate the losses in each branch before making a final pump selection.

Now let's look at the solution of this problem using TK!Solver. First, you need to formulate a few rules, for the entries in the Rule sheet. Most of the same rules were used in the manual solution, although you do not write them down on each time you size a pip-

Hydraulic Table — PRESSURE LOSS (PSI/FOOT LENGTH) IN PIPES AT AVERAGE FLOW VELOCITY (FT/SEC) and EQUIVALENT PIPE LENGTHS (FT.) FOR CIRCUIT COMPONENTS.

| SIZE (in) | PIPE / HOSE | O.D. (in) | I.D. (in) | WALL (in) | I.D. AREA (sq in) | 5 LOSS | 5 GPM | 7 LOSS | 7 GPM | 10 LOSS | 10 GPM | 15 LOSS | 15 GPM | 20 LOSS | 20 GPM | 25 LOSS | 25 GPM | 30 LOSS | 30 GPM | TEE (line) | TEE (branch) | ELBOW (90°) | ELBOW (45°) |
|---|---|---|---|---|---|---|---|---|---|---|---|---|---|---|---|---|---|---|---|---|---|---|---|
| 1/8 | PIPE-SCH 40 | .405 | .269 | .068 | .057 | 1.25 | .89 | 1.79 | 1.24 | 2.60 | 1.75 | 3.16 | 2.67 | 5.47 | 3.56 | 6.20 | 4.45 | 7.07 | 5.34 | | | | |
| 1/8 | PIPE-SCH 80 | .405 | .215 | .095 | .036 | 1.89 | .56 | 3.05 | .78 | 4.26 | 1.12 | 5.20 | 1.68 | 8.38 | 2.24 | 11.1 | 2.80 | 12.7 | 3.36 | | | | |
| 1/8 | HOSE | – | .125 | – | .012 | 5.96 | .186 | 8.37 | .260 | 11.9 | .372 | 18.0 | .558 | 24.0 | .744 | 30.0 | .930 | 35.7 | 1.11 | | | | |
| 1/4 | PIPE-SCH 40 | .540 | .364 | .088 | .104 | .67 | 1.62 | 1.05 | 2.27 | 1.64 | 3.24 | 1.92 | 4.96 | 2.97 | 6.48 | 3.23 | 8.10 | 3.73 | 9.72 | | | | |
| 1/4 | PIPE-SCH 80 | .540 | .302 | .119 | .072 | 1.11 | 1.12 | 1.49 | 1.57 | 2.11 | 2.24 | 2.84 | 3.36 | 4.15 | 4.48 | 5.08 | 5.60 | 6.30 | 6.72 | | | | |
| 1/4 | HOSE | – | .250 | – | .049 | 1.57 | .758 | 2.17 | 1.08 | 3.00 | 1.49 | 4.49 | 2.23 | 6.04 | 2.98 | 7.49 | 3.72 | 8.95 | 4.44 | | | | |
| 3/8 | PIPE-SCH 40 | .675 | .493 | .091 | .191 | .39 | 2.98 | .57 | 4.18 | .86 | 5.96 | 1.05 | 8.94 | 1.69 | 11.92 | 4.27 | 14.9 | 5.78 | 16.9 | .8 | 1.2 | 2.7 | .6 |
| 3/8 | PIPE-SCH 80 | .675 | .423 | .126 | .140 | .54 | 2.18 | .74 | 3.06 | 1.10 | 4.36 | 1.34 | 6.54 | 1.97 | 8.72 | 5.19 | 10.9 | 7.20 | 13.1 | | | | |
| 3/8 | HOSE | – | .375 | – | .110 | .685 | 1.71 | .97 | 2.43 | 1.34 | 3.35 | 2.02 | 5.03 | 2.68 | 6.71 | 3.33 | 8.36 | 3.99 | 10.0 | | | | |
| 1/2 | PIPE-SCH 40 | .840 | .622 | .109 | .304 | .24 | 4.74 | .36 | 6.65 | .49 | 9.48 | .68 | 14.22 | 2.09 | 18.98 | 3.38 | 23.7 | 4.28 | 28.4 | 1.05 | 1.5 | 3.5 | .75 |
| 1/2 | PIPE-SCH 80 | .840 | .546 | .147 | .234 | .30 | 3.65 | .45 | 5.12 | .71 | 7.30 | .78 | 10.9 | 2.47 | 14.6 | 3.61 | 18.2 | 5.00 | 21.9 | .9 | 1.4 | 2.9 | .68 |
| 1/2 | PIPE-SCH XX | .840 | .252 | .294 | .050 | 1.54 | .78 | 2.19 | 1.09 | 3.08 | 1.56 | 3.65 | 2.34 | 6.13 | 3.12 | 7.48 | 3.90 | 9.55 | 4.68 | | | | |
| 1/2 | HOSE | – | .500 | – | .196 | .387 | 3.03 | .547 | 4.30 | .755 | 5.94 | 1.13 | 8.90 | 2.4 | 11.9 | 3.15 | 15.3 | 4.5 | 17.7 | | | | |
| 3/4 | PIPE-SCH 40 | 1.050 | .824 | .113 | .533 | .14 | 8.32 | .22 | 11.7 | .27 | 16.6 | .78 | 25.0 | 1.47 | 33.3 | 2.19 | 41.6 | 2.90 | 49.9 | 1.4 | 2.1 | 4.5 | 1.0 |
| 3/4 | PIPE-SCH 80 | 1.050 | .742 | .154 | .432 | .16 | 6.74 | .26 | 9.45 | .37 | 13.5 | .87 | 20.2 | 1.71 | 27.0 | 2.48 | 33.7 | 3.52 | 40.4 | 1.2 | 1.6 | 4.0 | .8 |
| 3/4 | PIPE-SCH XX | 1.050 | .434 | .308 | .148 | .53 | 2.31 | .67 | 3.24 | 1.05 | 4.52 | 1.31 | 6.93 | 1.94 | 9.24 | 5.06 | 11.6 | 7.02 | 13.9 | | | | |
| 3/4 | HOSE | – | .750 | – | .442 | .171 | 6.82 | .248 | 9.92 | .336 | 12.4 | .502 | 20.1 | 1.33 | 26.8 | 2.02 | 33.4 | 2.90 | 41.3 | | | | |
| 1 | PIPE-SCH 40 | 1.315 | 1.049 | .133 | .863 | .10 | 13.5 | .13 | 18.9 | .34 | 26.9 | .39 | 40.4 | .78 | 53.8 | 1.64 | 67.3 | 2.24 | 80.7 | 1.7 | 2.6 | 5.7 | 1.2 |
| 1 | PIPE-SCH 80 | 1.315 | .957 | .179 | .719 | .11 | 11.2 | .15 | 15.7 | .24 | 22.4 | .44 | 33.6 | .85 | 44.8 | 1.84 | 56.1 | 2.93 | 67.3 | 1.6 | 2.5 | 5.2 | 1.1 |
| 1 | PIPE-SCH XX | 1.315 | .599 | .358 | .282 | .26 | 4.39 | .37 | 6.16 | .53 | 8.78 | .67 | 13.2 | 2.25 | 17.6 | 4.20 | 22.0 | 4.20 | 26.3 | 1.0 | 1.5 | 3.0 | .75 |
| 1 | HOSE | – | 1.00 | – | .785 | .097 | 12.2 | .136 | 17.1 | .194 | 24.4 | .610 | 36.6 | .987 | 48.8 | 1.51 | 61.2 | 2.02 | 73.4 | | | | |
| 1-1/4 | PIPE-SCH 40 | 1.660 | 1.380 | .140 | 1.496 | .05 | 23.4 | .08 | 31.7 | .25 | 46.7 | .39 | 70.1 | .78 | 93.4 | 1.18 | 117 | 1.47 | 140 | 2.4 | 3.7 | 7.5 | 1.6 |
| 1-1/4 | PIPE-SCH 80 | 1.660 | 1.278 | .191 | 1.280 | .07 | 20.0 | .09 | 28.1 | .26 | 39.9 | .44 | 58.9 | .85 | 79.8 | 1.27 | 99.8 | 1.80 | 120 | 2.1 | 3.5 | 7.0 | 1.5 |
| 1-1/4 | PIPE-SCH XX | 1.660 | .896 | .382 | .630 | .13 | 9.83 | .16 | 13.8 | .24 | 19.7 | .71 | 29.5 | 1.35 | 39.3 | 2.01 | 49.2 | 2.76 | 59.0 | 1.5 | 2.3 | 4.9 | 1.1 |
| 1-1/4 | HOSE | – | 1.25 | – | 1.23 | .062 | 19.1 | .087 | 26.8 | .125 | 38.2 | .436 | 57.3 | .738 | 76.4 | 1.08 | 95.5 | 1.52 | 115 | | | | |
| 1-1/2 | PIPE-SCH 40 | 1.900 | 1.610 | .145 | 2.046 | .04 | 31.8 | .06 | 44.5 | .19 | 63.5 | .33 | 95.3 | .64 | 127 | .96 | 159 | 1.26 | 191 | 2.8 | 4.3 | 9.0 | 2.0 |
| 1-1/2 | PIPE-SCH 80 | 1.900 | 1.500 | .200 | 1.767 | .04 | 27.6 | .06 | 38.6 | .21 | 55.1 | .42 | 82.7 | .71 | 110 | 1.06 | 138 | 1.36 | 165 | 2.6 | 4.2 | 8.2 | 1.8 |
| 1-1/2 | PIPE-SCH XX | 1.900 | 1.100 | .400 | .950 | .09 | 14.8 | .09 | 20.8 | .32 | 29.6 | .51 | 44.4 | 1.05 | 59.2 | 1.51 | 74.1 | 2.14 | 88.9 | 2.0 | 3.0 | 6.5 | 1.4 |
| 1-1/2 | HOSE | – | 1.50 | – | 1.77 | .044 | 27.7 | .061 | 38.6 | .180 | 55.1 | .353 | 82.7 | .59 | 110 | .86 | 138 | 1.21 | 166 | | | | |
| 2 | PIPE-SCH 40 | 2.375 | 2.067 | .154 | 3.355 | .03 | 52.3 | .08 | 73.4 | .14 | 105 | .24 | 159 | .48 | 209 | .69 | 262 | .85 | 324 | 3.5 | 5.5 | 11.0 | 2.5 |
| 2 | PIPE-SCH 80 | 2.375 | 1.939 | .218 | 2.953 | .03 | 46.0 | .09 | 64.6 | .15 | 92.0 | .26 | 138 | .52 | 184 | .73 | 230 | .98 | 275 | 3.4 | 5.0 | 10.8 | 2.4 |
| 2 | PIPE-SCH XX | 2.375 | 1.503 | .436 | 1.773 | .04 | 27.7 | .12 | 38.8 | .21 | 55.3 | .36 | 82.9 | .72 | 111 | 1.34 | 138 | 1.36 | 166 | 2.6 | 4.0 | 8.2 | 1.8 |
| 2 | HOSE | – | 2.00 | – | 3.14 | .024 | 48.9 | .034 | 68.6 | .123 | 97.3 | .256 | 147 | .41 | 196 | .60 | 245 | .80 | 293 | | | | |
| 2-1/2 | PIPE-SCH 40 | 2.875 | 2.469 | .203 | 4.788 | .03 | 74.8 | .07 | 105 | .11 | 149 | .20 | 224 | .37 | 299 | .53 | 374 | .72 | 449 | 4.2 | 6.5 | 14.0 | 3.0 |
| 2-1/2 | PIPE-SCH 80 | 2.875 | 2.323 | .276 | 4.238 | .03 | 66.1 | .07 | 92.6 | .12 | 132 | .21 | 198 | .39 | 264 | .57 | 331 | .87 | 397 | 4.0 | 6.1 | 13.0 | 2.9 |
| 2-1/2 | PIPE-SCH XX | 2.875 | 1.771 | .552 | 2.464 | .03 | 38.5 | .10 | 53.4 | .17 | 76.9 | .30 | 115 | .59 | 154 | .79 | 193 | 1.15 | 231 | 3.1 | 4.8 | 10.3 | 2.2 |
| 2-1/2 | HOSE | – | 2.50 | – | 4.91 | .016 | 76.5 | .045 | 107 | .09 | 153 | .18 | 229 | .30 | 306 | .43 | 382 | .617 | 459 | | | | |

Figure 17-2. Hydraulic Table.

Figure 17–3. Piping System Sizer.

ing system; you have an intuitive understanding of the rules, so there is no need to repeat them on each page of a real notebook. TK!Solver will need them, though, so let's look at the rules that you can use to construct the model.

A common equation for expressing the pressure drop in a pipe is the Darcy-Weisbach equation,

$$\Delta P = f\frac{L}{D}\ \frac{\varrho}{g_c}\ \frac{V^2}{2}$$

*Where:*   $\Delta P$ = pressure drop, in lb/ft$^2$

$f$ = friction factor, dimensionless

$L$ = flow length of pipe, in ft

$D$ = internal length of pipe, in ft

$\varrho$ = fluid density, in lbm/ft$^3$

$V$ = average velocity, in ft/s

$g_c$ = 32.2 ft − lbm/lbf − s$^2$

The friction factor is a function of the relative roughness of the pipe, $\epsilon/D$, that is, the ratio of the absolute roughness and the pipe diameter, and the Reynolds number, Re

$$\text{Re} = DV/v$$

*Where:*   $\epsilon$ = absolute roughness

$v$ = dynamic viscosity of the fluid

The friction factor is usually obtained from a Moody chart, available in a number of engineering text or reference books. In this first model, you will choose an average friction factor for turbulent flow in commercial steel pipe, which will be equivalent to using a set of hydraulic pressure drop tables, or a pressure loss chart.

The first step is to create the model shown in Figure 17–4. With this model, you can only calculate one unknown at a time, since you have only one rule. Although the model will calculate the pressure drop, given pipe diameter, velocity, and pipe length, the model isn't very useful when you haven't sized the pipe, yet. At this point, you can't save any time over manual methods using tables and charts. However, it is useful to check a model in steps to insure that it does work, and this model does agree with the Cameron Hydraulic Data charts, if you use the correct friction factor.

To obtain a model that will accept the information you normally have given, let's add another rule to the model, as well as some conversion factors on the Units sheet. The second rule relates the volume flow rate, in gallons per minute, to the pipe size and velocity. Now you can solve for two unknowns at a time, such as pressure drop and velocity, given volume flow rate and pipe length, and the constants. A solved model is shown in Figure 17–5. This model produces an accurate value for velocity, but does not agree with the hydraulic tables for pressure drop, as you are using a fixed value for the friction factor. In reality, the friction factor varies

```
(1c) Comment: PIPE SIZING, #1                                              204/!

==================== VARIABLE  SHEET ===========================================
St Input       Name    Output     Unit       Comment
-- -----       ----    ------     ----       -------
                                             PIPE SIZING, #1
                                             ***************
               dP                 fth2o      pressure drop in L feet of pipe
               f                             friction factor, dimensionless
               L                  ft         length of pipe
               rho                lbm/ft^3   fluid density
               V                  ft/s       fluid velocity
               D                  ft         inside diameter of pipe
               g                  ft*lbm/lb  acceleration of gravity


(1r) Rule: dP=(f*L*rho*V^2)/(D*g*2)                                        204/!

==================== RULE SHEET ================================================
S Rule
- ----
* dP=(f*L*rho*V^2)/(D*g*2)




(1f) From: ft                                                              204/!

==================== UNIT SHEET ================================================
From          To         Multiply By  Add Offset
----          --         -----------  ----------
ft            in         12
lbf/in^2      lbf/ft^2   144
lbf/in^2      fth2o      2.31
ft^3/s        gal/min    448.8
min           s          60
```

Figure 17–4. A Pipe Sizing Model—Step 1.

with pipe size in a complicated fashion, given the same absolute roughness.

To account for the variance in friction factors for the different pipe sizes, we need two more rules. We mentioned above that the friction factor can be obtained from a Moody chart, which is readily available, but it can also be calculated from an equation developed by C. F. Colebrook in 1939. The Colebrook equation is

$$\frac{1}{\sqrt{f}} = 1.14 + 2\log{(D/\epsilon)} - 2\log{[1 + \frac{9.3}{\text{Re}(\epsilon/D)\sqrt{f}}]}$$

*Where:*  $\text{Re} = VD/\nu$

$f$ = friction factor, dimensionless

$\epsilon$ = absolute roughness, in ft

```
(1c) Comment: PIPE SIZING, #2                                    203/!

================== VARIABLE SHEET ===================================
St Input      Name     Output   Unit     Comment
-- -----      ----     ------   ----     -------
                                         PIPE SIZING, #2
                                         ***************
              dP                fth2o    pressure drop in L feet of pipe
              f                          friction factor, dimensionless
              L                 ft       length of pipe
              rho               lbm/ft^3 fluid density
              V                 ft/s     fluid velocity
              D                 ft       inside diameter of pipe
              g                 ft*lbm/lb acceleration of gravity
              Q                 gal/min  volume flow rate of fluid in pipe




(1r) Rule: dP=(f*L*rho*V^2)/(D*g*2)                              203/!

=================== RULE SHEET ====================================
S Rule
- ----
* dP=(f*L*rho*V^2)/(D*g*2)
* Q=V*(D/2)^2*pi()
```

Figure 17–5. A Pipe Sizing Model—Step 2.

$D$ = inside pipe diameter, in ft

$V$ = average fluid velocity, in ft/sec

$\nu$ = kinematic viscosity, in ft²/sec

Now, a cursory examination of the Colebrook equation will reveal that it is very difficult to solve without using a computer, and most hydraulic tables were developed on mainframe computers several years ago. As you shall see, TK!Solver running on a microcomputer will handle the task of solving the Colebrook equation quite well. By adding the rule to calculate the Reynolds number, the Colebrook equation results in the model shown in Figure 17–6.

A few sample runs with this model reveal that it agrees with the hydraulic tables, within approximately 1 percent. However, the model is more complicated; the Direct Solver can no longer solve the model. If you attempt to use the Direct Solver, the first and fourth equations will remain unsatisfied. A glance at the Colebrook equation reveals the reason: the variable $f$ appears in the equation twice, and the equation cannot be rearranged so that it only appears once. You have to use the Iterative Solver to solve the model,

```
(3c) Comment: pressure drop in L feet of pipe                              202/!

==================== VARIABLE SHEET ===========================================
St Input        Name      Output      Unit      Comment
-- -----        ----      ------      ----      -------
                                                PIPE SIZING, #3
                                                ***************
                dP                    fth2o     pressure drop in L feet of pipe
                f                                friction factor, dimensionless
                L                     ft        length of pipe
                rho                   lbm/ft^3  fluid density
                V                     ft/s      fluid velocity
                D                     ft        inside diameter of pipe
                g                     ft*lbm/lb acceleration of gravity
                Q                     gal/min   volume flow rate of fluid in pipe
                Re                              Reynolds number
                v                     ft^2/s    kinematic viscosity
                epsilon               ft        absolute roughness


(1r) Rule: dP=(f*L*rho*V^2)/(D*g*2)                                        202/!

==================== RULE SHEET ===============================================
S Rule
- ----
* dP=(f*L*rho*V^2)/(D*g*2)
* Q=V*(D/2)^2*pi()
* Re=V*D/v
* 1/sqrt(f)=-2*log(epsilon/(3.7*D)+2.51/(Re*sqrt(f)))
*
```

```
(3c) Comment: pressure drop in L feet of pipe                              202/!

==================== VARIABLE SHEET ===========================================
St Input        Name      Output      Unit      Comment
-- -----        ----      ------      ----      -------
                                                PIPE SIZING, #3
                                                ***************
                dP                    fth2o     pressure drop in L feet of pipe
                f                                friction factor, dimensionless
                L                     ft        length of pipe
                rho                   lbm/ft^3  fluid density
                V                     ft/s      fluid velocity
                D                     ft        inside diameter of pipe
                g                     ft*lbm/lb acceleration of gravity
                Q                     gal/min   volume flow rate of fluid in pipe
                Re                              Reynolds number
                v                     ft^2/s    kinematic viscosity
                epsilon               ft        absolute roughness


(1r) Rule: dP=(f*L*rho*V^2)/(D*g*2)                                        202/!

==================== RULE SHEET ===============================================
S Rule
- ----
* dP=(f*L*rho*V^2)/(D*g*2)
* Q=V*(D/2)^2*pi()
* Re=V*D/v
* 1/sqrt(f)=-2*log(epsilon/(3.7*D)+2.51/(Re*sqrt(f)))
*
```

Figure 17–6.  The Pipe Sizing Model with the Colebrook Equation.

which involves setting one variable as a guess, and giving the Action command (!) from the Variable sheet or the Rule sheet, or the Solve command (/!) from any sheet.

You must enter an input for the variable $f$, and set it as a Guess; entering a Guess value for $dP$ will produce an error message and leave the model unsolved. A good guess for $f$ is .01 or .02; a guess of 0 for $f$ also seems to work, because TK!Solver works with real numbers, and a 0 is not always a 0. It would be advisable to use a value that is greater than 0.

Now we are ready to consider the overall method of sizing pipe with the model. We have a new type of tool, so our approach will be slightly different than the one we would use with hydraulic tables or slide rules. We want to let TK!Solver do as much of the work as possible.

Referring to Figure 17–1, you want to size each section of pipe and then calculate the pressure loss in the system, between the pump and the last outlet. For now, assume that the system is level, so that there is no change in potential energy—there is no lift involved. All you will consider is the loss of pressure due to friction between the fluid and the pipe. Now, if you were to use the hydraulic tables to size the system, you would first calculate the flow in each section of pipe by starting at the last outlet, labeled $F$, and work toward the pump, adding the flow at each outlet to the flow in the downstream section. The diagram is labeled with the flow leaving each outlet, and the flow in each section of pipe.

Next, you would consult the hydraulic table, looking for a pipe size that had an acceptable velocity and head loss (pressure loss) for the flow rate. Both hydraulic tables and friction charts give these values in loss per 100 feet. For example, when you size small piping for circulating chilled water in a building, you try to keep the velocity below 4 feet per second, to minimize noise. This process would be repeated for each section of pipe in the system, noting the pressure loss in tabular form as you proceed, as shown in Figure 17–7.

When you obtain a value for each section of pipe, you then calculate the equivalent length of each section by assigning an equivalent length to each fitting, valve, etc. When each section of pipe has been assigned a pressure loss, usually as lbs/in² per 100 feet, or feet of water per 100 feet of length, and an equivalent length, you can calculate the total pressure loss for each section of pipe. Lastly, total the individual pressure losses for the total loss in the system.

Actually, that description is a simplication, as the first attempt may produce a total pressure loss that is unacceptable; it

Form 9-53A (Rev. 10-72)

# HYDRAULIC CALCULATION SHEET

CONTRACT NO. _____

System No. _____ Sheet _____

NAME, ADDRESS OF PROPERTY

NOTES

DATE

Calculated by _____

| SPRINKLERS NOZZLES | | (q) Added Gpm | (Q) Total Gpm | Pipe Size In. | Pipe and Equivalent Length Fittings - Devices Feet | FRICTION LOSS | | (Elevation) Static Plus or Minus Psi | Required Pressure Psi | CALCULATN. REFERENCE | |
|---|---|---|---|---|---|---|---|---|---|---|---|
| Plan Ref. Pt. | Total Heads | Type | | | | | Psi/Ft. C = | Total Psi | | | Sheet No. | Point No. |
| | | | | | | PRESSURE AT HEAD → | | | | | |

Figure 17–7. Pipe Sizing Worksheet.

254

may lead to a poor pump selection, or may result in excessive pumping energy costs. In such an event, you must go through the procedure again, using a lower upper limit on velocity.

When you use the TK!Solver model, you are going to take a slightly different tack, because you have to be more explicit with the computer. The strategy will be to first calculate exact pipe sizes, based on an upper limit for velocity. Then you will let TK!Solver use a User Function to select actual pipe sizes. Finally, you will let TK!Solver make a second pass, calculating actual velocities and pressure losses, based on the actual pipe sizes. You can use the List Solver, so that you don't have to initiate the solution of the model for each pipe section.

Before you can use the List Solver, you must create the lists for the model, which you do by placing the cursor in the Status field of the Variable sheet and typing $L$, for each variable we want to associate with a list. In this first step, we will use a fixed value for $V$ (velocity), $v$ (kinematic viscosity), $g$ (acceleration), $L$ (length), and $\varrho$ (density). You will need an input list for $Q$ (flow rate), and output lists for $D$ (pipe diameter), $dP$ (pressure drop), and $V$ (velocity). You could also set up a list for $f$ (friction factor), but there is no need to do so, as you are not really interested in the individual friction factors for the sections of pipe. You don't actually need a list for Reynolds number, either, but we will create one, anyway. When the lists have been created, the complete model appears as shown in Figure 17–8.

To solve the model, fill the input list, $Q$ (flow rate), and enter a value in the input field so that TK!Solver will know that you wish to use the $Q$ list as the input list. You can avoid the necessity of entering a guess value for $f$ by diving to the $f$ Variable subsheet, and entering a value of .01 in the First Guess field. Then you give the List command, /L!, and TK!Solver will solve the model for each value of $Q$, filling the output lists with the corresponding values. The resulting lists are shown in Figure 17–9.

You could find the total pressure drop in the piping system, given the equivalent length of each pipe section, but the answer would be meaningless, because pipe is sold in standard sizes, not in the exact sizes you just calculated. Now you have to find a standard size for each section of pipe, such that the velocity will be less than the upper limit you chose. (In the larger pipe sizes, it is permissible to use an upper limit higher than the one you used, but you are using the one value to simplify matters.) What you need is a User Function that will transform the exact dimensions in the $D$ list to the dimensions of standard pipe sizes, and a Rule that will apply the function to the $D$ list. Each exact pipe size will be transformed to the dimension of the next larger standard pipe size.

```
(3c) Comment: pressure drop in L feet of pipe                        199/!

================== VARIABLE SHEET ==========================================
St Input        Name     Output     Unit      Comment
-- -----        ----     ------     ----      -------
                                              PIPE SIZING, #4
                                              ***************
L               dP                  fth2o     pressure drop in L feet of pipe
                f                              friction factor, dimensionless
                L                   ft        length of pipe
                rho                 lbm/ft^3  fluid density
                V                   ft/s      fluid velocity
L               D                   in        inside diameter of pipe
                g                   ft*lbm/lb acceleration of gravity
L               Q                   gal/min   volume flow rate of fluid in pipe
L               Re                            Reynolds number
                v                   ft^2/s    kinematic viscosity
                epsilon             ft        absolute roughness


(1r) Rule: dP=(f*L*rho*V^2)/(D*g*2)                                 199/!

================== RULE SHEET ==============================================
S Rule
- ----
* dP=(f*L*rho*V^2)/(D*g*2)
* Q=V*(D/2)^2*pi()
* Re=V*D/v
* 1/sqrt(f)=-2*log(epsilon/(3.7*D)+2.51/(Re*sqrt(f)))
*


(1n) Name: dP                                                      199/!

================== LIST SHEET =============================================
Name          Elements  Unit      Comment
----          --------  ----      -------
dP            6         fth2o
V                       ft/s
D             6         in
Q             6         gal/min
Re            6
id
inputid
sch40         23


(1f) From: ft                                                     199/!

================== UNIT SHEET =============================================
From          To        Multiply By  Add Offset
----          --        -----------  ----------
ft            in        12
lbf/in^2      lbf/ft^2  144
lbf/in^2      fth2o     2.31
ft^3/s        gal/min   448.8
min           s         60


(1n) Name: sch40id                                               199/!

================== USER FUNCTION SHEET ===================================
Name          Domain    Mapping Range   Comment
----          ------    ------- -----   -------
sch40id       sch40     Step    sch40   inside dimensions of sch 40 steel pipe
```

Figure 17-8.  The Pipe Sizing Model with Lists.

(1v) Value: 1.5350778293                                          192/!

================== LIST: dP =================================================
Comment:
Display Unit:                    fth2o
Storage Unit:                    lbf/ft^2
Element Value
------- -----
1        1.535077829
2        3.719287795
3        .7041818425
4        .7578401749
5        1.115586548
6        .120054452


(8o) Output: .285841234632                                        192/!

================== VARIABLE SHEET ===========================================
St Input    Name    Output    Unit    Comment
-- -----    ----    ------    ----    -------
                                      PIPESIZER - SCHEDULE 40 BLACK STEEL
                                      ************************************
L           dP      .64093061 fth2o   pressure drop in L feet of pipe
            f       .01724504         friction factor, dimensionless
L  1        L                 ft      length of pipe
            rho     62.4      lbm/ft^3 fluid density
    5       V                 ft/s    fluid velocity
L           D       .28584123 in      inside diameter of pipe
            g       32.174    ft*lbm/lb acceleration of gravity
L  1        Q                 gal/min volume flow rate of fluid in pipe
L           Re      9794.4502         Reynolds number
            v       .00001216 ft^2/s  kinematic viscosity
            epsilon .00015    ft      absolute roughness
L           Ds40    .364      in      standard sch 40 pipe id
L           dPs     .19139408 fth2o   pressure drop in sch40 pipe
L           Vs      3.0833086 ft/s    velocity in sch 40 pipe
L           Res     7691.3674         Reynolds number in sch 40 pipe
L           fs      .03920827         friction factor in sch 40 pipe
L           Dnom    'one_qtr          nominal pipe size


(1v) Value: 1.5350778293                                          192/!

================== LIST: dP =================================================
Comment:
Display Unit:                    fth2o
Storage Unit:                    lbf/ft^2
Element Value
------- -----
1        1.535077829
2        3.719287795
3        .7041818425
4        .7578401749
5        1.115586548
6        .120054452

Figure 17-9.  The Output Lists of the Pipe Sizing Model.

```
(1v) Value: 2.8584123464                                          192/!

=================== LIST: D ===============================================
Comment:
Display Unit:                  in
Storage Unit:                  ft
Element Value
------- -----
1       2.858412346
2       3.500825862
3       4.28761852
4       5.535291707
5       6.229765278
6       6.703571159


(1v) Value: 'three_in                                             192/!

=================== LIST: Dnom ============================================
Comment:
Display Unit:
Storage Unit:
Element Value
------- -----
1       'three_in
2       'three_$_half
3       'five_in
4       'six_in
5       'eight_in
6       'eight_in


(1v) Value: 91253.511669                                          192/!

=================== LIST: Res =============================================
Comment:
Display Unit:
Storage Unit:
Element Value
------- -----
1       91253.51167
2       118362.0802
3       124811.3713
4       173103.3226
5       166625.4135
6       192934.6894


(1v) Value: .020913415937                                         192/!

=================== LIST: fs ==============================================
Comment:
Display Unit:
Storage Unit:
Element Value
------- -----
1       .02091341594
2       .01993123073
3       .01911494871
4       .01801453597
5       .01769327469
6       .01734415782
```

Figure 17–9.  (continued).

```
(1v) Value: 1.0776507074                                         192/!

==================== LIST: dPs ==================================================
Comment:
Display Unit:            fth2o
Storage Unit:           lbf/ft^2
Element Value
------- -----
1       1.077650707
2       3.478518137
3       .3116011052
4       .4798728321
5       .3232772433
6       .0501906517


(1v) Value: 4.3401930975                                         192/!

==================== LIST: Vs ==================================================
Comment:
Display Unit:            ft/s
Storage Unit:           ft/s
Element Value
------- -----
1       4.340193098
2       4.867924112
3       3.608574461
4       4.164754631
5       3.04648294
6       3.527506562


(1v) Value: 97944.501998                                         192/!

==================== LIST: Re ==================================================
Comment:
Display Unit:
Storage Unit:
Element Value
------- -----
1       97944.502
2       119957.0265
3       146916.753
4       189668.7125
5       213465.0931
6       229700.2179


(1v) Value: 3.068                                                192/!

==================== LIST: Ds40 ==================================================
Comment:
Display Unit:            in
Storage Unit:           ft
Element Value
------- -----
1       3.068
2       3.548
3       5.047
4       6.065
5       7.981
6       7.981
```

Figure 17–9.  (continued).

You should recall at this point that User Functions can be mapped in three different ways: Table, Step, and Linear. The mapping of a function determines the manner in which you can interpolate between range and domain values. Remember that you cannot interpolate between range or domain values if a function is set for Table mapping. If the function is mapped as a Step function, you can interpolate between domain values, but not between range values. Finally, if the function is mapped as a linear function, you can interpolate either value.

The user function will take as an argument the exact pipe size, the domain value, and return the dimension of a standard pipe size, the range value. Since you want the function to return the dimension of a 4-inch pipe for any argument between the dimension of a 3-inch pipe and a 4-inch pipe, you want a Step function. When the function receives an argument that is not a standard pipe size, it will return the "previous" range value, which is the range value of the next larger size. The function for Schedule 40 steel pipe appears in Figure 17-10. It uses the same list for both the domain and the range, a list of standard inside pipe diameters, in this case for Schedule 40 steel pipe. The previous range value returned is always the next larger standard pipe size.

You must make an indelible mental note of one fact regarding user functions, and that is to keep the units consistent. The TK!Solver manual makes this cryptic note regarding user functions and the units associated with them: "Functional relationships are defined only for Display unit values." Remember that variables have Display units and Calculation units, and that lists have Storage units and Display units. For a user function to operate properly, units and all, the Display units of the Range list must be consistent with the Calculation units of the variable to which the user function returns a value.

For this model, that means you can enter the user function that converts the exact pipe sizes to standard sizes in inches. But, since you originally defined the variable $D$ (the calculated exact pipe size) in feet (ft), and since you want to pass values of $D$ to the user function as arguments, you must dive to the conversion list, sch40id, and set the display units to feet (ft), and the storage units to inches (in). Alternatively, you could enter the list in feet, by entering the size in inches divided by 12, i.e., 6.065/12 would enter the correct inside dimension of a 6-inch schedule 40 pipe, in feet.

Now, to develop the model into one that not only selects the correct standard pipe size, but also recalculates the velocity and pressure drop for the same flow rate in the standard size pipe, add four more equations, which are the first four equations with

```
USER FUNCTION: sch40id
Comment:                    inside dimensions of sch 40 steel pipe
Domain List:                sch40
Mapping:                    Step
Range List:                 sch40
Element Domain        Range
------- ------        -----
   1    1.8855        1.8855
   2    1.567833333   1.567833333
   3    1.406166667   1.406166667
   4    1.25          1.25
   5    1.09375       1.09375
   6    .9948333333   .9948333333
   7    .835          .835
   8    .6650833333   .6650833333
   9    .5054166667   .5054166667
  10    .4205833333   .4205833333
  11    .3355         .3355
  12    .2956666667   .2956666667
  13    .2556666667   .2556666667
  14    .20575        .20575
  15    .17225        .17225
  16    .1341666667   .1341666667
  17    .115          .115
  18    .0874166667   .0874166667
  19    .0686666667   .0686666667
  20    .0518333333   .0518333333
  21    .0410833333   .0410833333
  22    .0303333333   .0303333333
  23    .0224166667   .0224166667
```

Figure 17–10.  User Function for Sizing Schedule 40 Pipe.

subscripts added to the variable names: Ds40 for D, $f_s$ for f, $V_s$ for V, and so on. The entire model is shown in Figure 17–11.

TK!Solver solves the first group of equations first, because it has enough input values to do so. Then it converts the exact pipe sizes to standard sizes—it can't do the conversion until it has a value for *D,* the exact size. When it has a value for Ds40, the standard pipe size, it can solve the remaining group of equations, producing the velocity and pressure drop values for the desired flow rate in a standard size pipe, as well as the Reynolds number and friction factor.

Notice that the L value was changed to a list variable, so that you can enter a list of equivalent lengths, as well as a list of flow rates. The model will now size a piping path without attention from the operator: you enter the lists, having entered constants as rules and having set guess values for both *f* and *fs,* and then enter the /L! command. TK!Solver's List Solver does the rest of the work.

Although the model is now producing all the information you need, you can add some refinements—the output has not been con-

```
(1o) Output:                                                      192/!

================== VARIABLE SHEET =======================================
St Input     Name    Output      Unit      Comment
-- -----     ----    ------      ----      -------
                                           PIPESIZER - SCHEDULE 40 BLACK STEEL
                                           ***********************************
L            dP      .64093061   fth2o     pressure drop in L feet of pipe
             f       .01724504             friction factor, dimensionless
L    1       L                   ft        length of pipe
             rho     62.4        lbm/ft^3  fluid density
     5       V                   ft/s      fluid velocity
L            D       .28584123   in        inside diameter of pipe
             g       32.174      ft*lbm/lb acceleration of gravity
L    1       Q                   gal/min   volume flow rate of fluid in pipe
L            Re      9794.4502             Reynolds number
             v       .00001216   ft^2/s    kinematic viscosity
             epsilon .00015      ft        absolute roughness
L            Ds40    .364        in        standard sch 40 pipe id
L            dPs     .19139408   fth2o     pressure drop in sch40 pipe
L            Vs      3.0833086   ft/s      velocity in sch 40 pipe
L            Res     7691.3674             Reynolds number in sch 40 pipe
L            fs      .03920827             friction factor in sch 40 pipe
L            Dnom    'one_qtr              nominal pipe size
```

```
(1r) Rule: dP=(f*L*rho*V^2)/(D*g*2)                               192/!

=================== RULE SHEET =========================================
S Rule
- ----
* dP=(f*L*rho*V^2)/(D*g*2)
* Q=V*(D/2)^2*pi()
* Re=V*D/v
* 1/sqrt(f)=-2*log(epsilon/(3.7*D)+2.51/(Re*sqrt(f)))

* Ds40=sch40id(D)
* Dnom=noms40(Ds40)

* dPs=(f*L*rho*Vs^2)/(Ds40*g*2)
* Q=Vs*(Ds40/2)^2*pi()
* Res=Vs*Ds40/v
* 1/sqrt(fs)=-2*log(epsilon/(3.7*Ds40)+2.51/(Res*sqrt(fs)))
*
* rho=62.4
* g=32.174
* epsilon=.00015
* v=.00001216
```

```
(1f) From: ft                                                    192/!

================== UNIT SHEET ==========================================
From        To          Multiply By  Add Offset
----        --          -----------  ----------
ft          in          12
lbf/in^2    lbf/ft^2    144
lbf/in^2    fth2o       2.31
ft^3/s      gal/min     448.8
min         s           60
```

Figure 17–11.  The Pipesizer Model.

```
(1n) Name:  dP                                                      192/!

==================== LIST SHEET ==========================================
Name         Elements   Unit        Comment
----         --------   ----        -------
dP           6          fth2o
V                       ft/s
D            6          in
Q            6          gal/min
Re           6
id
inputid
sch40        23         ft
Ds40         6          in
dPs          6          fth2o
Vs           6          ft/s
Res          6
fs           6
L            6          ft
noms40       23
Dnom         6
```

Figure 17–11. (continued).

sidered. You could have the program print out the various lists, but
a table would have a much more professional appearance.

Recall that you create tables using the Table sheet, shown
again in Figure 17–12. However, before you make the entries that
TK!Solver will use to create the table, you need to add another user
function. The model selects a standard pipe using the sch40id user
function, which returns an inside dimension that corresponds to a
standard size. A person who works with pipe dimensions day in and
day out will eventually learn all the dimensions, and will know at a
glance that a pipe with an inside diameter of 6.065 inches is a
schedule 40 6-inch pipe, and that a pipe with an inside diameter of
5.761 inches is a schedule 80 6-inch pipe, but most people need to
refer to handbook tables for that type of information. Your table
will be much more understandable if you add a user function that
will return a nominal pipe size in addition to the actual inside
diameter.

You set up this user function by displaying the User Function
sheet on the screen with the =F command, and enter the name of
the function and the two lists it will use. You have already entered
one of the lists you need—the list of actual inside diameters you
used to create the sizing function, sch40id. The second list is a list
of literal constants. You use the literal constants so that you can
display the pipe size in familiar terms, i.e., "6 inch sch 40," or "3-1/2
in. XS." There is no need for interpolation in this function, so you
leave the mapping field alone; TK!Solver selects the Table option

```
(1n) Name: sch40id                                                 192/!

=================== USER FUNCTION SHEET =====================================
Name         Domain       Mapping Range       Comment
----         ------       ------- -----       -------
sch40id      sch40        Step    sch40        inside dimensions of sch 40 steel pipe
noms40       sch40        Table   noms40       nominal schedule 40 pipe sizes



(s) Screen or Printer: Printer                                     192/!

=================== TABLE SHEET =============================================
Screen or Printer:          Printer
Title:                      Schedule 40 Pipe Sizes
Vertical or Horizontal:     Vertical
List       Width First Header
----       ----- ----- ------
Q          10    1     GPM
Dnom       10    1     Size
Ds40       10    1     I D, in.
Vs         10    1     Vel, fps
fs         10    1     f
L          10    1     lgth, ft
dPs        10    1     delta P, ft
```

Figure 17–12.  The Table Sheet.

```
USER FUNCTION: noms40
Comment:                    nominal schedule 40 pipe sizes
Domain List:                sch40
Mapping:                    Table
Range List:                 noms40
Element Domain         Range
------- ------         -----
   1    1.8855         'twenty_fou
   2    1.567833333    'twenty_in
   3    1.406166667    'eighteen_i
   4    1.25           'sixteen_in
   5    1.09375        'fourteen_i
   6    .9948333333    'twelve_in
   7    .835           'ten_in
   8    .6650833333    'eight_in
   9    .5054166667    'six_in
  10    .4205833333    'five_in
  11    .3355          'four_in
  12    .2956666667    'three_$_ha
  13    .2556666667    'three_in
  14    .20575         'two_$_half
  15    .17225         'two_in
  16    .1341666667    'one_$_half
  17    .115           'one_$_qtr
  18    .0874166667    'one_in
  19    .0686666667    'three_qtr
  20    .0518333333    'one_half
  21    .0410833333    'three_eigt
  22    .0303333333    'one_qtr
  23    .0224166667    'one_eigth
```

Figure 17–13.  The Sizename Function.

as the default, and that is what you want. The list is shown in Figure 17-13 and the Table sheet is shown in Figure 17-14.

You must enter another rule so that TK!Solver will fill the list of nominal pipe sizes. This rule is simply to call to the user function you just created. On each pass the List Solver makes, it will look up the nominal pipe size that corresponds to the schedule 40 inside diameter it selects after calculating the exact pipe size.

Finally, you have the output of the program in the form of a table which would be acceptable for any engineering report, shown in Figure 17-15. The input for the model comprises a maximum acceptable velocity in the pipe, a list of flow rates in gallons per minute, and a list of corresponding equivalent lengths for the sections of the piping system. You initiate the solution of the model by entering a trivial value as an input for the $Q$ list and the $L$ list, so that TK!Solver will know how to use those lists as input lists, and then entering the command that invokes the List Solver—/L!. Although the solution is not reached instantaneously, it does run unattended, so speed is of secondary importance.

Before leaving this model, a few words of explanation are in order. You have created a model that duplicates the performance of the informal technique you would use with pipe sizing charts and tables, and you made the same assumptions you would have made if you have been using either of those tools. You assumed the flow would be turbulent, and made no attempt to check for laminar or transitional flow. For the average application, this is a reasonable assumption.

You could improve the model somewhat—you could easily add the ability to select schedule 80 pipe sizes, copper tube sizes, duc-

```
(s) Screen or Printer: Printer                                          192/!

==================== TABLE SHEET =========================================
Screen or Printer:          Printer
Title:                      Schedule 40 Pipe Sizes
Vertical or Horizontal:     Vertical
List        Width First Header
----        ----- ----- ------
Q            10     1    GPM
Dnom         10     1    Size
Ds40         10     1    I D, in.
Vs           10     1    Vel, fps
fs           10     1    f
L            10     1    lgth, ft
dPs          10     1    delta P, ft
```

Figure 17-14. The Table Sheet for the PIPESIZR Model.

tile iron pipe sizes, and so on. You could have the model check for transitional flow or laminar flow. However, let's leave the model for now, and move on to another one.

Schedule 40 Pipe Sizes

| GPM | Size | I D, in. | Vel, fps | f | lgth, ft | delta P, f |
|-----|------|----------|----------|---|----------|------------|
| 100 | three_in | 3.068 | 4.34019310 | .020913416 | 45 | 1.07765071 |
| 150 | three_$_ha | 3.548 | 4.86792411 | .019931231 | 140 | 3.47851314 |
| 225 | five_in | 5.047 | 3.60857446 | .019114949 | 34 | .311601105 |
| 375 | six_in | 6.065 | 4.16475463 | .018014536 | 50 | .479872832 |
| 475 | eight_in | 7.981 | 3.04648294 | .017693275 | 85 | .323277243 |
| 550 | eight_in | 7.981 | 3.52750656 | .017344158 | 10 | .050190652 |

(2c) Comment: ***************                                         203/!

=================== VARIABLE SHEET ===============================================

| St | Input | Name | Output | Unit | Comment |
|----|-------|------|--------|------|---------|
| -- | ----- | ---- | ------ | ---- | ------- |
|    |       |      |        |      | DUCT SIZING, #1 |
|    |       |      |        |      | *************** |
|    |       | rho  |        | lbm/ft^3 | mass density of fluid |
|    |       | v1   |        | ft/s | velocity at section 1 |
|    |       | gc   |        | ft*lbm/lb | dimensional constant |
|    |       | pz1  |        | lbf/ft^2 | atmospheric pressure at elev z1 |
|    |       | p1   |        | lbf/ft^2 | static pressure (gage) at section 1 |
|    |       | gamma |       | lbf/ft^3 | specific weight |
|    |       | z1   |        | ft | elevation 1 |
|    |       | v2   |        | ft/s | velocity at section 2 |
|    |       | pz2  |        | lbf/ft^2 | atmospheric pressure at elev z2 |
|    |       | p2   |        | lbf/ft^2 | static pressure (gage) at section 2 |
|    |       | z2   |        | ft | elevation 2 |
|    |       | dPt  |        | lbf/ft^2 | total pressure loss due to fluid resis |

Figure 17-15. The Output of the PIPESIZR Model.

# CHAPTER 18

# Duct Design

Ducts are used to convey air and other gases from one point to another. A duct system will often include a fan, which increases or decreases the pressure in the duct system, and so induces a flow of the gas through the duct system. On other occasions, the high or low pressure is not produced by a mechanical device, such as a fan or a pump, but by some process, such as combustion. In either case, the duct system must be sized to convey a quantity of gas, or in some cases, a quantity of gas in which some form of particulate matter is suspended, with constraints on the pressure lost to friction, velocity, maximum pressure in the system, etc.

Duct design is a common activity, as we find duct systems all around us—in houses, cars, aircraft, office buildings, factories, etc. Unfortunately, duct design is not a trivial matter. The physical processes involved are complex and not easily observed. Some of the equations used to describe the processes involved can be derived from basic principles, but many more have been derived experimentally, and cannot be applied universally.

As with all complex problems we encounter as engineers, the attention we can devote to a duct design problem is influenced by economics. Where performance is critical and funds are available, we can apply complex solution methods and obtain accurate results. Where funds are not available, we use simpler methods, and accept the loss of performance. Thus, the air conditioning system in a space vehicle is very well designed, at a very high cost, and the air conditioning system in the average house is poorly designed.

Because TK!Solver can solve problems faster than we can manually, and because it runs unattended, we can consider applying more accurate design methods to our design problems, at no increase in cost. Once a model, or series of models, is developed, we can use it over and over. So, without further delay, let's proceed to the development of some duct design tools.

# THE BERNOULLI EQUATIONS

The basic equations for analyzing fluid flow are the Bernoulli equations, which can be developed by setting the forces on an element of a stream tube in fluid flow to the change in momentum of the element produced by those forces. A diagram of the forces involved is shown in Figure 18–1. We won't go through the complete derivation of Bernoulli's equations; we will simply present them in three common forms.

$$\frac{V^2}{2} + \frac{P}{\varrho} + gz = \text{constant} \tag{a}$$

$$\frac{\varrho V_1{}^2}{2} + P_1 + \varrho g z_1 = \varrho \frac{V_2{}^2}{2} + P_2 + \varrho g z_2 = \text{constant total pressure} \tag{b}$$

$$\frac{V_2{}^2}{2g} + \frac{P_1}{g\varrho} + z_1 = \frac{V_2{}^2}{zg} + \frac{P_2}{g\varrho} + z_2 = \text{constant total head} \tag{c}$$

*Where:*  $V$ = velocity
$P$ = pressure

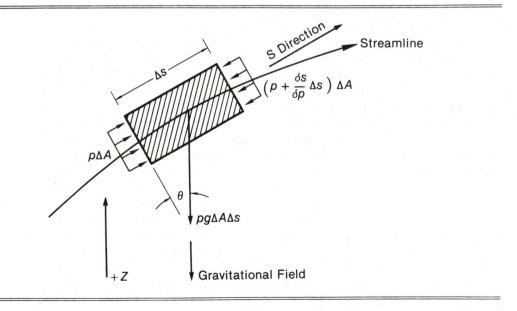

Figure 18–1. Forces on an Element of a Fluid Stream.

$\varrho$ = density

$g$ = gravitational constant

$z$ = elevation

The first term in equation (a) represents the dynamic pressure, the second term represents the static pressure, and the last term represents the potential pressure, and the sum of the three pressures is constant in a given system. The constant sum is known as *Bernoulli's constant.* The second form of the equation is normally used in gas flow problems, and the last form of the equation is normally used in liquid flow problems.

Bernoulli's equations were developed for the ideal flow of an incompressible fluid without friction losses. However, by adding a term in the equation to account for friction losses, you can use the equations for the analysis of real fluids in real duct systems. You can begin to build the first model with an equation that expresses the Bernoulli relationship between two points in a duct system. Assuming that the program has been reset, make the entries shown in Figure 18-2.

This model is not very useful for a duct design problem, but it does illustrate some basic facts. If you take static pressure and velocity measurements on the section of duct shown in Figure 18-3, and enter them as inputs to the model, you can calculate the losses due to friction with the duct, and friction within the fluid stream due to viscosity. It is these losses that you will be concerned with when you begin to size duct systems, as they affect not only the duct sizes, but also the size of air moving devices and the energy required to move the desired amount of air through the ducts.

```
(1r) Rule: rho*v1^2/2*gc+pz1+p1+gamma*z1=rho*v2^2/2*gc+pz2+p2+gamma*z2+dPt 203/!

===================== RULE SHEET ================================================
S Rule
- ----
* rho*v1^2/2*gc+pz1+p1+gamma*z1=rho*v2^2/2*gc+pz2+p2+gamma*z2+dPt
* gc=32.174
```

Figure 18-2. The First Step of the DUCTDESN Model—
Model DUCT1 on TKMODELS1 Disk.

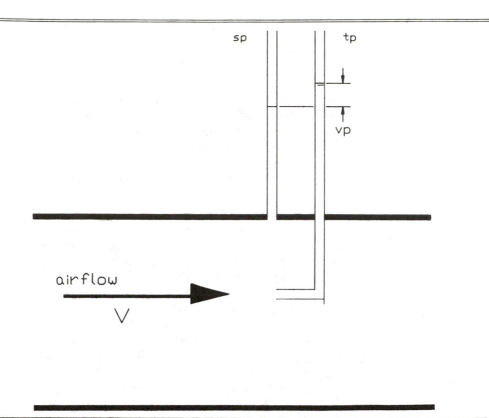

Figure 18–3.  Pressure Readings in a Duct.

# THE STACK EFFECT

However, with the addition of seven more rules, two of which are simply constants, you can investigate a phenomenon known as the *stack effect*. A chimney, or stack, works because of the difference in the density of the air in the chimney and the density of the air surrounding the chimney. The air entering the chimney expands as it is heated by the fire in the fireplace, and as it is heated, becomes less dense. Since it is then lighter than the surrounding air, it tends to rise due to buoyant effects.

We are not analyzing chimneys in this session, but the principle is relevant to fluid flow in ducts that have changes in elevation—duct risers, as shown in Figure 18-4. When you duct air from one floor of a building to another, as you do when there is an air

```
(1c) Comment: ***   STACKEFF   ***                                    201/!

==================== VARIABLE SHEET ===================================
St Input      Name      Output    Unit       Comment
-- -----      ----      ------    ----       -------
                                             ***   STACKEFF   ***
                                             *******************
              rho                 lbm/ft^3   mass density of fluid flowing in duct
              v1                  ft/s       velocity at section 1
              gc                  ft*lbm/lb  dimensional constant
              pz1                 lbf/in^2   atmospheric pressure at elev z1
              p1                  lbf/in^2   static pressure (gage) at section 1
              gamma               lbf/ft^3   specific weight
              z1                  ft         elevation 1
              v2                  ft/s       velocity at section 2
              pz2                 lbf/in^2   atmospheric pressure at elev z2
              p2                  lbf/in^2   static pressure (gage) at section 2
              z2                  ft         elevation 2
              dPt                 lbf/in^2   total pressure loss due to fluid resis
              pa                  lbf/in^2   reference atmospheric pressure
              g                   ft/s^2     acceleration due to gravity
              rho#a               lbm/ft^3   reference air density
              pse                 lbf/in^2   stack effect
              sv                  ft^3/lbm   specific volume


(2r) Rule: g=32.174                                                   201/!

=================== RULE SHEET ========================================
S Rule
- ----
* rho*v1^2/2*gc+pa+pz1-pz1+p1+gamma*z1=rho*v2^2/2*gc+pa+pz2-pz2+p2+gamma*z2+dPt
* g=32.174
* gc=32.174
* pz1=pa-(g/gc)*rho#a*z1
* pz2=pa-(g/gc)*rho#a*z2
* pse=(g/gc)*(rho#a-rho)*(z2-z1)
* gamma=rho*g/gc
* sv=1/rho
```

Figure 18–4. A Duct System with Risers.

handling unit on one floor that serves several floors, you are conveying air at a temperature that is either lower or higher than the surrounding air. If the temperatures differ, so do the densities, and when the densities differ, the stack effect appears. Although the stack effect is not significant in low-rise buildings, it is significant in high-rise buildings. The stack effect tends to increase or decrease the air flow, depending upon the direction of the air flow, up or down, and the direction of the temperature difference. If the air flow is upward, and the air flowing in the duct is hotter than the surrounding air, the stack effect contributes to the air flow; if the air flow is upward, and the air flowing in the duct is colder than the surrounding air, the stack effect hinders the air flow.

The model, as shown in Figure 18–5, presents the stack effect in terms of pressure. In fact, most of the effects we will consider are

```
(5o) Output:                                                         201/!

=================== VARIABLE SHEET ====================================
St Input       Name     Output     Unit       Comment
-- -----       ----     ------     ----       -------
                                              ***  STACKEFF  ***
                                              ******************
               rho                 lbm/ft^3   mass density of fluid flowing in duct
               v1                  ft/s       velocity at section 1
               gc                  ft*lbm/lb  dimensional constant
               pz1                 lbf/in^2   atmospheric pressure at elev z1
               p1                  lbf/in^2   static pressure (gage) at section 1
               gamma               lbf/ft^3   specific weight
               z1                  ft         elevation 1
               v2                  ft/s       velocity at section 2
               pz2                 lbf/in^2   atmospheric pressure at elev z2
               p2                  lbf/in^2   static pressure (gage) at section 2
               z2                  ft         elevation 2
               dPt                 lbf/in^2   total pressure loss due to fluid resis
               pa                  lbf/in^2   reference atmospheric pressure
               g                   ft/s^2     acceleration due to gravity
               rho#a               lbm/ft^3   reference air density
               pse                 lbf/in^2   stack effect
               sv                  ft^3/lbm   specific volume


(1r) Rule: rho*v1^2/2*gc+pa+pz1-pz1+p1+gamma*z1=rho*v2^2/2*gc+pa+pz2-pz2+p 201/!

=================== RULE SHEET =======================================
S Rule
- ----
* rho*v1^2/2*gc+pa+pz1-pz1+p1+gamma*z1=rho*v2^2/2*gc+pa+pz2-pz2+p2+gamma*z2+dPt
* g=32.174
* gc=32.174
* pz1=pa-(g/gc)*rho#a*z1
* pz2=pa-(g/gc)*rho#a*z2
* pse=(g/gc)*(rho#a-rho)*(z2-z1)
* gamma=rho*g/gc
* sv=1/rho
```

Figure 18–5. STACKEFF—A Stack Effect Model.

caused by differences in pressure, or cause differences in pressure. Fans move air by increasing the total pressure in the fan housing. Restrictions in duct systems cause a loss of total pressure, hindering air flow downstream. Certain types of air flow control devices require a minimum upstream pressure for proper operation. The stack effect will produce an effective increase or decrease in total pressure and, if not recognized, can be the cause of great puzzlement when the system is started and tested.

To use the STACK1 model, you must input two air density values, rho#a, which is the density of the ambient or surrounding air, rho, which is the density of the air in flowing in the duct, and the two elevations between which you want to calculate the stack effect. If you wish to calculate the pressure loss in the duct due to friction, you must also enter $v1$ and $v2$, the velocities at both sections, and *pa*, the ambient air pressure. Notice that *p*1 and *p*2 are gauge pressures, which you can measure with an ordinary pressure gauge.

You can enter the specific volume of air at the various conditions instead of the density, and let the program calculate both specific weight and density. Specific volume as a function of temperature is available from a psychrometric chart, as shown in Figure 18–6. You could eliminate the need for the psychrometric chart by expanding the model to include a rule relating specific volume or density to temperature, which is normally available in such problems. However, the psych chart is also readily available, so we will move on to other aspects of duct design, rather than expand this model further. A sample input and output is shown in Figure 18–7.

Evaluating the stack effect is not an everyday task, even for those who do nothing but duct design. We started this session with that model to illustrate a point. The stack effect is not significant in most duct design projects, but it is often overlooked in projects where it is significant, because of the time factor, and in some cases because of the lack of a ready tool. TK!Solver provides the ready tools and, therefore, the time for this kind of task.

## THE EQUAL FRICTION METHOD

The STACKEFF model is interesting, and useful on occasion, but the design of a duct system requires other methods. Duct design is like sizing pipe, in that a system will consist of a network of sections of varying size. Typically, the system starts at an air moving device and spreads out like a tree, just as a piping system does. A typical supply duct system is shown in Figure 18–8.

As was mentioned earlier, the fan in the duct system produces a high pressure in its housing, and it is this high pressure zone that causes the air to move through the duct. If there were no resistance in the system—no viscosity in the air, no friction with the sides of the duct, and no changes in direction of the air flow—the total pressure would remain constant along the length of the duct. The total pressure, as noted above, consists of velocity pressure, static pressure, and potential pressure. (Piping designers may read "head" for "pressure," if desired.) In a horizontal duct, there is no change in elevation, so the total pressure consists only of velocity and static pressure. If you were to measure the static pressure, the pressure exerted normal to the side of a duct, you would find it con-

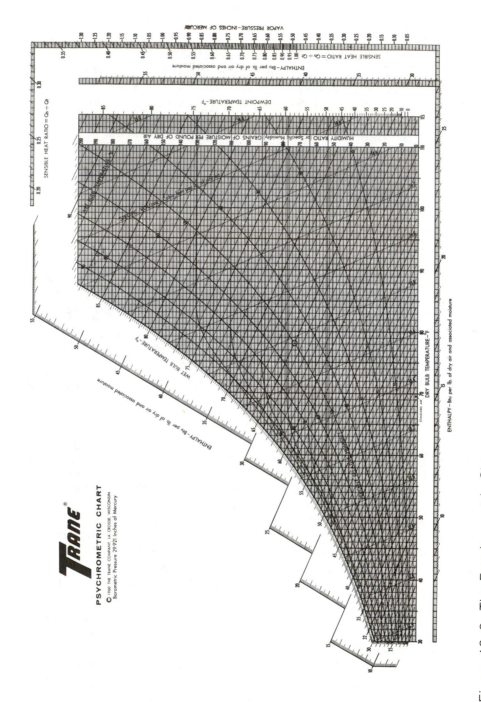

Figure 18–6. The Psychrometric Chart.

```
(19i) Input:                                                    201/!

==================== VARIABLE SHEET ====================================
St Input        Name    Output     Unit      Comment
-- -----        ----    ------     ----      -------
                                             ***  STACKEFF  ***
                                             ******************
   .068         rho                lbm/ft^3  mass density of fluid flowing in duct
   25           v1                 ft/s      velocity at section 1
                gc      32.174     ft*lbm/lb dimensional constant
                pz1     14.7       lbf/in^2  atmospheric pressure at elev z1
   .011         p1                 lbf/in^2  static pressure (gage) at section 1
                gamma   .068       lbf/ft^3  specific weight
   0            z1                 ft        elevation 1
   25           v2                 ft/s      velocity at section 2
                pz2     14.588889  lbf/in^2  atmospheric pressure at elev z2
   .007         p2                 lbf/in^2  static pressure (gage) at section 2
   200          z2                 ft        elevation 2
                dPt     -.0904444  lbf/in^2  total pressure loss due to fluid resis
   14.7         pa                 lbf/in^2  reference atmospheric pressure
                g       32.174     ft/s^2    acceleration due to gravity
   .08          rho#a              lbm/ft^3  reference air density
                pse     .01666667  lbf/in^2  stack effect
                sv      14.705882  ft^3/lbm  specific volume
```

Figure 18–7.  Input and Output for the STACKEFF Model.

stant along a section of constant velocity. If the velocity were to change, as it would if the cross-sectional area of the duct were to change, the velocity pressure would increase with the velocity, and the static pressure would decrease.

Unfortunately, air does have viscosity, and there is friction between the air stream, and there is some resistance associated with all the components of a duct system. The velocity pressure is a function of the air velocity, potential pressure is a function of elevation, and total pressure is still the sum of velocity, static, and potential pressure.

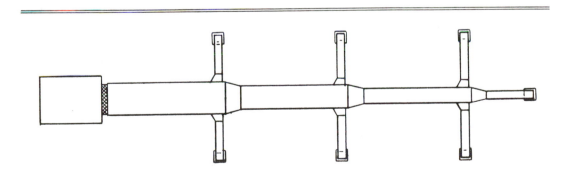

Figure 18–8.  A Typical Supply Duct System.

If you were to measure the static pressure at the beginning and the end of a section of the duct system (one with no branches), you would find that the static pressure is not the same at both ends. You could verify that the air velocity is the same at both ends, and if you insure that the duct is horizontal, you could rule out changes in potential pressure. You can conclude that the total pressure has decreased along the length of the duct; that decrease is due to friction, turbulence, etc., which duct designers lump under the single term *duct friction*. The relationships are shown in Figure 18–9.

The simplest and most commonly used duct design method is the *Equal Friction method*. The basic assumptions of the method are that all the pressure losses in the duct system can be considered frictional losses, and that the various resistances in the system can be expressed as an equivalent length of straight duct. Of course, these assumptions are not strictly true, but for simple duct systems, the method produces workable designs.

To use the equal friction method, you first lay the duct system out to convey the air to the areas served. You calculate the individual air quantities, and note them on the duct layout. Then tabulate the cumulative air flow quantities, working back from the end of each branch until you reach the fan.

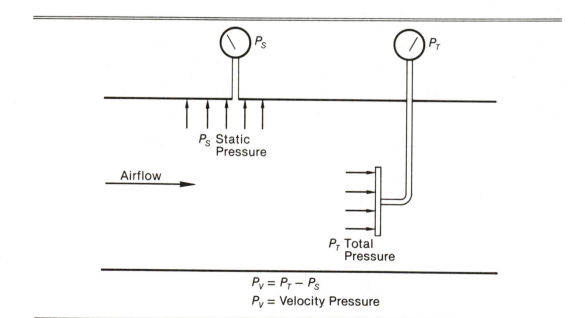

Figure 18–9. Pressure Relationships in a Duct System.

Once you have the air flow quantities for each section of duct, you have a choice to make. You can size the duct based on the amount of pressure the fan is capable of producing, or you can size the duct system and then select a fan capable of producing the pressure and flow required. For commercial installations, the second method is probably most common, so it will be considered first.

Since you are going to select a fan after the duct is sized, you can begin by selecting a pressure loss per unit length, and sizing the duct to produce that loss. After you have sized the duct sections, total the equivalent lengths of the sections along the longest path, and multiply the unit pressure loss by the total equivalent length. Given the friction loss, usually expressed as inches of water column per 100 feet of duct length, you can look up the duct diameter using charts available from a number of sources, or you can use slide rules based on the same equations.

## THE EQUFRICT MODEL

If you consider all pressure losses as frictional losses, you can calculate the loss with the Darcy-Weisbach equation, as you did with the piping model in the last chapter. You will use a slightly different form, which will produce results in the preferred units, inches of water column:

$$\Delta P_{fr} = f_o(Cf\,L/D)\,P_v$$

Where:  $\Delta P_{fr}$ = friction losses

$f_o$ = friction factor

$L$ = duct length

$D$ = duct diameter

$Cf$ = conversion factor

If you could be assured of laminar flow, you could calculate the friction factor directly from the Reynolds number, but for most applications you will need the Colebrook equation. Again, you will use these formulas in forms that will produce values in units of inches of water column:

$$Re = cf\,\varrho\,D\,V/\mu$$

$$f_D = 64/Re$$

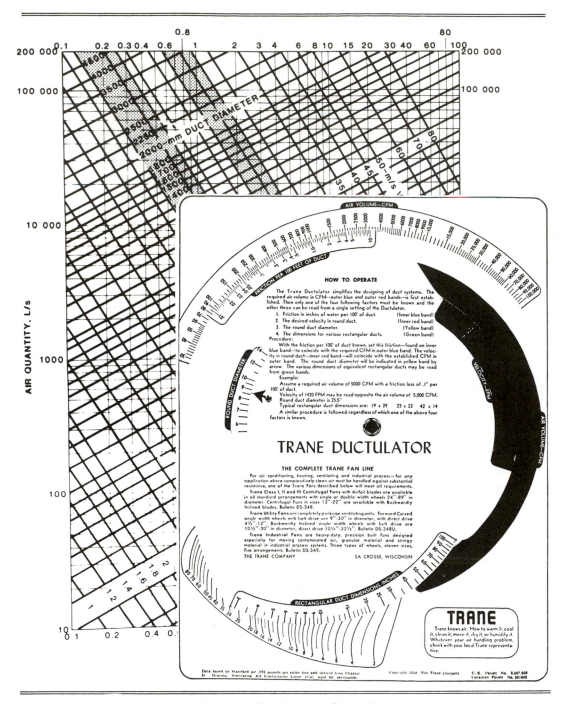

Figure 18–10.  Friction Charts and Slide Rules.

$$\frac{1}{\sqrt{f_o}} = -2\log_{10}[\ \frac{cf\epsilon}{3.7\ D}\ +\ \frac{2.51}{Re\sqrt{fD}}\ ]$$

Since it is possible for air flow to be laminar in a duct, both friction factor relationships will be incorporated in this model. Remember that if you enter too many rules in a model, you can have an overdefined model, which will produce an error in the status field of the overdefined variable. If you were to enter both rules for calculating the friction factor—the Colebrook equation and the simpler function of the Reynolds number only—the friction factor would be overdefined. You need a way to select the proper relationship and ignore the other.

TK!Solver does not require us to be programmers, and indeed does not offer some of the decision-making tools that programmers use to control the sequence in which their programs operate. Instead, the sequence in which a TK!Solver model operates is controlled with additional rules and with the TK!Solver functions. The first step of the EQUFRICT model, shown in Figure 18–11, illustrates the technique.

Set up two indicator variables, lam and trans, using the STEP($x$1,$x$2) function. This function returns 1 if the first argument is larger than or equal to the second argument, and 0 if the first argument is less than the second. In the model, either lam or trans can be true (1) and the other must be false (0). If the rule for Re produces a value less than or equal to 2000, lam will be true, and both sides of the equation relating friction factor to Reynolds number will be multiplied by 1. Both sides of the Colebrook equation will be multiplied by 0, so only the first rule produces a value for $f$, friction factor. If the Reynolds number is greater than 2000, the situation is reversed, and the Colebrook equation produces the value for $f$. Figure 18–12 shows a sample solution of EQUFRIC1.

## THE EQUAL VELOCITY METHOD

At this point, you have a working model that duplicates the performance of the slide rules and charts that are commonly used to size duct systems using the equal friction method. In fact, this model will also carry out another method, the equal velocity method, that is often used to size duct systems where the velocity must be relatively constant, such as dust collection systems. These systems are sized to maintain a constant conveying velocity in the duct so

```
(1c) Comment: ***   EQUFRIC1   ***                                        201/!

================== VARIABLE SHEET ========================================
St Input      Name      Output    Unit        Comment
-- -----      ----      ------    ----        -------
                                              ***   EQUFRIC1   ***
                                              ******************
              Pv                  in_wc       velocity pressure
              V                   ft/s        velocity of fluid in duct
              Q                   ft^3/min    air flow rate
              A                   ft^2        cross-sectional area of duct
              dPf                 in_wc       friction loss
              f                               friction factor, dimensionless
              L                   ft          length of duct
              D                   in          duct diameter
              rho                 lbm/ft^3    air density
              Re                              Reynolds number
              mu                  lbf*s/ft^   fluid dynamic viscosity
              lam                             1 if Re <= 2000, 0 otherwise
              trans                           1 if Re > 2000, 0 otherwise
              epsilon             ft          absolute roughness of duct wall




(1r) Rule: Pv=rho*(V/1097)^2                                              201/!

================== RULE SHEET ============================================
S Rule
- ----
* Pv=rho*(V/1097)^2
* V=144*Q/A
* A=pi()*(D/2)^2
* dPf=f*(12*L/D)*Pv
* Re=rho*D*V/(mu*12*32.174)   "D is in inches, V in ft/sec, mu approx 3.8x10^-6
* lam=step(2000,Re)
* trans=step(Re,2001)
* lam* f=(64/Re)  *lam
* trans*  1/sqrt(f)=-2*log((12*epsilon/3.7*D)+2.51/(Re*sqrt(f)))  *trans
* mu=3.8E-6                   "fluid dynamic viscosity at 75'F
* rho=.075                    "standard air density
```

Figure 18–11. The EQUFRICT Model, Phase 1.

that the material to be conveyed remains entrained in the air stream. Noise is usually not a problem, and the fan is sized to match the total pressure requirement. You can use the EQUFRIC1 model for velocity sizing by entering the flow rate, $Q$, and the velocity, $V$, instead of the flow rate and pressure drop. If you enter the total duct length, rather than a unit length as you normally would with the equal friction method, TK!Solver will provide both the proper duct size and the pressure drop for that section of duct.

You could use EQUFRIC1 just as you would use a friction chart or a slide rule, but you would probably not save much time doing so. You need to enhance the model so that it sizes each sec-

```
(7i) Input:                                                    201/

==================== VARIABLE SHEET ============================================
St Input        Name    Output      Unit     Comment
-- -----        ----    ------      ----     -------
                                             ***  EQUFRIC1  ***
                                             ******************
                Pv      .24929180   in_wc    velocity pressure
   2000         V                   ft/s     velocity of fluid in duct
   4000         Q                   ft^3/min air flow rate
                A       288         ft^2     cross-sectional area of duct
                dPf     2.6814712   in_wc    friction loss
                f       .17164659            friction factor, dimensionless
    100         L                   ft       length of duct
                D       19.149229   in       duct diameter
                rho     .075        lbm/ft^3 air density
                Re      1957819.6            Reynolds number
                mu      .0000038    lbf*s/ft^ fluid dynamic viscosity
                lam     0                    1 if Re <= 2000, 0 otherwise
                trans   1                    1 if Re > 2000, 0 otherwise
   .001         epsilon             ft       absolute roughness of duct wall
```

Figure 18–12. Sample Solution of EQUFRIC1.

tion of the duct system, unattended, just as the piping model does. In fact, it would be very useful if you could also specify a total friction loss for the longest branch, so that you can size duct systems to match packaged air handling units are furnished with standard fans and motors. While you are at it, you may as well upgrade the model so that it can size rectangular duct also.

Set up the model to size the entire duct run by designating the $Pv$, $V$, $A$, $dPf$, $L$, and $D$ variables as list variables, much as you did with the pipe sizing model. You add a feature you did not incorporate in the pipe sizing model, and that feature is a series of rules that total the $L$ and $dPf$ lists. These rules provide the total equivalent length and the total friction loss for the path you are sizing. These rules use the sum('listname) function, which calculates the sum of all the elements in a list.

The revised model, EQUFRIC2, which is shown in Figure 18–13, calculates the inside dimension of a square duct, simply by taking the square root of the area. It will also calculate the inside dimensions of a rectangular duct if you input the aspect ratio, the ratio of the width to the height of the duct. This calculation needs to be refined, because a square or rectangular duct has more surface area than a round duct of the same cross-sectional area, and consequently more friction. But, you're getting closer to the model you want.

The final refinements to EQUFRIC2 consist of a rule that allows you to input a total allowable friction loss for the branch,

```
(1i) Input:                                                                    195/!

=================== VARIABLE SHEET ===========================================
St Input        Name    Output      Unit      Comment
-- -----        ----    ------      ----      -------
                                              ***  EQUFRIC2  ***
                                              ******************
L               Pv      .14022664   in_wc     velocity pressure
LG 1000         V                   ft/min    velocity of fluid in duct
L  4000         Q                   ft^3/min  air flow rate
L               A       2.6666667   ft^2      cross-sectional area of duct
L               dPf     .32911968   in_wc     friction loss
                f       .01862833             friction factor, dimensionless
L  250          L                   ft        length of duct
L               D       22.111626   in        duct diameter
                rho     .075        lbm/ft^3  air density
                Re      1163286.5             Reynolds number
                mu      .0000038    lbf*s/ft^ fluid dynamic viscosity
                lam     0                     1 if Re <= 2000, 0 otherwise
                trans   1                     1 if Re > 2000, 0 otherwise
     .001       epsilon             ft        absolute roughness of duct wall
L               W       27.712813   in        duct width
L               H       13.856406   in        duct height
                Tlength 870         ft        total equivalent length of duct path


St Input        Name    Output      Unit      Comment
-- -----        ----    ------      ----      -------
                                              ***  EQUFRIC2  ***
                                              ******************
L               Pv      .14022664   in_wc     velocity pressure
LG 1000         V                   ft/min    velocity of fluid in duct
L  4000         Q                   ft^3/min  air flow rate
L               A       2.6666667   ft^2      cross-sectional area of duct
L               dPf     .32911968   in_wc     friction loss
                f       .01862833             friction factor, dimensionless
L  250          L                   ft        length of duct
L               D       22.111626   in        duct diameter
                rho     .075        lbm/ft^3  air density
                Re      1163286.5             Reynolds number
                mu      .0000038    lbf*s/ft^ fluid dynamic viscosity
                lam     0                     1 if Re <= 2000, 0 otherwise
                trans   1                     1 if Re > 2000, 0 otherwise
     .001       epsilon             ft        absolute roughness of duct wall
L               W       27.712813   in        duct width
L               H       13.856406   in        duct height
                Tlength 870         ft        total equivalent length of duct path
                TPf     1.5         in_wc     total friction loss in path
     1.5        maxsp               in_wc     allowable friction loss
     2          ar                            aspect ratio of a rectangular duct
                IDsq    14.280790   in        inside dimension of square duct

S Rule
- ----
  dPf/L=maxsp/(sum('L))

  Pv=rho*(V/1097)^2
  V=144*Q/A
  A=pi()*(D/2)^2
  dPf=f*(12*L/D)*Pv
  Re=rho*D*V/(mu*12*32.174)  "D is in inches, V in ft/sec, mu approx 3.8x10^-6
```

Figure 18-13. The EQUFRIC2 Model.

```
lam=step(2000,Re)
trans=step(Re,2001)
lam* f=(64/Re)  *lam
trans* 1/sqrt(f)=-2*log((12*epsilon/(3.7*D))+2.51/(Re*sqrt(f)))  *trans
mu=3.8E-6                 "fluid dynamic viscosity at 75'F
rho=.075                  "standard air density

IDsq=sqrt(A)
Tlength=sum('L)           "Total equivalent length of duct path
TPf=sum('dPf)            "Total friction loss in path

A=ar*H^2
W=ar*H
```

### DUCT SIZING

| Q | L | V | A | D | W | H | dPf |
|---|---|---|---|---|---|---|---|
| 100 | 100 | 686.700202 | .145623956 | 5.16716962 | 6.47608673 | 3.23804336 | .172413793 |
| 300 | 140 | 896.677692 | .334568377 | 7.83210949 | 9.81609355 | 4.90804677 | .241379310 |
| 375 | 200 | 946.267620 | .396293810 | 8.52402961 | 10.6832868 | 5.34164341 | .344827586 |
| 450 | 40 | 988.730413 | .455129117 | 9.13489618 | 11.4488945 | 5.72444726 | .068965517 |
| 600 | 50 | 1059.47989 | .566315612 | 10.1897980 | 12.7710178 | 6.38550891 | .086206897 |
| 1000 | 80 | 1197.28584 | .835222439 | 12.3747770 | 15.5094830 | 7.75474149 | .137931035 |
| 1200 | 30 | 1250.53234 | .959591338 | 13.2641512 | 16.6241483 | 8.31207413 | .051724138 |
| 1500 | 50 | 1318.81450 | 1.13738513 | 14.4407609 | 18.0988098 | 9.04940492 | .086206897 |
| 1800 | 100 | 1377.26389 | 1.30693908 | 15.4797513 | 19.4009911 | 9.70049555 | .172413793 |
| 2000 | 80 | 1412.17342 | 1.41625666 | 16.1141458 | 20.1960867 | 10.0980433 | .137931035 |

| Name | Elements | Unit | Comment |
|---|---|---|---|
| ---- | -------- | ---- | ------- |
| Pv | 10 | in_wc | |
| V | 10 | ft/s | |
| Q | 10 | ft^3/min | |
| A | 10 | ft^2 | |
| dPf | 10 | in_wc | |
| L | 10 | ft | |
| D | 10 | in | |
| W | 10 | | |
| H | 10 | | |

Figure 18–13. (Continued).

and the entries in the Table sheet that allow you to print out a table showing the various parameters for each section of the duct path you are analyzing. This model, EQUFRIC3, is shown in Figure 18–14.

## STATIC REGAIN

The equal friction method is satisfactory for duct systems which are of the constant volume type, particularly if they are small or medium sized. However, the recent emphasis on energy conservation has led to widespread use of variable volume air conditioning systems, where the system provides only as much air as is required

```
St Input       Name    Output    Unit      Comment
-- -----       ----    ------    ----      -------
                                           ***   EQUFRIC3   ***
                                           ******************
L              Pv      .29404220 in_wc     velocity pressure
L              V       2172.1051 ft/min    velocity of fluid in duct
L     1        Q                 ft^3/min  air flow rate
L              A       1.8415315 ft^2      cross-sectional area of duct
L              dPf     .86206897 in_wc     friction loss
               f       .01914769           friction factor, dimensionless
L     1        L                 ft        length of duct
L              D       18.374938 in        duct diameter
               rho     .075      lbm/ft^3  air density
               Re      1328936.5           Reynolds number
               mu      .0000038  lbf*s/ft^ fluid dynamic viscosity
               lam     0                   1 if Re <= 2000, 0 otherwise
               trans   1                   1 if Re > 2000, 0 otherwise
      .001     epsilon           ft        absolute roughness of duct wall
L              W       23.029570 in        duct width
L              H       11.514785 in        duct height
               Tlength 870       ft        total equivalent length of duct path
               TPf     20.541688 in_wc     total friction loss in path
      3        maxsp             in_wc     allowable friction loss
      2        ar                          aspect ratio of a rectangular duct
L              IDsq    16.284365 in        inside dimension of square duct


S Rule
- ----
* dPf=maxsp*L/sum('L)

* Pv=rho*(V/1097)^2
* V=144*Q/A
* A=pi()*(D/2)^2
* dPf=f*(12*L/D)*Pv
* Re=rho*D*V/(mu*12*32.174)   "D is in inches, V in ft/sec, mu approx 3.8x10^-6
* lam=step(2000,Re)
* trans=step(Re,2001)
* lam* f=(64/Re)    *lam
* trans*  1/sqrt(f)=-2*log((12*epsilon/(3.7*D))+2.51/(Re*sqrt(f)))  *trans
* mu=3.8E-6                   "fluid dynamic viscosity at 75'F
* rho=.075                    "standard air density

* IDsq=sqrt(A)
* Tlength=sum('L)             "Total equivalent length of duct path
* TPf=sum('dPf)               "Total friction loss in path

* A=ar*H^2
* W=ar*H
```

Figure 18–14. The EQUFRIC3 Model.

to handle the air conditioning requirement. The thermostat controls a variable air volume box, which reduces the air flow as the requirement for air conditioning decreases, by closing a damper in the box. If all the variable air volume boxes close their dampers, even partially, a controller senses the decreased demand for air and "throttles" the fan, so that it supplies less air. A diagram of a variable air volume system is shown in Figure 18–15.

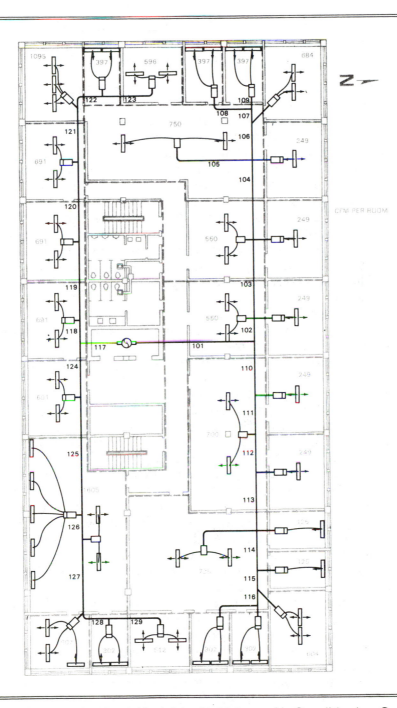

Figure 18–15. A Variable Air Volume Air Conditioning System.

The equal friction method produces duct designs in which the static pressure varies throughout the system. An air outlet close to the fan will have more pressure available and will consequently supply more air than one of the same size near the end of the longest path. The designer is faced with a choice between making the shorter paths smaller than normal to increase the frictional losses, or to add dampers in the shorter branches, which has the same effect. Either method increases the fan horsepower and operating costs over that which is absolutely required. The answer to the dilemma is the static regain method of sizing duct.

Recall that the total pressure at any point in the duct system is the sum of the velocity and static pressures, disregarding potential pressure. If the velocity decreases, the velocity pressure decreases, and the static pressure must increase. Of course, the total pressure does decrease along the direction of the flow, due to frictional losses, but you can increase the static pressure at any point in the duct system by increasing the duct size, which decreases the velocity.

You should note that the velocity decreases in a main duct, immediately following a branch duct take-off, if the cross-sectional area of the main duct remains constant. If you were to measure the static pressures as shown in Figure 18–16, you would find that the static pressure is greater downstream of the take-off than it is upstream of the take-off.

The static regain method exploits this fact by sizing each section of duct so that the increase in static pressure at each branch take-off offsets the frictional losses in the succeeding section. The end result is that the static pressure is the same at the entrance of each branch duct in the system. Proper use of the static regain method produces duct systems that are easily balanced, and simplifies the selection of the variable air volume boxes, which depend on the proper static pressure for proper operation.

You must still lay the duct system out to serve the desired areas, and you must still tabulate the air flows in each section of the duct system. However, that part of the task is minor compared to the velocity pressure calculations you must perform with the static regain method, which you will turn over to TK!Solver.

You can use much of the equal friction models for the static regain model—after all, you're not changing the laws of physics, just the method of calculation. You retain one simplification that can be debated, and that is that all losses can be expressed in terms of friction and duct resistance can be expressed as equivalent duct lengths. But, this is a book on TK!Solver, not on air conditioning.

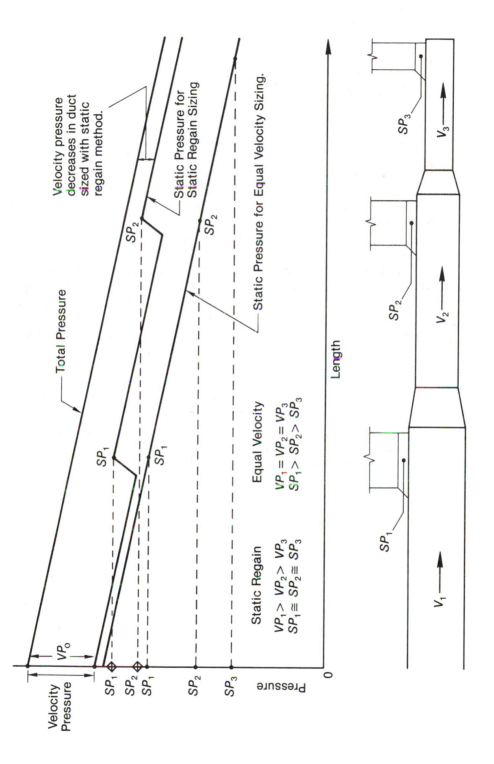

Figure 18–16. Static Pressures at Branch Take-Offs.

287

The first approach to the problem of static regain duct sizing will be to input a static pressure for the last branch entrance, and to increase the velocity in each section as you proceed toward the fan, by the friction loss you have just calculated for the preceding section. Begin with the EQUFRIC3 model, deleting the rule that you used to prorate the total friction losses to the various sections of duct. Then you will enter a rule, or rules, that will carry the friction loss calculated for one section over to the next section as an input. Finally, you have to make provisions for the first section, as there will be no previous value for friction loss.

Although you could use a technique similar to the one used to determine which formula to use for the friction factor, TK!Solver provides you with a much simpler method—use the build-in ELEMENT function. In order to make things clear long after you create the model, use the first form of the ELEMENT function, ELEMENT( ), to return a value to the variable, current. Then, use the third form of the function, ELEMENT('listname,element__number,default), to select the value of the previous element, using the expression, current−1, to point to the desired element. The function returns the default value if the desired element does not exist, and, indeed, the zeroth element does not. Therefore, when TK!Solver's List Solver makes its first pass, it uses the value of $P$min + $dPf$ for $P$total, the total pressure entering the section of duct. The final static regain model is shown in Figure 18–17.

As with the previous models, STATREGx works from a list of duct sizes and lengths, and produces several lists as outputs. These lists are presented in the form of a table, a sample of which is shown in Figure 18–18.

```
St Input      Name      Output     Unit       Comment
-- -----      ----      ------     ----       -------
                                              ***   STATREGN  ***
                                              static regain duct sizing model
                                              ********************************
L             Pv        .08        lbf/ft^2   velocity pressure
L             V         499.8      ft/min     velocity of fluid in duct
L   500       Q                    ft^3/min   air flow rate
L             A         1.0004002  ft^2       cross-sectional area of duct
L             dPf       .26186464  lbf/ft^2   friction loss
L             f         .03694272             friction factor, dimensionless
L   100       L                    ft         length of duct
L             D         1.1286049  ft         duct diameter
              rho       .075       lbm/ft^3   air density
L             Re        5767.1274             Reynolds number
              mu        .0000038   lbf*s/ft^  fluid dynamic viscosity
              lam       0                     1 if Re <= 2000, 0 otherwise
              trans     1                     1 if Re > 2000, 0 otherwise
    .001      epsilon              ft         absolute roughness of duct wall
L             W         16.973958  in         duct width
L             H         8.4869789  in         duct height
              Tlength   200        ft         total equivalent length of duct path
              TPf       .11530388  in_wc      total friction loss in path
    2         ar                              aspect ratio of a rectangular duct
L             IDsq      12.002401  in         inside dimension of square duct
L             prev
    .09617234 Pmin                 in_wc      minimum static pressure
L             D1        162.51911  in         duct diameter in inches
              first     0
L             Qcurren   8.3333333  ft^3/min
              Ps        3          lbf/ft^2
L             Pt        2.6795     lbf/ft^2
              gc        32.174     lbm*ft/lb  conversion factor
L             Lt        80         ft         running length

S Rule
- ----
* prev=element()-1
* first=step(1,element())
* first*    V=8.33  *first
* Lt=L+element('Lt,prev,0)
* gc=32.174
* Ps=3
* first*  Pv=.08  *first
* (1-first)*  Pv=element('Pv,prev,0)+element('dPf,prev,0)  *(1-first)
* Pt=Pv+Ps
* (1-first)*    V=sqrt(Pv*2*gc/rho)    *(1-first)
* A=Q/V
* D=2*sqrt(A/pi())
* rho=.075                    "standard air density
* mu=3.8E-6                   "fluid dynamic viscosity at 75'F
* Re=rho*D*V/(mu*gc)   "D is in ft, V in ft/sec, mu approx 3.8x10^-6
* lam=step(2000,Re)
* trans=step(Re,2001)
* lam*  f=(64/Re)  *lam
* trans*  1/sqrt(f)=-2*log((epsilon/(3.7*D))+2.51/(Re*sqrt(f)))   *trans
* dPf=f*(L/D)*Pv
* D1=D
* IDsq=sqrt(A)
* Tlength=sum('L)              "Total equivalent length of duct path
* TPf=sum('dPf)               "Total friction loss in path
* H=sqrt(A/ar)
* W=ar*H
```

Figure 18–17.  The STATREGx Model.

```
(1n) Name: Pv                                                              188/!

================== LIST SHEET ==================================================
Name        Elements   Unit          Comment
----        --------   ----          -------
Pv          10         lbf/ft^2
V           10         ft/s
Q           10         ft^3/s
A           10         ft^2
dPf         10         lbf/ft^2
L           10         ft
D           10         ft
W           10
H           10
f           10
Re          10
IDsq        10         in
D1          10
Qcurrent
Ps
Pt          10         lbf/ft^2
Lt          10
prev        10
```

```
(1f) From: ft                                                              188/!

================== UNIT SHEET ==================================================
From        To         Multiply By  Add Offset
----        --         -----------  ----------
ft          in         12
ft^2        in^2       144
in_wc       lbf/ft^2   5.199
min         s          60
ft/s        ft/min     60
ft^3/s      ft^3/min   60
```

```
(s) Screen or Printer: Printer                                             188/!

================== TABLE SHEET =================================================
Screen or Printer:              Printer
Title:
Vertical or Horizontal:         Vertical
List        Width First Header
----        ----- ----- ------
Pv          10      1   Pv
V           10      1   V
Q           10      1   Q
A           10      1   A
dPf         10      1   dPf
L           10      1   L
D           10      1   D
Re          10      1   Re
D1          10      1   D1
Pt          10      1   Pt
f           10      1   f
Lt          10      1   Lt
prev        10      1   prev
```

Figure 18–17. (Continued).

| prev | Pv | V | Q | A | dPf | L |
|---|---|---|---|---|---|---|
| 0 | .08 | 8.33 | 8.33333333 | 1.00040016 | .052372929 | 20 |
| 1 | .132372929 | 10.6570373 | 20 | 1.87669420 | .054433226 | 20 |
| 2 | .186806155 | 12.6599645 | 25 | 1.97472908 | .071392122 | 20 |
| 3 | .258198277 | 14.8837911 | 40 | 2.68748734 | .078290619 | 20 |
| 4 | .336488896 | 16.9911300 | 65 | 3.82552543 | .079441970 | 20 |
| 5 | .415930866 | 18.8906747 | 100 | 5.29361717 | .078441329 | 20 |
| 6 | .474372195 | 20.5951004 | 200 | 9.71104759 | .062923892 | 20 |
| 7 | .557296087 | 21.8665311 | 250 | 11.4329977 | .063404903 | 20 |
| 8 | .620700990 | 23.0769343 | 350 | 15.1666593 | .058758274 | 20 |
| 9 | .679459264 | 24.1445217 | 500 | 20.7086314 | .052709101 | 20 |

| D | Re | D1 | Pt | f | Lt |
|---|---|---|---|---|---|
| 1.12860491 | 5767.12741 | 1.12860491 | 3.08 | .036942716 | 20 |
| 1.54579470 | 10105.5722 | 1.54579470 | 3.13237293 | .031782402 | 40 |
| 1.58565544 | 12314.4187 | 1.58565544 | 3.18680616 | .030299672 | 60 |
| 1.84981490 | 16889.4088 | 1.84981490 | 3.25819828 | .028044950 | 80 |
| 2.20699122 | 23003.5826 | 2.20699122 | 3.33648890 | .026052529 | 100 |
| 2.59615922 | 30085.0961 | 2.59615922 | 3.41593087 | .024480773 | 120 |
| 3.51631765 | 44424.7122 | 3.51631765 | 3.49437220 | .022377916 | 140 |
| 3.81535645 | 51178.5078 | 3.81535645 | 3.55729609 | .021704109 | 160 |
| 4.39440444 | 62208.6458 | 4.39440444 | 3.62070099 | .020799678 | 180 |
| 5.13488543 | 76053.9849 | 5.13488543 | 3.67945926 | .019916955 | 200 |

Figure 18–18.  The Output of STATREGx.

# CHAPTER 19

# Electrical Networks

Electrical engineering is a broad field, ranging from electronics to power generation to controls, etc. We will consider several models to illustrate the variety of problems that TK!Solver can solve. In this session, we will consider the analysis of electrical networks, or circuits, using two basic tools, loop current analysis, and node voltage analysis.

Recall from your studies of basic electricity that electric current, resistance, and voltage are related by Ohm's Law

$$I = \frac{E}{R}$$

*Where:* $I$ = current

$E$ = voltage (emf)

$R$ = resistance

In an electric circuit, you can also apply Kirchhoff's Laws:

**Voltage**—In any closed electric circuit, the algebraic sum of the voltage drops must equal the algebraic sum of the applied emfs.

**Current**—The algebraic sum of the currents entering a point in an electric circuit must equal the algebraic sum of the currents leaving that point.

These laws are applicable in both simple and complex circuits, but in a complex one, their application is tedious.

## LOOP CURRENT ANALYSIS

The subject of analyzing complicated circuits is often referred to as *network analysis,* and many techniques have been developed to make the process as quick and painless as possible. Rather than consider a complicated electrical circuit, we will consider the simple one shown in Figure 19–1, since the principles are the same, regardless of complexity.

The first technique we will consider is loop current analysis, which follows directly from Kirchhoff's voltage law. This method does have some limitations; for instance, it can only be applied to planar networks, but it is useful nevertheless. Begin by insuring that the network contains voltage sources with series impedances, rather than current sources, which may involve the application of Thevenin's theorem. Then set up an equation for each loop of the network, based on Kirchhoff's voltage law.

You assume a separate current for each loop, flowing in a clockwise direction. Each loop current is a variable, and the resistances (impedances) are constants. You proceed around each loop, adding the voltage drops, which are the products of the loop currents and the resistances. In the network shown above, each loop current appears in every loop. When you are finished writing equations, you will have as many linear equations as you have loop currents, or unknowns.

If you did not have TK!Solver at hand, you would construct a coefficient matrix and a constant vector for the system of si-

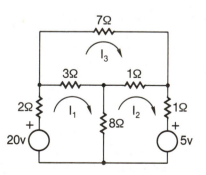

Figure 19–1. A Direct-Current Electrical Circuit.

```
(3s) Status:  Guess                                                    204/!

==================== VARIABLE SHEET ========================================
St Input      Name    Output    Unit    Comment
-- -----      ----    ------    ----    -------
                                        DIRECT CURRENT CIRCUIT MODEL
                                        *******************************
G 0           I1                        LOOP CURRENT #1
G 0           I2                        LOOP CURRENT #2
              I3                        LOOP CURRENT #3



==================== RULE SHEET ============================================
S Rule
- ----
*  13*I1-8*I2-3*I3 = 20
*  -8*I1+10*I2-I3 = -5
*  -3*I1-I2+11*I3 = 0
```

Figure 19–2. TK!Solver Model for Figure 19–1.

multaneous equations. Then you would solve the system of equations with Cramer's Rule, Gaussian elimination, Gauss-Jordan elimination, or any of the other methods invented BT (Before TK!Solver) to solve simultaneous equations.

But, since you do have TK!Solver, you simply enter the loop equations in the Rule sheet and solve the model. Of course, if you attempt to solve the model for this network with the Direct Solver, it will not work.

The reason the Direct Solver will not work is that all the equations contain the same number of unknowns. Recall that TK!Solver's Direct Solver must find an equation from which it can solve for one of the unknowns. Then it can eliminate that unknown from all the equations, and find another equation with only one unknown.

You could take the same approach you would take without TK!Solver, that is, manipulate the system of equations to arrive at a triangular coefficient matrix—one with zeroes above the diagonal. Then the Direct Solver would be able to solve the first equation, followed by the second, etc. But, that process would involve more work than was required to build the model in the first place. To solve this system of simultaneous equations without manipulating the equations, you must invoke the Iterative Solver by entering guesses for all but one unknown. With the Iterative Solver, TK!Solver can solve the system of equations.

If you wish to solve this type of model for differing resistances, you can enter the rules as shown in Figure 19–3. Then

```
     St Input        Name      Output      Unit      Comment
     -- -----        ----      ------      ----      -------
                                                     LOOP CURRENT ANALYSIS (5 LOOPS)
                                                     ***********************************
                                                     APPLIED EMF IN LOOPS
                     V1
                     V2
                     V3
                     V4
                     V5
                                                     LOOP CURRENTS
                     I1
                     I2
                     I3
                     I4
                     I5
                                                     RESISTANCES IN LOOP 1
                     R11
                     R12
                     R13
                     R14
                     R15
                                                     RESISTANCES IN LOOP 2
                     R21
                     R22
                     R23
                     R24
                     R25
                                                     RESISTANCES IN LOOP 3
                     R31
                     R32
                     R33
                     R34
                     R35
                                                     RESISTANCES IN LOOP 4
                     R41
                     R42
                     R43
                     R44
                     R45
                                                     RESISTANCES IN LOOP 5
                     R51
                     R52
                     R53
                     R54
                     R55


     S Rule
     - ----
     * V1=R11*I1+R12*I2+R13*I3+R14*I4+R15*I5
     * V2=R21*I1+R22*I2+R23*I3+R24*I4+R25*I5
     * V3=R31*I1+R32*I2+R33*I3+R34*I4+R35*I5
     * V4=R41*I1+R42*I2+R43*I3+R44*I4+R45*I5
     * V5=R51*I1+R52*I2+R53*I3+R54*I4+R55*I5
```

Figure 19-3.  The Loop Current Model Generalized.

you must enter input values for the applied emfs and the resistances, in addition to the guesses for the loop currents. However, if you have standard circuits, or if you wish to evaluate the circuit's performance with varying resistances, you can easily do so.

If you save the generalized loop current model, you can avoid a lot of typing and editing. However, if you enter a generalized model large enough to accept any planar network, you must input a lot of zeroes in the unused loops. Notice that the model of Figure 19–3 will solve a five-loop network, but you must have INPUT values of 0 for loops 4 and 5, and guesses for two of the other three loops. You must also input 0 for all the unused resistances in all five equations.

You can avoid the necessity for all the zero inputs by building and storing models for various numbers of loops: LOOP3, LOOP4, LOOP5, LOOP6, etc. Then load the model with the correct number of loops for the problem at hand.

You can also evaluate the performance of the circuit model with varying values by associating one or more variables with lists and using the List Solver. This type of procedure should be restricted to critical components, however, as it could become quite lengthy because of the number of possible combinations.

## NODE VOLTAGE ANALYSIS

You apply Kirchhoff's current law in the method of node voltage analysis, the converse of loop current analysis. The model has an appearance similar to that of the loop current model, but instead of working with an impedance matrix (the Rxx values in the coefficient matrix), you work with an admittance matrix. The admittance of a circuit element is simply the reciprocal of the impedance.

The circuit of Figure 19–1 is shown in Figure 19–4, converted to the correct form by the application of Norton's theorem. Notice that all voltage sources with series impedances have been replaced with current sources with shunt impedances, shown as admittances.

You number the nodes of the circuit, rather than the loops, as $V1$, $V2$, etc., except for one node, the reference node. In Figure 19–4, the reference node is the line across the bottom of the circuit diagram. These node voltages are the unknowns, the number of numbered nodes determines the number of equations you will use in the model, or in other words, the size of the coefficient matrix. The values of the current sources form the constant vector, and you use the admittances of the circuit to fill the coefficient matrix.

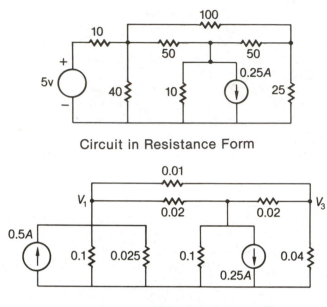

**Circuit in Resistance Form**

**Circuit Converted in Admittance Form**

Figure 19–4. Circuit Diagram in Admittance Form.

    Fill the admittance matrix, beginning with the diagonal of the matrix—$A11$, $A22$, $A33$, etc. Each element of the diagonal is the sum of the admittances connected directly to the corresponding node. Then fill the remainder of the admittance matrix. Each remaining element is the sum of the admittances between the two nodes corresponding to the row and column subscripts. Notice that the matrix is symmetrical about the diagonal—$A21 = A12$, $A45 = A54$. The node voltage model is shown in Figure 19–5.

```
                         ****************************************
                         CURRENT SOURCES
        I1
        I2
        I3
        I4
        I5
                         NODE VOLTAGES (REL TO REF NODE)
        V1
        V2
        V3
        V4
        V5
                         ADMITTANCE MATRIX ROW 1
        A11              ** DIAGONAL ELEMENT **
        A12
        A13
        A14
        A15
                         ADMITTANCE MATRIX ROW 2
        A21
        A22              ** DIAGONAL ELEMENT **
        A23
        A24
        A25
                         ADMITTANCE MATRIX ROW 3
        A31
        A32
        A33              ** DIAGONAL ELEMENT **
        A34
        A35
                         ADMITTANCE MATRIX ROW 4
        A41
        A42
        A43
        A44              ** DIAGONAL ELEMENT **
        A45
                         ADMITTANCE MATRIX ROW 5
        A51
        A52
        A53
        A54
        A55              ** DIAGONAL ELEMENT **


        S Rule
        - ----
        * I1=A11*V1+A12*V2+A13*V3+A14*V4+A15*V5
        * I2=A21*V1+A22*V2+A23*V3+A24*V4+A25*V5
        * I3=A31*V1+A32*V2+A33*V3+A34*V4+A35*V5
        * I4=A41*V1+A42*V2+A43*V3+A44*V4+A45*V5
        * I5=A51*V1+A52*V2+A53*V3+A54*V4+A55*V5
```

Figure 19-5. Node Voltage Analysis Model.

# CHAPTER 20

# Electrical Power

Whenever you deal with electrical circuits, you must be concerned with power—the instantaneous product of voltage and current. In a steady-state, direct current circuit, calculation of power is straightforward. In alternating current circuits with resistive, inductive, and capacitive elements matters are a little more complicated.

When you solve the integrodifferential equations that describe the behavior of electrical circuits, the solutions generally include terms of the form, $E_{0e}jwt$, or of the form $E_0 (\cos wt + j \sin wt)$. Regardless of the form used, the terms include both real and imaginary parts. Remember that TK!Solver cannot work with imaginary numbers.

However, you can still work with imaginary or complex numbers using TK!Solver if you treat the real and imaginary parts of complex numbers separately and just interpret them as real or imaginary. Recall that the real and imaginary components of a number cannot really be added; if you add two complex numbers, you add the real parts together, and the imaginary parts together. If you multiply or divide one complex number by another, you treat it as polynomial. The product of two imaginary numbers is a real number, and the product of a real number and an imaginary number is an imaginary number. It takes a little bookkeeping, but you can use TK!Solver to solve models that include complex numbers.

## PHASORS

Alternating current circuits are often studied in the steady-state—the applied voltage is a steady-state sinusoidal waveform,

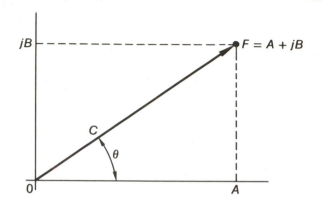

Figure 20–1. The Complex Plane.

and so is the resulting current. Nevertheless, such circuits are described by differential equations.

If the waveforms are steady-state, you can describe, analyze, and design the circuits without resorting to the use of differential equations, if you use phasor notation. Phasor notation allows you to represent the complex numbers generated in the solution of periodic functions, by plotting them on the complex plane, as illustrated in Figure 20–1.

The horizontal axis is the real axis and the vertical axis is the imaginary axis. A real number lies on the real axis and a pure imaginary number lies on the imaginary axis. A complex number lies between the real and imaginary axes, in any quadrant, depending on the respective signs of the real and imaginary parts.

The location of a complex number in the complex plane can be described in either of two ways: rectangular coordinates or polar coordinates. The number, $F$, shown in Figure 20–1 is described in rectangular form as

$$F = A + jB$$

and in polar form as

$$F = C \angle \theta$$

These two forms are related by the equations

$$C = \sqrt{A^2 + B^2}$$

$$A = C \cos \Theta$$

$$B = C \sin \Theta$$

$$\tan \Theta = B/A$$

Although the number $F$ is a point in the complex plane, it is commonly represented as a vector—a line segment from the origin of the complex plane to the point, as shown in Figure 20-1. Finally, the phasor representing a current, voltage, or impedance may be represented as the length of the vector and the angle between the real axis and the vector, i.e.,

$$C \angle \Theta$$

========================================= POWER  FACTOR

Power is the instantaneous product of voltage and current. At time $t$, given voltage as $V \angle \Theta$, and current as $I \angle \phi$, the power is

$$p(t) = [V_{max} \cos(wt + \Theta)] \times [I_{max} \cos(wt + \phi)]$$

This equation can be rewritten as

$$p(t) = \frac{V_{max} I_{max}}{2} [\cos(\Theta - \phi) + \cos(2wt + \Theta + \phi)]$$

The second cosine term is zero, and the first is a constant, so the expression for average power can be written

$$P = V_{RMS} I_{RMS} \cos(\Theta - \phi)$$

since

$$V_{RMS} = \frac{V_{MAX}}{V2}$$

Of course, you can represent the power in phasor notation as

$$P = V_{\text{RMS}} \, I_{\text{RMS}} \, \angle \, (\Theta - \phi)$$

and you can, in fact, arrive at the power directly from phasor notation.

The power factor is the quantity, $\cos(\Theta - \phi)$, or, alternatively, $\cos(\phi - \Theta)$. The power factor angle is simply the difference between the voltage and current angles.

You can use TK!Solver to analyze the circuit of Figure 20–2 in regard to power factor. This circuit is typical of a power circuit, with a 230 v RMS 60 Hz voltage source, and resistive, capacitive, and inductive loads.

First, you need to find the impedance of the various circuit elements, and then, of the entire circuit. You don't need to enter a rule to find the real component of the impedance, since it is the actual value of the resistive element.

You do need to enter rules to find the impedance of the reactive elements, the capacitor and the inductor. Since these two values are imaginary, you cannot add them to the real portion. In fact, it is handier to work with the reactances of the inductive and capacitive elements, which do not involve the imaginary number $j$.

The TK!Solver model for the circuit shown in Figure 20–2 is shown in Figure 20–3. With the first three rules, you find the reactances. The fourth rule is simply a reminder that you are ultimately interested in impedance, even though you cannot display total circuit impedance in textbook form. In the next three rules, you convert the impedance to phasor notation and, from them on, you work strictly with phasor notation.

You could refine this model somewhat to reflect real power circuits. You could associate lists with the three variables, $R$, $C$, and

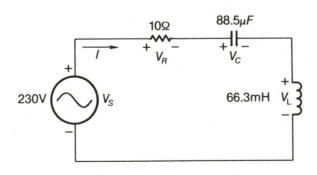

Figure 20–2. Power Circuit.

```
St Input       Name    Output     Unit    Comment
-- -----       ----    ------     ----    -------
                                          POWER FACTOR MODEL
                                          ********************
   60          F                  HZ      FREQUENCY
   .0663       L                  HENRIES INDUCTANCE
   .0000885    C                  FARADS  CAPACITANCE
   10          R                  OHMS    RESISTANCE
   230         VS                 VOLTS   SOURCE VOLTAGE (RMS)
   0           VTHETA             RADIANS VOLTAGE PHASOR ANGLE IN RADIANS

               XL      24.994511  OHMS    INDUCTIVE REACTANCE
               XC      -29.97268  OHMS    CAPACITIVE REACTANCE
               XT      -4.978171  OHMS    TOTAL REACTANCE
               ZR      10         OHMS    RESISTIVE IMPEDANCE
               ZMAG    11.170595          MAGNITUDE OF IMPEDANCE PHASOR
               ZTHETA  -.4618998  RADIANS IMPEDANCE PHASOR ANGLE IN RADIANS
               ZANGLE  -26.46491  DEGREES IMPEDANCE PHASOR ANGLE IN DEGREES
               IMAG    20.589772          MAGNITUDE OF CURRENT PHASOR
               ITHETA  .46189978  RADIANS CURRENT PHASOR ANGLE IN RADIANS
               IANGLE  26.464908  DEGREES CURRENT PHASOR ANGLE IN DEGREES
               PF      .89520748          POWER FACTOR
               S       4735.6476  VA      APPARENT POWER IN VA
               P       4239.3871  WATTS   AVERAGE POWER IN WATTS
               Q       -2110.439  VAR     REACTIVE POWER IN VA
```

```
S Rule
- ----
  XL = 2*PI()*F*L            "INDUCTIVE REACTANCE
  XC = -1/(2*PI()*F*C)       "CAPACITIVE REACTANCE
  XT = XL + XC               "TOTAL REACTANCE

  ZR = R

  ZMAG = SQRT(ZR^2 + XT^2)
  ZTHETA = ATAN2(XT,ZR)
  ZANGLE = ZTHETA*(180/PI())

  IMAG = VS/ZMAG
  ITHETA = VTHETA - ZTHETA
  IANGLE = ITHETA*(180/PI())

  PF = COS(VTHETA - ITHETA)
  S = VS*IMAG
  P = ABS(S)*COS(VTHETA - ITHETA)
  Q = ABS(S)*SIN(VTHETA - ITHETA)
```

Figure 20–3. The Power Factor Model.

*L*, so that you could enter the various components from a building wiring diagram, for instance. Of course, you would have to give some thought to the rules for finding the total resistance from a list of individual resistances, as you cannot simply add them with the SUM function. (They are wired in parallel). The same thought must be given to the total inductive and capacitive reactances.

# POWER FACTOR CORRECTION

It is important to be able to determine the power factor of a circuit, but it is even more important to be able to do something about it. Most industrial plants, for instance, have low inductive power factors, because of a large number of motors. In this case, a low power factor can be corrected by placing a capacitance in parallel with the load, i.e., a shunt capacitance across the load terminals. The justification for correcting the power factor is economic—many power companies levy a charge against customers with low power factors.

The addition of a shunt capacitance to an inductive load causes a new line current which has a reactive component that cancels the reactive component of the current drawn by the existing load. If the power factor is known, you still should determine whether the power factor is inductive or capacitive. However, in most building power applications, you can safely assume that the power factor is inductive—the voltage leads the current. Therefore, shunt capacitance is required.

Figure 20–4 is the TK!Solver model used to calculate the correcting reactance, given real power in KVA, and apparent power in KW. You can enter the corrected power factor and TK!Solver will calculate the correcting current, the corrected current, the correcting reactance, and the KVAR rating of the correcting reactance. It's a handy model to keep around.

# POWER GENERATION

In any power application other than residential services, there is three-phase power. Three phase power is more economical and produces smoother torque. In fact, most power companies have an upper limit on the horsepower rating of single-phase motors that customers are at least encouraged to follow.

Three phase power can be supplied in either of two types of services—three wire or four wire. A three wire service comprises three power lines, and the four wire service comprises the three power lines and a neutral line. Each of the three power lines carries the same average, or RMS, voltage, and the peak voltages are the same, but the peaks are 120′ apart. The three phases are plotted in Figure 20–5, compliments of TK!Solver.

It is desirable from all points of view to have a balanced configuration, regardless of whether the service and load are three wire (delta), or four wire (wye), or a combination of the two. For instance,

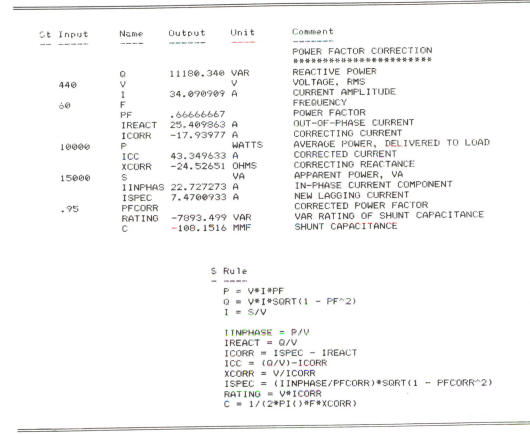

```
St Input      Name    Output     Unit     Comment
-- -----      ----    ------     ----     -------
                                          POWER FACTOR CORRECTION
                                          ***************************
              Q       11180.340  VAR      REACTIVE POWER
    440       V                   V       VOLTAGE, RMS
              I       34.090909   A       CURRENT AMPLITUDE
    60        F                            FREQUENCY
              PF      .66666667           POWER FACTOR
              IREACT  25.409863   A       OUT-OF-PHASE CURRENT
              ICORR   -17.93977   A       CORRECTING CURRENT
    10000     P                   WATTS   AVERAGE POWER, DELIVERED TO LOAD
              ICC     43.349633   A       CORRECTED CURRENT
              XCORR   -24.52651   OHMS    CORRECTING REACTANCE
    15000     S                   VA      APPARENT POWER, VA
              IINPHAS 22.727273   A       IN-PHASE CURRENT COMPONENT
              ISPEC   7.4700933   A       NEW LAGGING CURRENT
    .95       PFCORR                      CORRECTED POWER FACTOR
              RATING  -7893.499   VAR     VAR RATING OF SHUNT CAPACITANCE
              C       -108.1516   MMF     SHUNT CAPACITANCE


              S Rule
              - ----
                P = V*I*PF
                Q = V*I*SQRT(1 - PF^2)
                I = S/V

                IINPHASE = P/V
                IREACT = Q/V
                ICORR = ISPEC - IREACT
                ICC = (Q/V)-ICORR
                XCORR = V/ICORR
                ISPEC = (IINPHASE/PFCORR)*SQRT(1 - PFCORR^2)
                RATING = V*ICORR
                C = 1/(2*PI()*F*XCORR)
```

Figure 20-4. Power Factor Correction.

if a wye connection is balanced, no current flows in the neutral line. Another benefit of working with balanced configurations is that a single line can be analyzed with the assurance that it will be representative of the other two lines as well.

With the model shown in Figure 20–6, you can perform single line analysis on a system comprising a generator and a load. You enter the line to line voltage, the total load in KW, and the power factor. The model provides the KVA rating of the generator, the in-phase and out-of-phase components of the current, and the compensating reactance required to bring the power factor to unity.

## POWER TRANSFORMERS

You find power transformers at every stage of the power generation, distribution, and consumption process. Power is generated at

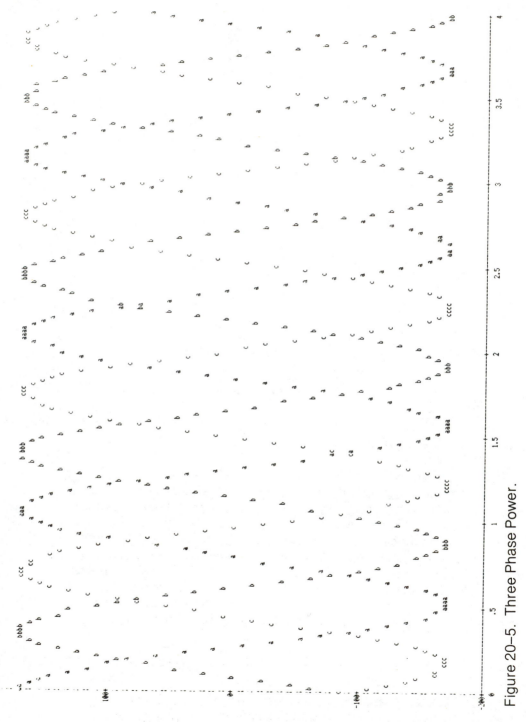

Figure 20-5. Three Phase Power.

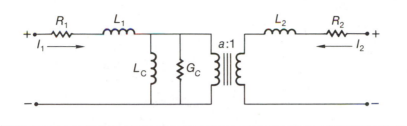

Figure 20–6. Schematic Model of a Power Transformer.

one voltage, distributed over transmission lines at another, distributed from substations at yet another, and used in the plant at even another. A schematic model of a power transformer is shown in Figure 20–6.

The power transformer model can be described in terms of a number of parameters, which are indicated on the schematic diagram, and listed below:

R1   primary winding resistance

L1   primary winding leakage inductance

Lc   core inductance

Gc   core conductance

$a_{ps}$   primary to secondary turns ratio Np/Ns

L2   secondary winding leakage inductance

R2   secondary winding resistance

These parameters can be calculated from measurements of voltage, current, and power taken during short circuit and open circuit tests.

You can speed the process of calculating the transformer model parameters by entering the equations that describe the circuit model in TK!Solver models. Two models are shown below; Figure 20–7 is the open circuit test model, and Figure 20–8 is the short circuit model. Both models are shown solved with sample inputs.

## TRANSMISSION LINES

Transmission lines are an integral part of both power systems and communications systems. When you study some elementary elec-

```
St Input       Name     Output       Unit     Comment
-- -----       ----     ------       ----     -------
                                              POWER TRANSFORMER, OPEN CIRCUIT TEST
                                              ************************************
               S        440          VA       APPARENT POWER
      440       V1                    V        PRIMARY VOLTAGE
      1         I1                    A        PRIMARY AMPERAGE
               Q        428.48571    VA       REACTIVE POWER
      100       P                     W        AVERAGE POWER
               GC       5.1653E-4    MHOS     CORE INDUCTANCE
               BC       -.0022133    MHOS     SUSCEPTANCE
               APS      .4                    PRIMARY - SECONDARY TURNS RATIO
      1100      V2                    V        SECONDARY VOLTAGE, OPEN CIRCUIT


S Rule
- ----
  S = V1*I1
  Q^2 = S^2 - P^2
  GC = P/V1^2
  BC = -Q/V1^2
  APS = V1/V2
```

Figure 20–7.  The Open Circuit Transformer Model.

tric and electronic circuits, you can almost ignore the wiring between components. At the worst, you calculate the amperage in some parts of a circuit and size the wire accordingly.

However, you encounter long distances, high voltages, and high frequencies in transmission line problems, and factors other than the ampacity of a wire become important. Not only do you have to consider the resistance of a transmission line, but you also have to consider the inductance along the line, and the capacitance between lines, and between the line and ground. In some problems, you may even have to consider the conductance from line to line.

One interesting phenomenon encountered in the study of transmission lines is the skin effect. Conductors carrying direct current or alternating current of very low frequencies present a resistance that is essentially uniform over the entire cross-section of the conductor. As the frequency of the signal increases, the time required for an external field to penetrate the conductor begins to affect the impedance. Each time the direction of the current reverses, the magnetic field associated with the current must reverse also. The end result is that at very high frequencies, metallic conductors carry the majority of the current in the outer layer of the conductor, rather than uniformly through the entire conductor.

```
St Input      Name    Output      Unit    Comment
-- -----      ----    ------      ----    -------
                                          POWER TRANSFORMERS
                                          *****************
              APS     4.9984              PRIMARY TO SECONDARY TURNS RATIO
              NP                          PRIMARY TURNS
              NS                          SECONDARY TURNS
              IRATED  11.363636   A       RATED CURRENT
   15000      VARTG               VA      KVA RATING OF XFORMER
   1320       VRTG                V       VOLTAGE RATING OF XFORMER
              SSC     1136.3636   VA      APPARENT POWER AT TEST CURRENT
   100        VSC                 V       PRIMARY TEST VOLTAGE
              QSC     853.71091   VAR     REACTIVE POWER, SHORT CIRCUIT
   750        PSC                 W       AVERAGE POWER, SHORT CIRCUIT
              R1      2.904       OHMS    PRIMARY WINDING RESISTANCE
              R2      .11623438   OHMS    SECONDARY WINDING RESISTANCE
              ISC1    11.363636   A       PRIMARY CURRENT
              X1      3.3055686   OHMS    PRIMARY REACTANCE
              X2      .13230741   OHMS    SECONDARY REACTANCE
   56.8       ISC2                A       SECONDARY CURRENT


S Rule
- ----
* APS = NP/NS
  IRATED = VARTG/VRTG
  ISC1=IRATED
  SSC = IRATED * VSC
  QSC = SQRT(SSC^2 - PSC^2)
  R1 = APS^2*R2
  R1 = PSC/(2*ISC1^2)
  X1 = QSC/(2*ISC1^2)
  X1 = APS^2*X2
  APS = ISC2/IRATED
```

Figure 20-8.  The Short Circuit Transformer Model.

As the skin effect reduces the thickness of the conducting layer, the impedance increases. You can calculate the skin depth, the thickness of the conducting layer, as a function of frequency, resistivity, and permittivity. Since impedance is generally a complex number, you must calculate the impedance in two components, real and imaginary. You can handle this situation very simply with two rules, as shown in the model of Figure 20-9.

You can also calculate the skin effect for a range of frequencies and plot the skin effect versus frequency by associating lists with the variables delta and $f$. A model is shown in Figure 20-10, but notice that you plotted delta versus the logarithm of frequency, so that you could cover a wider frequency band.

A coaxial cable is a special case of the general transmission line, so you enter special rules using variables whose names are easily associated with the coaxial line.

| St Input | Name | Output | Unit | Comment |
|----|------|--------|------|---------|
| | | | | TRANSMISSION LINES 1 |
| | | | | ******************** |
| | delta | 2.6584143 | MILS | SKIN DEPTH |
| 1000000 | f | | HZ | FREQUENCY |
| 1.2566E-6 | mu | | H/M | PERMEABILITY |
| 1.8E-8 | rho | | OHM*M | RESISTIVITY |
| | Aeff | | M^2 | EFFECTIVE CROSS SECT AREA AT f |
| | r | | MILS | RADIUS |
| | Rf | | OHMS | RESISTANCE AT FREQUENCY f |
| | Ro | | OHMS | DC RESISTANCE |
| | A | | M^2 | CROSS SECTIONAL AREA |
| | Zlwr | 2.6657E-4 | OHMS | IMPEDANCE/UNIT LENGTH, REAL COMPONENT |
| | Zlwj | 2.6657E-4 | j*OHMS | IMPEDANCE/UNIT LENGTH, IMAG COMPONENT |
| | Ll_e | | H/M | EXTERNAL INDUCTANCE, PARALLEL, 0 WIRES, |
| 1 | D | | CM | DISTANCE BETWEEN PARALLEL CONDUCTORS |
| | Lcoax | 4.6052E-7 | H/M | EXTERNAL INDUCTANCE, COAXIAL LINE |
| 2 | IDo | | CM | OUTER CONDUCTOR I. D. |
| .2 | ODi | | CM | INNER CONDUCTOR O. D. |
| | Xl_e | | OHMS/M | EXTERNAL REACTANCE |
| | XL | 2.8937803 | OHMS/M | TOTAL INDUCTIVE REACTANCE |
| | Xcoax | 2.8935138 | OHMS/M | INDUCTIVE REACTANCE, COAX |
| | Cll | 8.786E-13 | F/M | CAPACITANCE, PARALLEL WIRES |
| | epsilon | 8.842E-13 | | PERMITTIVITY |
| 16 | r1 | | MILS | RADIUS, WIRE # 1 |
| 16 | r2 | | MILS | RADIUS, WIRE #2 |
| | Ccoax | 2.413E-12 | F/M | CAPACITANCE, COAXIAL CONDUCTOR |
| 1 | epsilon | | | PERMITTIVITY RATIO |

```
S Rule
- ----
  "TRANSMISSION LINES 1
  "********************
  "SKIN EFFECT
* A = pi()*r^2
  delta = 1/(SQRT(PI()*f*mu/rho))
* Aeff = pi()*r^2-pi()*(r-delta)^2
* Rf = Ro*A/Aeff
  Zlwr = SQRT(PI()*f*mu*rho)
  Zlwj = SQRT(PI()*f*mu*rho)

  "EXTERNAL INDUCTANCE
* Ll_e = (mu/pi())*ln(D/r)
  Lcoax = 2E-7*ln(IDo/ODi)
* Xl_e = 2*pi()*f*Ll_e
  Xcoax = Lcoax*2*pi()*f
  XL = Zlwr+Xcoax

  "CAPACITANCE
  Cll = 2*pi()*epsilon/ln((D-ri)*(D-r2)/(ri*r2))
  Ccoax = 2*pi()*epsilon/ln(IDo/ODi)
  epsilon = (1/(36*pi()*10E9))*epsilonr
```

Figure 20-9.  Transmission Line Model.

CHARACTERISTICS OF MULTILAYER, STEEL REINFORCED BARE ALUMINUM CONDUCTORS

| CODENAME | AREA | ALUMSTR | STEELSTR | OD | RDC | RAC20C | RAC50C | GMR | XL1FT | XC1FT |
|---|---|---|---|---|---|---|---|---|---|---|
| WAXWING | 266800 | 18 | 1 | .609 | .0646 | .3488 | .3831 | .0198 | .476 | .109 |
| PARTRIDGE | 266800 | 26 | 7 | .642 | .064 | .3452 | .3792 | .0217 | .465 | .1074 |
| OSTRICH | 300000 | 26 | 7 | .68 | .0569 | .307 | .3372 | .0229 | .458 | .1057 |
| MERLIN | 336400 | 18 | 1 | .684 | .0512 | .2767 | .3037 | .0222 | .462 | .1055 |
| LINNET | 336400 | 26 | 7 | .721 | .0507 | .2737 | .3006 | .0243 | .451 | .104 |
| ORIEL | 366400 | 30 | 7 | .741 | .0504 | .2719 | .2987 | .0255 | .445 | .1032 |
| CHICKADEE | 397500 | 18 | 1 | .743 | .0433 | .2342 | .2572 | .0241 | .452 | .1031 |
| IBIS | 397500 | 26 | 7 | .783 | .043 | .2323 | .2551 | .0264 | .441 | .1015 |
| PELICAN | 477000 | 18 | 1 | .814 | .0361 | .1957 | .2148 | .0264 | .441 | .1004 |
| FLICKER | 477000 | 24 | 7 | .846 | .0359 | .1943 | .2134 | .0284 | .432 | .0992 |
| HAWK | 477000 | 26 | 7 | .858 | .0357 | .1931 | .212 | .0289 | .43 | .0988 |
| HEN | 477000 | 30 | 7 | .883 | .0355 | .1919 | .2107 | .0304 | .424 | .098 |
| OSPREY | 556500 | 18 | 1 | .879 | .0309 | .1679 | .1843 | .0284 | .432 | .0981 |
| PARAKEET | 556500 | 24 | 7 | .914 | .0308 | .1669 | .1832 | .0306 | .423 | .0969 |
| DOVE | 556500 | 26 | 7 | .927 | .0307 | .1663 | .1826 | .0314 | .42 | .0965 |
| ROOK | 636000 | 24 | 7 | .977 | .0269 | .1461 | .1603 | .0327 | .415 | .095 |
| GROSBEAK | 636000 | 26 | 7 | .99 | .0268 | .1454 | .1596 | .0335 | .412 | .0946 |
| DRAKE | 795000 | 26 | 7 | 1.108 | .0215 | .1172 | .1284 | .0373 | .399 | .0912 |
| TERN | 795000 | 45 | 7 | 1.063 | .0217 | .1188 | .1302 | .0352 | .406 | .0925 |
| RAIL | 954000 | 45 | 7 | 1.165 | .0181 | .0977 | .1092 | .0386 | .395 | .0897 |
| CARDINAL | 954000 | 54 | 7 | 1.196 | .018 | .0988 | .1082 | .0402 | .39 | .089 |
| ORTOLAN | 1033500 | 45 | 7 | 1.213 | .0167 | .0924 | .1011 | .0402 | .39 | .0885 |
| BLUEJAY | 1113000 | 45 | 7 | 1.259 | .0155 | .0861 | .0941 | .0415 | .386 | .0874 |
| FINCH | 1113000 | 54 | 19 | 1.293 | .0155 | .0856 | .0937 | .0436 | .38 | .0866 |
| BITTERN | 1272000 | 45 | 7 | 1.382 | .0136 | .0762 | .0832 | .0444 | .378 | .0855 |
| BOBOLINK | 1431000 | 45 | 7 | 1.427 | .0121 | .0684 | .0746 | .047 | .371 | .0837 |
| PLOVER | 1431000 | 54 | 19 | 1.465 | .012 | .0673 | .0735 | .0494 | .365 | .0829 |
| LAPWING | 1590000 | 45 | 7 | 1.502 | .0109 | .0623 | .0678 | .0498 | .364 | .0822 |
| FALCON | 1590000 | 54 | 19 | 1.545 | .0108 | .0612 | .0667 | .0523 | .358 | .0814 |
| BLUEBIRD | 2156000 | 84 | 19 | 1.762 | .008 | .0476 | .0515 | .0586 | .344 | .0776 |

Figure 20-10. Power Transmission Line Model, Conductor Pairs.

| St | Input | Name | Output | Unit | Comment |
|----|-------|------|--------|------|---------|
| | | | | | POWER TRANSMISSION LINES |
| | | | | | ************************** |
| L | | 'BLUEBIRD CODENAM | | | CODENAME OF CONDUCTOR |
| L | | AREA | | CMILS | AREA OF ALUMINUM IN CIRCULAR MILS |
| L | | ALUMSTR | | | # STRANDS ALUMINUM |
| L | | STEELST | | | # STRANDS STEEL |
| L | | OD | | INCHES | OUTSIDE DIAMETER |
| L | | RDC | | MOHMS/FT | RESISTANCE TO DIRECT CURRENT, 20' C |
| L | | RAC20C | | OHMS/MILE | RESISTANCE TO ALT CURRENT, 20' C, 60HZ |
| L | | RAC50C | | OHMS/MILE | RESISTANCE TO ALT CURRENT, 50' C, 60HZ |
| L | | GMR | | FT | GEOMETRIC MEAN RADIUS |
| L | | XL1FT | | OHMS/MILE | INDUCTIVE REACTANCE, 1 FT SPACING, 60H |
| L | | XC1FT | | MOHMS*MIL | CAPACITIVE REACTANCE, 1 FT SPACING, 60 |
| L | | XL | .67272799 | OHMS/MILE | INDUCTIVE REACTANCE |
| | 60 | f | | HZ | FREQUENCY |
| | 15 | D | | FT | LINE TO LINE SPACING |
| | | VGMR | .0586 | FT | GEOMETRIC MEAN RADIUS FROM TABLE |
| | | KL | 1.9545402 | | CORRECTIVE FACTOR, FOR XL |
| | | VXL1FT | .344 | OHMS/MILE | VALUE OF XL1FT FROM TABLE |
| | | VXL | .67236183 | OHMS/MILE | CORRECTED VALUE OF XL |
| | | VOD | 1.762 | INCHES | VALUE OF OD FROM TABLE |
| | | VRDC | .008 | MOHMS/FT | DC RESISTANCE FROM TABLE |
| | | VRAC20C | .0476 | OHMS/MILE | RESISTANCE TO ALT CURRENT FROM TABLE |
| | | VXC1FT | .0776 | MOHMS*MIL | CAPACITIVE REACTANCE FROM TABLE |
| | | KC | 2.0369298 | | CORRECTIVE FACTOR, FOR XC |
| | | r | .881 | INCHES | CONDUCTOR RADIUS |
| | | VXC | .15806575 | MOHMS*MIL | VALUE OF XC FROM TABLE |
| | | R | .07341667 | FT | |

S Rule
-- ----
VGMR = FGMR(CODENAME)
XL = 2.022E-3*f*LN(D/VGMR)
KL = 1 + LN(D)/LN(1/VGMR)
VXL1FT = FXL1FT(CODENAME)
VXL = KL*VXL1FT
VOD = FOD(CODENAME)
VRDC = FRDC(CODENAME)
VRAC20C = FRAC20C(CODENAME)

VXC1FT = FXC1FT(CODENAME)
r = FOD(CODENAME)/24
KC = 1+LN(D)/LN(1/r)
VXC1FT = FXC1FT(CODENAME)
VXC = KC*VXC1FT
R = FOD(CODENAME)/24

Figure 20–10. (Continued).

# POWER TRANSMISSION LINES

Power transmission lines differ from the general transmission line in that they include both conductors selected for their current carrying properties and conductors selected for structural reasons. The conductors are stranded and the core has a different composition than the outer layers. Furthermore, they occur in sets of three or four, for the distribution of three-phase power. Our simple model would not be adequate for calculating the parameters of resistance, inductance, and capacitance.

However, there is no reason to calculate these parameters, even though TK!Solver could handle the calculations, given the correct rules. All the parameters you could possibly need have been calculated and are available in handbooks, usually at nominal costs from manufacturers' associations.

You can still use TK!Solver's power to advantage, by entering the handbook data into a TK!Solver model and defining various user functions to perform table look-up. Of course, you can include other rules in the same model to make calculations from the values returned by the look-up functions.

The model of Figure 20–10 illustrates the technique of retrieving values from look-up tables and then performing calculations based on those values. This model calculates the parameters of inductive and capacitive reactance for two conductor pairs of multilayer, steel reinforced bare aluminum conductors.

# CHAPTER 21

# Mechanics of Materials

Mechanics of materials, or strength of materials, is an important phase of every field of engineering. Although the electronic engineer may prefer to work with circuitry, and the mechanical engineer may prefer to size piping systems, both must consider the structural aspects of the systems they design. TK!Solver is particularly valuable for the engineer who must solve certain types of problems only occasionally; the electrical engineer who must size a structural support for a piece of electrical equipment is only one example.

## STRESS–STRAIN CURVES

Stress is force per unit area; the equation for stress is

$$\sigma = F/A$$

*Where:*   $\sigma$ is stress

$F$ is force

$A$ is cross-sectional area

When you apply opposing axial forces to metallic material, creating stress, the sample changes length in the direction of the force. The ratio of the change in length to the original length is the strain, given by the equation

$$\epsilon = e/L$$

*Where:*   $\epsilon$ is strain

$e$ is elongation

$L$ is length

The relation of stress and strain for a given material usually is presented as a curve. The data for the curve cannot be calculated, but is obtained by placing test specimens under tension and measuring the elongation and tension simultaneously. A few typical stress strain curves are shown in Figure 21–1.

Once you have obtained test data on a sample and constructed the stress-strain curve, you can infer several values that can be used to predict the performance of the material. The stress-strain curve is a model, in graphical form.

Figure 21–2 illustrates the points of interest on the general stress-strain curve. Notice that the initial portion of the curve is almost a straight line, indicating that the material is elastic over that range of stress. If you apply a force to a sample that does not cause the stress to exceed the stress value of point 1, and then remove the force, the sample will return to its original length along the same curve. Point 1 is called the *elastic limit*.

If you apply a force that causes the stress to exceed the elastic limit, and then remove the force, the stress will return to zero along a curve to the right of the original curve. Although the stress is again zero, the strain apparently remains—the material is permanently deformed.

If you continue to apply force, increasing the stress, you eventually cause the material to fail, indicated by point 4 on the curve. The highest point on the curve, point 3, is the ultimate tensile strength.

Test data is normally obtained from an automatic testing machine, which applies the tensile force and records both the ten-

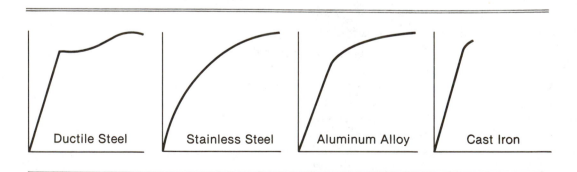

Figure 21–1. Stress-Strain Curves for Various Metals.

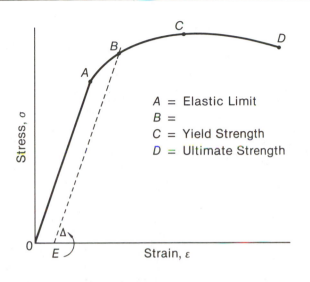

A = Elastic Limit
B =
C = Yield Strength
D = Ultimate Strength

Figure 21-2. The General Stress-Strain Curve.

sile force and the elongation as functions of time. Once you have obtained the test data, you can calculate the various parameters using a TK!Solver model, as shown in Figure 21-3.

The first step, of course, is to enter the test data as two lists, $F$ and $e$. At this point, you can perform one interesting task by entering two rules and completing the Plot sheet; you can plot the stress-strain curve, as shown in Figure 21-4.

You calculate the Modulus of Elasticity, $E$, from the first two sets of values for $\sigma$ and $\epsilon$. The Modulus of Elasticity is the slope of the stress-strain curve at the origin. By using the ELEMENT function, you can calculate the value of $E$ on the second pass of the List Solver. There is no need to associate a list with $E$, as its value is constant, although you could calculate a list of values and fit a straight line to the linear part of the curve, using techniques discussed earlier.

The next property of interest is the *yield* strength, which is defined as the point at which the sample retains a permanent set. Because not all metallic materials exhibit the same stress-strain curve shape, you commonly calculate the yield strength as the point where the sample retains a permanent set of 0.2 percent. If the test force is removed at this point, the sample will return to a length that is 100.2 percent of the original length.

```
 St  Input       Name     Output     Unit     Comment
 --  -----       ----     ------     ----     -------
                                              STRESS-STRAIN CURVE ANALYSIS
                                              *****************************
 L               sigma               LB/IN^2  STRESS
 L   1           F                   LB       TENSILE FORCE APPLIED TO SAMPLE
     .19634954   A                   IN^2     CROSS-SECTIONAL AREA OF SAMPLE
 L               epsilon                      STRAIN
 L   1           e                   IN       ELONGATION
     2           L                   IN       ORIGINAL LENGTH
 L               depsilo                      INCREMENT OF epsilon
 L               dsigma                       INCREMENT OF sigma
 L               E        9982198.0           MODULUS OF ELASTICITY
 L               SET                 IN        PERMANENT SET IN SAMPLE
 L               eps_ult                      ELONGATION AT FRACTURE
                 sigmaTU  45836.624  LB/IN^2  ULTIMATE STRENGTH
                 YS       37113.300            


 S  Rule
 -  ----
    sigma = F/A
    epsilon = e/L

    depsilon = ELEMENT('epsilon,ELEMENT(),0)-ELEMENT('epsilon,ELEMENT()-1,0)
    dsigma = ELEMENT('sigma,ELEMENT(),0)-ELEMENT('sigma,ELEMENT()-1,0)

    E = ELEMENT('dsigma,1,0)/ELEMENT('depsilon,1,1)
    SET = epsilon - sigma/E

    eps_ult = ELEMENT('e,COUNT('e),0)/L
    sigmaTU = MAX('F)/A

    YS = FSET(.002)/A
```

Figure 21-3. The Stress-Strain Model.

   The simplest way to find the yield strength is to define
another variable, SET, which is the difference between the strain
calculated from the measured elongation and the value of the
stress divided by the modulus of elasticity. SET is defined as a list
variable and you can easily identify the interval in which the yield
point occurs. You can find two other parameters, elongation at
fracture and ultimate strength, very easily.
   You cannot find an exact value of yield strength because the
point of intersection of the two curves may fall between data
points. However, you can interpolate between $F$ values for a point
that corresponds to an exact value of SET, with the user function,
FSET. However, FSET uses the SET list as the domain, and the
list is empty when you invoke the List Solver the first time. Therefore, the program beeps on each pass and presents an error
message until the SET list contains a value greater than 0.002.

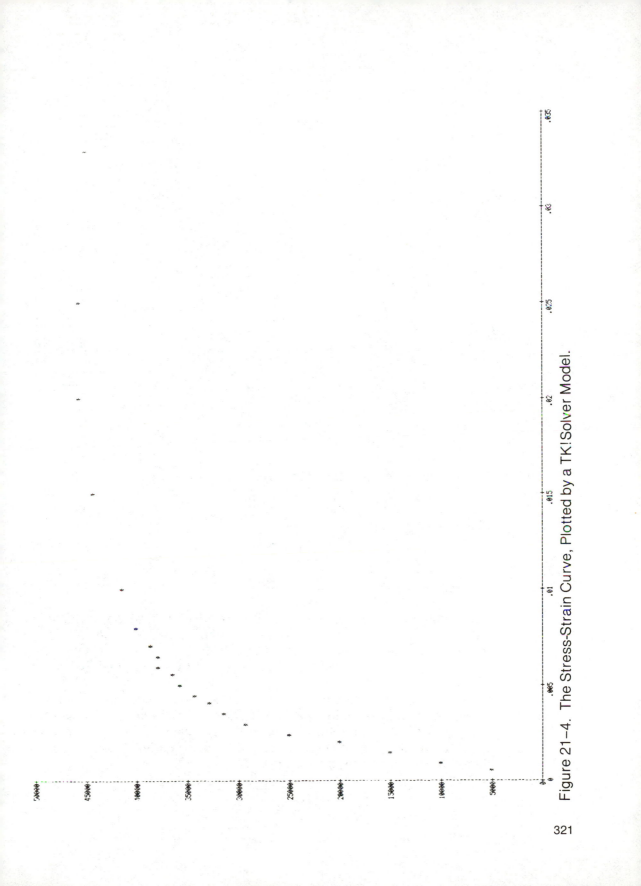

Figure 21–4.  The Stress-Strain Curve, Plotted by a TK!Solver Model.

STRESS-STRAIN TABLE

| Point | e | F | epsilon | sigma | depsilon | dsigma | SET |
|---|---|---|---|---|---|---|---|
| 1 | .001 | 980 | .0005 | 4991.09902 | .0005 | 4991.09902 | 0 |
| 2 | .002 | 1960 | .001 | 9982.19803 | .0005 | 4991.09902 | 0 |
| 3 | .003 | 2940 | .0015 | 14973.2971 | .0005 | 4991.09902 | 1.E-14 |
| 4 | .004 | 3920 | .002 | 19964.3961 | .0005 | 4991.09902 | 4.E-14 |
| 5 | .005 | 4900 | .0025 | 24955.4951 | .0005 | 4991.09902 | -3.E-14 |
| 6 | .006 | 5700 | .003 | 29029.8616 | .0005 | 4074.36654 | 9.18367E-5 |
| 7 | .007 | 6200 | .0035 | 31576.3407 | .0005 | 2546.47909 | 3.36735E-4 |
| 8 | .008 | 6500 | .004 | 33104.2282 | .0005 | 1527.88745 | 6.83673E-4 |
| 9 | .009 | 6800 | .0045 | 34632.1156 | .0005 | 1527.88745 | .001030612 |
| 10 | .01 | 7000 | .005 | 35650.7073 | .0005 | 1018.59164 | .001428571 |
| 11 | .011 | 7200 | .0055 | 36669.2989 | .0005 | 1018.59164 | .001826531 |
| 12 | .012 | 7400 | .006 | 37687.8905 | .0005 | 1018.59164 | .002224490 |
| 13 | .013 | 7500 | .0065 | 38197.1863 | .0005 | 509.295818 | .002673469 |
| 14 | .014 | 7600 | .007 | 38706.4822 | .0005 | 509.295818 | .003122449 |
| 15 | .016 | 7800 | .008 | 39725.0738 | .001 | 1018.59164 | .004020408 |
| 16 | .02 | 8150 | .01 | 41507.6092 | .002 | 1782.53536 | .005841837 |
| 17 | .03 | 8650 | .015 | 44054.0883 | .005 | 2546.47909 | .010586735 |
| 18 | .04 | 8950 | .02 | 45581.9757 | .005 | 1527.88745 | .015433673 |
| 19 | .05 | 9000 | .025 | 45836.6236 | .005 | 254.647909 | .020408163 |
| 20 | .066 | 8800 | .033 | 44818.0320 | .008 | -1018.5916 | .028510204 |

Figure 21-4. (Continued).

From that point on, the yield strength value appears for each element of the input lists. Although it would be nice to construct only "beep-free" models, you are still primarily interested in models that produce answers.

# POISSON'S RATIO

Once you have calculated the properties $\sigma$, $\epsilon$, and $E$, you can calculate another property of a metallic material—the bulk modulus, $K$. The bulk modulus relates the change in volume of a sample to a uniform tri-axial stress, such as would be applied by hydrostatic pressure.

If you subject a sample to uni-axial stress, as shown in Figure 21-5, you would be able to measure not only the elongation, as in the previous example, but a contraction in the two lateral directions (perpendicular to the uni-axial stress). The lateral strains produced by the uni-axial stress are

$$\epsilon_y = mu_y * \epsilon_X$$

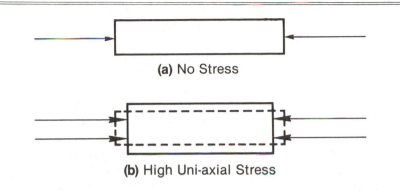

**(a)** No Stress

**(b)** High Uni-axial Stress

Figure 21–5.  Deformation of Sample under Uni-Axial Stress.

and

$$\epsilon_Z = mu_Z{}^*\epsilon_X$$

The terms, $mu_Y$ and $mu_Z$, are known as Poisson's ratio; if the material is isotropic, that is, if its properties are the same regardless of direction, $mu_Y = mu_Z$. You can calculate Poisson's ratio with a single rule

$$mu = epsilon/(eZ/Z)$$

and the bulk modulus, $K$, with another

$$K = E/(3{*}(1 - 2{*}mu))$$

## THERMAL EXPANSION

When materials of interest in engineering problems change temperature they also change size—linear dimensions increase as the temperature increases. The expansion is proportional to the increase (or decrease) in temperature, and the proportional constant is called the *coefficient of thermal expansion.*

   If you subject a sample to both an increase in temperature and an axial force, the effects are additive. A tensile force and an increase in temperature will cause an increase in length equal to the sum of the two effects calculated separately.

You can calculate another parameter that is often very important. If you subject a member to an increase in temperature, but restrain its increase in length, you create a thermal stress. Steam heating systems provide a good example of the phenomenon. A steam line is normally anchored at intervals. If the piping system were merely anchored, to a building structure, for instance, substantial forces would develop at the anchor points. Therefore, it is common practice to install expansion loops between the anchor points.

The model shown in Figure 21–6 illustrates all three types of calculations—thermal expansion, combined thermal and superimposed deflections, and forces due to restrained thermal expansion. In an effort to minimize the use of a handbook to locate the coefficients of thermal expansion, the coefficients were entered in a list and the respective materials were entered in another list. Then you define a user function so that you can enter the name of the material, and let the TK!Solver program do the retrieval for you.

## SHEAR STRESSES

When we discussed tensile stress above, we were considering a normal stress—the force acted perpendicular to the plane of interest. Compressive forces are also normal forces.

If you subject a sample to tensile or compressive forces that act parallel to a given plane, but that are not aligned axially, you produce a shear stress in the plane. Of course, most engineering problems involve forces that produce a combination of shear and normal stresses, but shear stress will be considered separately.

If you consider the test sample shown in Figure 21–7, which cracks along a plane 60 degrees from the centerline, you know that both shear and normal forces are involved. You can resolve the total stress into shear and normal components relative to any plane, using the rules shown in the model of Figure 21–8. However, you are interested in the stresses relative to the plane in which the failure occurred. Therefore, enter the angle between the failure plane and the plane normal to the axial force and calculate the shear and normal stresses relative to that plane.

## BEAMS

A beam is a structural member designed to support loads applied along its length. Although the loads may be applied at arbitrary

```
St Input      Name     Output      Unit     Comment
-- -----      ----     ------      ----     -------
                                            THERMAL DEFLECTIONS & STRESSES
                                            ********************************
              dL@t    .029856      IN       CHANGE OF LENGTH DUE TO TEMP CHANGE
              alpha   .00001244    IN/IN-F  COEFFICIENT OF THERMAL EXPANSION
   12         L                    IN       LENGTH
   200        dt                   oF       CHANGE IN TEMPERATURE
              dL                   IN       CHANGE IN LENGTH, COMBINED FORCES
              F                    LB       AXIAL FORCE
   .25        A                    IN^2     CROSS-SECTIONAL AREA
   10000000   E                    LB/IN^2  MODULUS OF ELASTICITY
              Fconstr -4136.667    LB       FORCE IN CONSTRAINED MEMBER
   .01        dLbar                IN       KNOWN DEFLECTION
   'ALUMINUM  MATL                          MATERIAL


USER FUNCTION: Falpha                       S Rule
Element Domain         Range                - ----
-------- ------        -----                  dL@t = alpha*L*dt
   1       'ALUMINUM   .00001244          *   dL = dL@t + F*L/(A*E)
   2       'ANTIMONY   .00000755              Fconstr =(dLbar - alpha*L*dt)*A*E/L
   3       'BARIUM     0                       alpha = Falpha(MATL)
   4       'BISMUTH    .00000077
   5       'BORON      0
   6       'BRASS1     .00001
   7       'BRASS2     .00001
   8       'BRASS3     .00001
   9       'BRASS4     .00001
   10      'BRONZE     .00001
   11      'CADMIUM    0
   12      'CALCIUM    0
   13      'CHROMIUM   0
   14      'COBALT     .00000683
   15      'COPPER     .000009
   16      'GOLD       .00000778
   17      'IRIDIUM    .00000361
   18      'IRON_CAST  .00000655
   19      'IRON_WROUG .00000661
   20      'LEAD       .0000163
   21      'MAGNESIUM  .00001444
   22      'MANGANESE  .00001294
   23      'MERCURY    0
   24      'MOLYBDENUM .00000294
   25      'NICKEL     .000007
   26      'PLATINUM   .00000496
   27      'POTASSIUM  0
   28      'SILVER     .00001025
   29      'SODIUM     0
   30      'STEEL_CARB .00000633
   31      'TANTULUM   .00000361
   32      'TELLURIUM  .0000093
   33      'TIN        .00001496
   34      'TITANIUM   .0000049
   35      'TUNGSTEN   .00000239
   36      'URANIUM    0
   37      'VANADIUM   .0000043
   38      'ZINC       .000017
```

Figure 21–6.  Thermal and Combined Expansion.

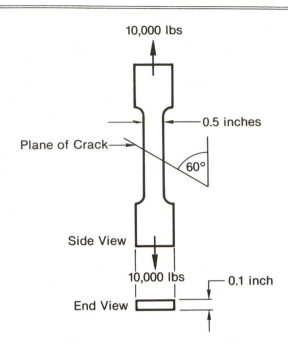

10,000 lbs

0.5 inches

Plane of Crack

60°

Side View

10,000 lbs    0.1 inch

End View

Figure 21-7.  Crack in Test Specimen from Shear Stress.

```
(7c) Comment: ANGLE BETWEEN PLANE AND PLANE NORMAL TO FORCE          203/
=================== VARIABLE SHEET ================================
St Input      Name   Output   Unit     Comment
-- -----      ----   ------   ----     -------
                                       ***********
              sigma  200000   LB/IN^2  STRESS
   10000      F               LB       FORCE
   .05        A               IN^2     CROSS-SECTIONAL AREA
              sig@the 150000  LB/IN^2  STRESS NORMAL TO PLANE
   30         theta           DEG      ANGLE BETWEEN PLANE AND PLANE NORMAL T
              tau@the 86602.540 LB/IN^2 SHEAR STRESS
              tauMAX  100000   LB/IN^2  MAXIMUM SHEAR STRESS IN SAMPLE
=================== RULE SHEET ====================================
S Rule
- ----
  sigma = F/A
  sig@theta = sigma*COS(theta)^2
  tau@theta = (1/2)*sigma*SIN(PI()/2-theta)
  tauMAX = sigma/2
```

Figure 21-8.  Shear Stress Model.

angles to the axis of the beam, the ability of the beam to resist the components perpendicular to the axis is more critical than its ability to resist axial forces. These normal loads produce shear and bending in the beam.

A beam is normally subjected to two kinds of loads—concentrated and distributed. Figure 21–9 illustrates both kinds of loads applied to a single beam, as well as the reactions at the beam supports. You can normally expect to be given the loads to be applied to a beam; you must find the reactions and the stresses in the beam.

You begin the analysis of the beam by deciding that you will use the List Solver. You construct a list of the forces acting on the beam with the list elements corresponding to the distance from one end of the beam. If there is a uniform distributed load, you have three choices of ways to enter the data in the list: you can type the entries, one at a time; you can type the entry once and copy it to successive fields with the /C command; or you can enter the value in the first field and the last field and then use the Fill List option of the Action command. The last choice is the quickest by far; it will fill the list with the value in seconds.

If you have a load that increases linearly from one end of the beam to the other, you can still use the Fill List option of the Action command. But, if the distributed load increases and then decreases, you cannot use the Fill List option.

If the distributed load varies according to one or more functions that are piecewise continuous over the length of the beam, you can use the List Solver to fill the list. If more than one rule is needed to describe the loading, you can use the Block option of the List Solver to fill the list in a piecewise manner.

Once you have entered the distributed loads, you can enter the concentrated loads, diving to the list if necessary and entering the values of the concentrated loads at the correct elements. Of course, you have to remember to enter the sum of the existing value and the concentrated load, remembering that the TK!Solver program will do the arithmetic. A list of the loads for the beam of Figure 21–9 is shown in Figure 21–10.

After you have entered all the loads, you find the reactions at the ends of the beam. Beginning at one end, sum the moments of the forces about that end, and divide the sum by the length; that value is the reaction at the opposite end. Then, sum the loads and the opposite end reaction (its sign is opposite to that of the loads) to obtain the reaction at the origin (we must keep track of signs). The model for this calculation is shown in Figure 21–11.

There is no reason to associate lists with the reactions, $R1$ and $R2$, as the variable sheet can be printed after the List Solver has

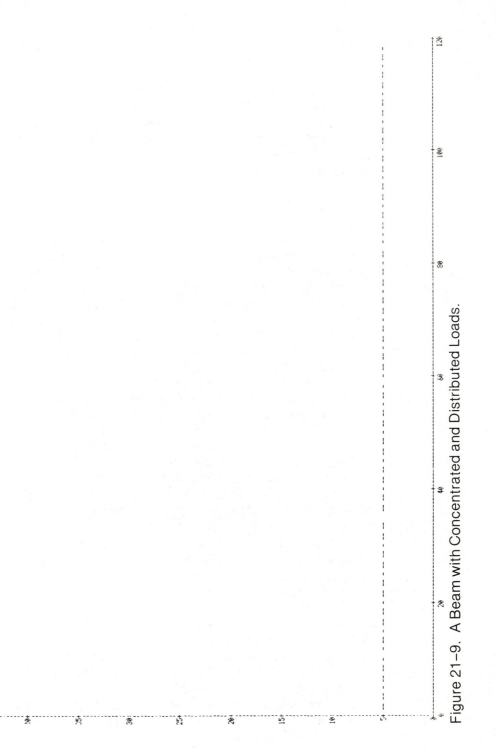

Figure 21–9.  A Beam with Concentrated and Distributed Loads.

```
LIST: LOAD
Element Value
------- -----
1    0        52   5        76   5
2    5        53   5        77   5
3    5        54   5        78   5
4    5        55   5        79   5
5    5        56   5        80   5
6    5        57   5        81   5
7    5        58   5        82   5
8    5        59   5        83   5
9    5        60  35        84  40
10   5        61   5        85   5
11   5        62   5        86   5
12  20        63   5        87   5
13   5        64   5        88   5
14   5        65   5        89   5
15   5        66   5        90   5
16   5        67   5        91   5
17   5        68   5        92   5
18   5        69   5        93   5
19   5        70   5        94   5
20   5        71   5        95   5
21   5        72   5        96   5
22   5        73   5        97   5
23   5        74   5        98   5
24   5        75   5        99   5
25   5                     100   5
26   5                     101   5
27   5                     102   5
28   5                     103   5
29   5                     104   5
30   5                     105   5
31   5                     106   5
32   5                     107   5
33   5                     108   5
34   5                     109   5
35   5                     110   5
36  50                     111   5
37   5                     112   5
38   5                     113   5
39   5                     114   5
40   5                     115   5
41   5                     116   5
42   5                     117   5
43   5                     118   5
44   5                     119   5
45   5                     120   0
46   5
47   5
48   5
49   5
50   5
51   5
```

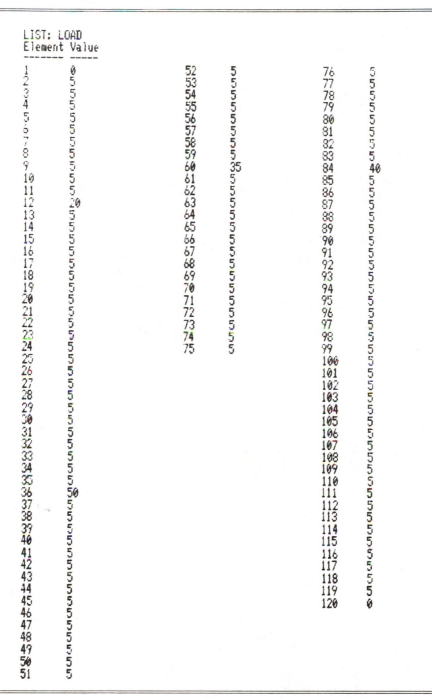

Figure 21-10.  The List of Loads for the Beam of Figure 21-9.

```
(6i) Input: 1                                                            193/!

=================== VARIABLE SHEET ============================================
St Input        Name        Output      Unit    Comment
-- -----        ----        ------      ----    -------
                                                BEAM ANALYSIS
                                                *************

L               MOMENT                  IN-LB   MOMENT ABOUT ORIGIN
L    1          LOAD                    LB      LOADS
L    1          XCOORD                  IN      DISTANCE FROM ORIGIN
                PREV        120                 PREVIOUS ELEMENT
                R2          -343.0208   LB      REACTION AT OPPOSITE END
                R1          -371.9792   LB      REACTION AT ORIGIN END


(1r) Rule: PREV = ELEMENT()-1                                            193/!

=================== RULE SHEET ================================================
S Rule
- ----
  PREV = ELEMENT()-1
  MOMENT = LOAD*(XCOORD+ELEMENT('XCOORD,PREV,0))/2
  R2 = -SUM('MOMENT)/MAX('XCOORD)
  R1 = -(SUM('LOAD)+R2)
```

LOADS AND MOMENTS

| XCOORD | LOAD | MOMENT | | | |
|--------|------|--------|------|------|--------|
| 0  | 0  | 0     | 31 | 5  | 152.5 |
| 1  | 5  | 2.5   | 32 | 5  | 157.5 |
| 2  | 5  | 7.5   | 33 | 5  | 162.5 |
| 3  | 5  | 12.5  | 34 | 5  | 167.5 |
| 4  | 5  | 17.5  | 35 | 50 | 1725  |
| 5  | 5  | 22.5  | 36 | 5  | 177.5 |
| 6  | 5  | 27.5  | 37 | 5  | 182.5 |
| 7  | 5  | 32.5  | 38 | 5  | 187.5 |
| 8  | 5  | 37.5  | 39 | 5  | 192.5 |
| 9  | 5  | 42.5  | 40 | 5  | 197.5 |
| 10 | 5  | 47.5  | 41 | 5  | 202.5 |
| 11 | 20 | 210   | 42 | 5  | 207.5 |
| 12 | 5  | 57.5  | 43 | 5  | 212.5 |
| 13 | 5  | 62.5  | 44 | 5  | 217.5 |
| 14 | 5  | 67.5  | 45 | 5  | 222.5 |
| 15 | 5  | 72.5  | 46 | 5  | 227.5 |
| 16 | 5  | 77.5  | 47 | 5  | 232.5 |
| 17 | 5  | 82.5  | 48 | 5  | 237.5 |
| 18 | 5  | 87.5  | 49 | 5  | 242.5 |
| 19 | 5  | 92.5  | 50 | 5  | 247.5 |
| 20 | 5  | 97.5  | 51 | 5  | 252.5 |
| 21 | 5  | 102.5 | 52 | 5  | 257.5 |
| 22 | 5  | 107.5 | 53 | 5  | 262.5 |
| 23 | 5  | 112.5 |    |    |       |
| 24 | 5  | 117.5 |    |    |       |
| 25 | 5  | 122.5 |    |    |       |
| 26 | 5  | 127.5 |    |    |       |
| 27 | 5  | 132.5 |    |    |       |
| 28 | 5  | 137.5 |    |    |       |
| 29 | 5  | 142.5 |    |    |       |
| 30 | 5  | 147.5 |    |    |       |

Figure 21-11. A Model to Calculate Reactions.

```
         LOADS AND MOMENTS                    LOADS AND MOMENTS

   XCOORD      LOAD       MOMENT        XCOORD     LOAD      MOMENT

     54         5         267.5          108        5        537.5
     55         5         272.5          109        5        542.5
     56         5         277.5          110        5        547.5
     57         5         282.5          111        5        552.5
     58         5         287.5          112        5        557.5
     59        35        2047.5          113        5        562.5
     60         5         297.5          114        5        567.5
     61         5         302.5          115        5        572.5
     62         5         307.5          116        5        577.5
     63         5         312.5          117        5        582.5
     64         5         317.5          118        5        587.5
     65         5         322.5          119        0        0
     66         5         327.5          120
     67         5         332.5
     68         5         337.5
     69         5         342.5
     70         5         347.5
     71         5         352.5
     72         5         357.5
     73         5         362.5
     74         5         367.5
     75         5         372.5
     76         5         377.5
     77         5         382.5
     78         5         387.5
     79         5         392.5
     80         5         397.5
     81         5         402.5
     82         5         407.5
     83        40        3300
     84         5         417.5
     85         5         422.5
     86         5         427.5
     87         5         432.5
     88         5         437.5
     89         5         442.5
     90         5         447.5
     91         5         452.5
     92         5         457.5
     93         5         462.5
     94         5         467.5
     95         5         472.5
     96         5         477.5
     97         5         482.5
     98         5         487.5
     99         5         492.5
    100         5         497.5
    101         5         502.5
    102         5         507.5
    103         5         512.5
    104         5         517.5
    105         5         522.5
    106         5         527.5
    107         5         532.5
```

Figure 21–11. (Continued).

completed its processing. You are only interested in the final values of $R1$ and $R2$, but the List Solver calculates a value for $R1$ and $R2$ for each element of the input lists.

On occasions (of which this appears to be one), the TK!Solver program simply cannot perform all the calculations you desire in a single model. You can apply various techniques described in the chapter on controlling TK!Solver, but it's often quicker to circumvent the issue completely by using two models.

You can construct a list of the forces in the free-body diagram as you calculate the reactions, with the rule

$$\text{FORCE} = \text{LOAD}$$

When the List Solver has completed processing, you must enter the reactions, $R1$ and $R2$, in the first and last elements of the force list.

At this point, you must remove the first set of rules from the Rule sheet, lest the TK!Solver program wipe out some critical lists in the second pass of the List Solver. You can do this in one of several ways. The first and most obvious way is to simply delete the rules and enter the new ones. The second way is to convert the rules to comments by adding double quotes with the Edit command. The third way is to add logical variables, such as PASS1, PASS2, etc. A fourth way is to disassociate the variables from the lists you wish to protect from the second pass.

A foolproof way to make two passes is to build and store the model with both sets of rules, but with the second set entered as comments in the Rule sheet, and to pay close attention to which variables are associated with lists.

After you modify the model for the second pass, you again invoke the List Solver and build the lists for the shear diagram and the moment diagram. You derive the shear diagram from the free body diagram and the moment diagram from the shear diagram. The shear diagram is a curve whose value at any point is the area under the curve of the forces in the free-body diagram, to the left of the same coordinate, and the moment diagram is a curve whose value at any point is the area under the shear diagram curve, to the left of the same coordinate.

You could not construct the shear and moment diagrams on the first pass because you need the reactions to construct the shear diagram. However, at any point, the moment diagram depends only on the area under the shear diagram, to the left the corresponding point, so you can construct the shear and moment diagrams simultaneously.

A word of caution is in order, though. The rules for the shear and moment diagrams both retrieve values previously calculated for their respective lists. The shear diagram rule adds a force to the previous element of the shear diagram list to calculate the current element. The moment diagram rule operates in the same way. To be safe always start with empty lists, in such a case.

After making the second pass with the List Solver, you have the lists necessary to plot the shear and moment diagrams. Figure 21–12 shows plots of the load diagram, the shear diagram, and the moment diagram all on the same plot. The shape of the moment curve is exaggerated because the units are inch-pounds; you could convert the curve to foot-pounds (a much more likely unit for a 10-foot long beam) by changing the display units, and defining the correct conversion factor.

# COLUMNS

Whenever you solve problems involving beams, you are certain to solve related problems involving columns. Columns are subjected to axial, compressive stresses but, before a column will fail from strictly compressive stress, it will bow because it is long compared to a characteristic transverse dimension. A column supporting a load will assume a shape similar to a sine wave when disturbed.

The average stress in a column is calculated with the same relation used to calculate tensile stress

$$\sigma = F/A$$

When you consider columns, you have to consider the peak local stress, as well as the average stress. The peak local stress is

$$\sigma_c = F/A + Mc/I$$

where $M$ is the moment associated with the bow of the column, $c$ is the distance from the neutral axis to the outermost fiber, and $I$ is the moment of inertia about the centroidal axis.

The *Euler load* is defined as the load at which an initially straight column becomes neutrally stable. Above the Euler load, the force, $F$, is independent of the deflection, $\delta$, up to the point where the local stress, $\sigma_c$, reaches the elastic limit. The equation for the Euler load is shown in the TK!Solver model of Figure 21–13, along with the rules for the average stress and the peak local stress.

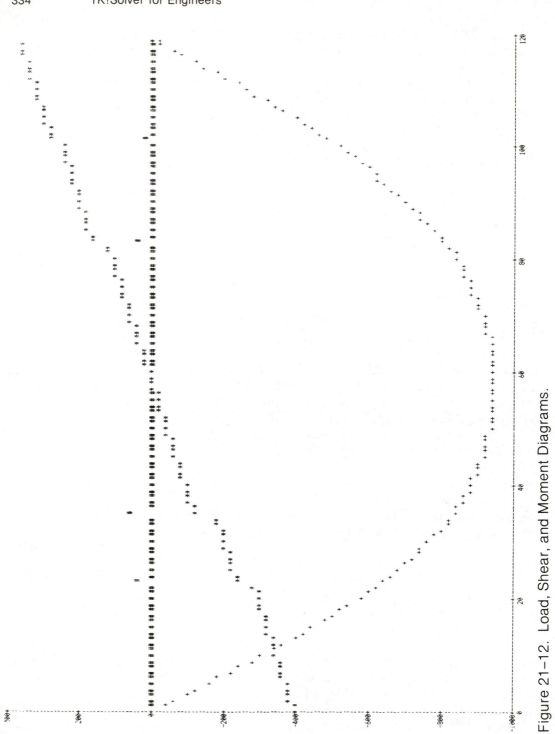

Figure 21–12. Load, Shear, and Moment Diagrams.

(s) Screen or Printer: Printer                                                182/!

```
=================== PLOT SHEET =================================================
Screen or Printer:          Printer
Title:                      LOAD, SHEAR, AND MOMENT DIAGRAMS
Display Scale ON:           Yes
X-Axis:                     XCOORD
Y-Axis      Character
------      ---------
LOAD        #
SHEAR       *
MOMENTD     +
```

(13u) Unit: LB                                                                182/!

```
=================== VARIABLE SHEET =============================================
St Input    Name      Output     Unit      Comment
-- -----    ----      ------     ----      -------
                                           BEAM ANALYSIS
                                           *************
            MOMENT               IN-LB     MOMENT ABOUT ORIGIN
            LOAD                 LB        LOADS
L  1        XCOORD               IN        DISTANCE FROM ORIGIN
            PREV      120                  PREVIOUS ELEMENT
            R2        -364.2083  LB        REACTION AT OPPOSITE END
            R1        -395.7917  LB        REACTION AT ORIGIN END
            TOTALLO   760        LB        TOTAL LOAD ON BEAM
L  1        FORCE                LB        FORCES IN FREE-BODY DIAGRAM
L           MOMENTD              FT-LB     MOMENT DIAGRAM
L           SHEAR                LB        SHEAR DIAGRAM
```

(8r) Rule: MOMENTD = ELEMENT('MOMENTD,PREV,0)+(XCOORD-ELEMENT('XCOORD,PREV 182/!

```
=================== RULE SHEET ================================================
S Rule
- ----
  PREV = ELEMENT()-1
  MOMENT = LOAD*(XCOORD+ELEMENT('XCOORD,PREV,0))/2
  R2 = -SUM('MOMENT)/MAX('XCOORD)
  R1 = -(SUM('LOAD)+R2)
  TOTALLOAD = SUM('LOAD)
  FORCE = LOAD
  SHEAR = FORCE + ELEMENT('SHEAR,PREV,0)
  MOMENTD = ELEMENT('MOMENTD,PREV,0)+(XCOORD-ELEMENT('XCOORD,PREV,0))*(SHEAR)
```

Figure 21–12. (Continued).

```
St Input     Name      Output    Unit     Comment
-- -----     ----      ------    ----     -------
                                          BEAM ANALYSIS
                                          *************
             MOMENT              IN-LB    MOMENT ABOUT ORIGIN
             LOAD                LB       LOADS
   L   1     XCOORD              IN       DISTANCE FROM ORIGIN
             PREV      120                PREVIOUS ELEMENT
             R2        -364.2083 LB       REACTION AT OPPOSITE END
             R1        -395.7917 LB       REACTION AT ORIGIN END
             TOTALLO   760       LB       TOTAL LOAD ON BEAM
   L   1     FORCE               LB       FORCES IN FREE-BODY DIAGRAM
   L         MOMENTD             FT-LB    MOMENT DIAGRAM
   L         SHEAR               LB       SHEAR DIAGRAM
```

```
S Rule
- ----
  PREV = ELEMENT()-1
  MOMENT = LOAD*(XCOORD+ELEMENT('XCOORD,PREV,0))/2
  R2 = -SUM('MOMENT)/MAX('XCOORD)
  R1 = -(SUM('LOAD)+R2)
  TOTALLOAD = SUM('LOAD)
  FORCE = LOAD
  SHEAR = FORCE + ELEMENT('SHEAR,PREV,0)
  MOMENTD = ELEMENT('MOMENTD,PREV,0)+(XCOORD-ELEMENT('XCOORD,PREV,0))*(SHEAR)
```

Figure 21-12. (Continued).

The slenderness ratio, $L/k$, determines whether the local stress in a column will reach the yield stress before the Euler load is reached. Allowable stress in a column can be plotted as a function of the slenderness ratio, for both the Euler load and short column action. The point at which the two curves intersect is the critical slenderness ratio, and location of this point is a function of the compressive yield stress. The critical slenderness ratio decreases as the compressive yield strength increases. The curve of allowable column stress vs. slenderness ratio is determined experimentally for a given type of column. You can generally use the Euler stress to calculate working stress ($\sigma E$/safety factor) if it is less than one-half the yield strength. You can enhance the TK!Solver model for column with the rules shown in Figure 21-14 so that it will calculate the working stress from the lower of the Euler stress and one-half the yield stress.

If all columns had frictionless, pinned ends, the models above would be adequate. However, frictionless, pinned ends are not so common. Instead of using the length of the column, you need to use the effective length, which you find with the formula

$$L' = L/\sqrt{C}$$

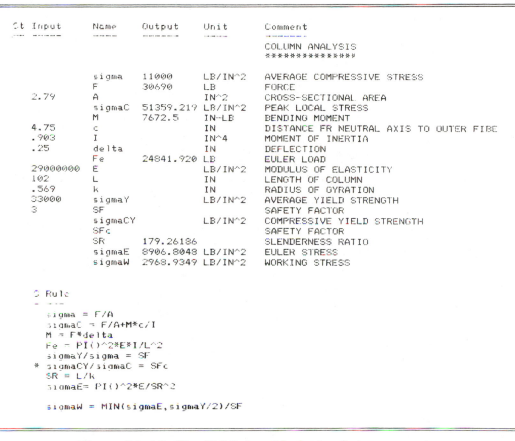

```
St Input       Name    Output     Unit      Comment
--- -------    ------  --------   -------    ---------

                                            COLUMN ANALYSIS
                                            ****************
               sigma   11000      LB/IN^2   AVERAGE COMPRESSIVE STRESS
               F       30690      LB        FORCE
    2.79       A                  IN^2      CROSS-SECTIONAL AREA
               sigmaC  51359.219  LB/IN^2   PEAK LOCAL STRESS
               M       7672.5     IN-LB     BENDING MOMENT
    4.75       c                  IN        DISTANCE FR NEUTRAL AXIS TO OUTER FIBE
    .903       I                  IN^4      MOMENT OF INERTIA
    .25        delta              IN        DEFLECTION
               Fe      24841.920  LB        EULER LOAD
    29000000   E                  LB/IN^2   MODULUS OF ELASTICITY
    102        L                  IN        LENGTH OF COLUMN
    .569       k                  IN        RADIUS OF GYRATION
    33000      sigmaY             LB/IN^2   AVERAGE YIELD STRENGTH
    3          SF                           SAFETY FACTOR
               sigmaCY            LB/IN^2   COMPRESSIVE YIELD STRENGTH
               SFc                          SAFETY FACTOR
               SR      179.26186            SLENDERNESS RATIO
               sigmaE  8906.8048  LB/IN^2   EULER STRESS
               sigmaW  2968.9349  LB/IN^2   WORKING STRESS


S Rule
- ----
    sigma = F/A
    sigmaC = F/A+M*c/I
    M = F*delta
    Fe = PI()^2*E*I/L^2
    sigmaY/sigma = SF
  * sigmaCY/sigmaC = SFc
    SR = L/k
    sigmaE= PI()^2*E/SR^2

    sigmaW = MIN(sigmaE,sigmaY/2)/SF
```

Figure 21–13. The TK!Solver Model for Columns.

where $C$ is the restraint coefficient. $C$ varies from 1.0 for pinned ends, to 4.0 for built-in ends. If you wish to make the models precise yet simple to use, you can add a user function to retrieve a value of $C$, given a symbolic value describing the end conditions. (Actually, the models are safe as presented; perfectly fixed ends are not achievable in practice, and values of $C$ greater than 2 should not be used.)

## BEAM FORMULAS

In general, the exact solution of beam problems involves the use of integrodifferential equations. The TK!Solver program, as has been noted several times, cannot perform symbolic differentiation of in-

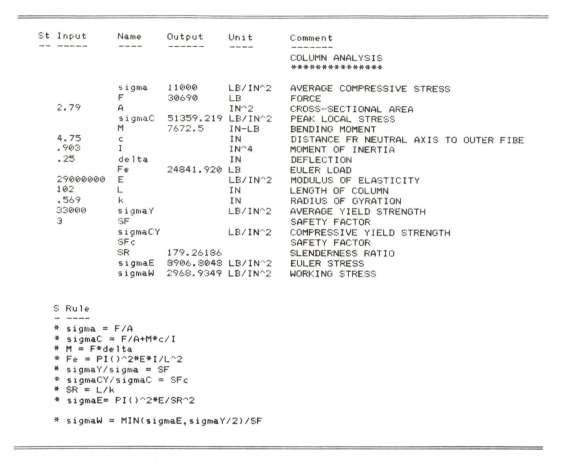

```
St Input       Name     Output      Unit      Comment
-- -----       ----     ------      ----      -------
                                              COLUMN ANALYSIS
                                              ***************
               sigma    11000       LB/IN^2   AVERAGE COMPRESSIVE STRESS
               F        30690       LB        FORCE
   2.79        A                    IN^2      CROSS-SECTIONAL AREA
               sigmaC   51359.219   LB/IN^2   PEAK LOCAL STRESS
               M        7672.5      IN-LB     BENDING MOMENT
   4.75        c                    IN        DISTANCE FR NEUTRAL AXIS TO OUTER FIBE
   .903        I                    IN^4      MOMENT OF INERTIA
   .25         delta               IN        DEFLECTION
               Fe       24841.920   LB        EULER LOAD
   29000000    E                    LB/IN^2   MODULUS OF ELASTICITY
   102         L                    IN        LENGTH OF COLUMN
   .569        k                    IN        RADIUS OF GYRATION
   33000       sigmaY               LB/IN^2   AVERAGE YIELD STRENGTH
   3           SF                             SAFETY FACTOR
               sigmaCY              LB/IN^2   COMPRESSIVE YIELD STRENGTH
               SFc                            SAFETY FACTOR
               SR       179.26186             SLENDERNESS RATIO
               sigmaE   8906.8048   LB/IN^2   EULER STRESS
               sigmaW   2968.9349   LB/IN^2   WORKING STRESS


S Rule
- ----
*  sigma = F/A
*  sigmaC = F/A+M*c/I
*  M = F*delta
*  Fe = PI()^2*E*I/L^2
*  sigmaY/sigma = SF
*  sigmaCY/sigmaC = SFc
*  SR = L/k
*  sigmaE= PI()^2*E/SR^2

*  sigmaW = MIN(sigmaE,sigmaY/2)/SF
```

Figure 21-14. TK!Solver Model for Columns, # 2.

tegration. However, you were no better off before the advent of the TK!Solver program, in that there is seldom enough time to solve each and every structural problem as you might in the classroom. Instead, you use standard formulas which have already been solved for a number of configurations.

Unfortunately, you cannot store formulas somewhere and retrieve them as you need them. However, you can enter all the standard formulas you normally use in a single model and use the techniques of Chapter 11 to select the appropriate formulas. Figure 21-15 is a partial model which uses the case-type construction. To select the set of rules that will be used, enter a 1 in the input of CASEx; any rule that contains CASEx will be solved and the others will not. Notice that if you enter a 1 in more than one CASE? variable, you may produce the Overdefined error condition.

Of course, you can also associate lists with the variables, and produce output lists from which you can generate moment diagrams and deflection curves.

```
(20i) Input:                                                     195/!

================== VARIABLE SHEET ================================
St Input      Name    Output    Unit    Comment
-- -----      ----    ------    ----    -------
                                        *************

     1        CASE1
              CASE2
              CASE3
              CASE4
              CASE5
              CASE6
              CASE7
              CASE8

              M       2000
   1000       F
     2        x
              w
              L
              y
              E
              I


(2r) Rule: CASE1* M = F*x *CASE1                                 195/!

================== RULE SHEET ====================================
S Rule
- ----
  "MOMENT RULES"
* CASE1* M = F*x *CASE1
* CASE2* M = .5*w*x^2 * CASE2
* CASE3* M = w*x^3/(6*L) *CASE3
* CASE4* M = -.5*F*x *CASE4
* CASE5* M = (w*x/2)*(x-L)
* CASE6* M = (-w*x/(6*L))*(L^2-x^2)
* CASE7* M = .5*F*((L/4)-x)
* CASE8* M = (w*L^2/2)*((1/6)-(x/L)+(x/L)^2)
  "DEFLECTION"
* CASE1* y = (F/(6*E*I))*(2*L^3-3*L^2*x+x^3) *CASE1
* CASE2* y = (w/(24*E*I))*(3*L^4-4*L^3*x+x^4)
* CASE3* y = (w/(120*E*I*L))*(4*L^5-5*L^4+x^5)
* CASE4* y = (F*x/(48*E*I))*(3*L^2-4*x^2)
* CASE5* y = (w*x/(24*E*I))*(L^3-2*L*x^2+x^3)
* CASE6* y = (w*x/(360*E*I*L))*(7*L^4-10*L^2*x^2+3*x^4)
* CASE7* y = (F*x^2/(48*E*I))*(3*L-4*x)
* CASE8* y = (w*x^2/(24*E*I))/(L-x)^2
```

Figure 21–15.  The Beam Formula Model.

```
St Input      Name     Output     Unit      Comment
-- -----      ----     ------     ----      -------
                                            BEAM FORMULAS
                                            *************

   1          CASE1
              CASE2
              CASE3
              CASE4
              CASE5
              CASE6
              CASE7
              CASE8

              M        2000
   1000       F
   2          x
              w
              L
              y
              E
              I


S Rule
- ----
   "MOMENT RULES
   "************
   CASE1* M = F*x *CASE1
*  CASE2* M = .5*w*x^2 * CASE2
*  CASE3* M = w*x^3/(6*L) *CASE3
*  CASE4* M = -.5*F*x *CASE4
*  CASE5* M = (w*x/2)*(x-L)
*  CASE6* M = (-w*x/(6*L))*(L^2-x^2)
*  CASE7* M = .5*F*((L/4)-x)
*  CASE8* M = (w*L^2/2)*((1/6)-(x/L)+(x/L)^2)

   "DEFLECTION
   "**********
*  CASE1* y = (F/(6*E*I))*(2*L^3-3*L^2*x+x^3) *CASE1
*  CASE2* y = (w/(24*E*I))*(3*L^4-4*L^3*x+x^4)
*  CASE3* y = (w/(120*E*I*L))*(4*L^5-5*L^4+x^5)
*  CASE4* y = (F*x/(48*E*I))*(3*L^2-4*x^2)
*  CASE5* y = (w*x/(24*E*I))*(L^3-2*L*x^2+x^3)
*  CASE6* y = (w*x/(360*E*I*L))*(7*L^4-10*L^2*x^2+3*x^4)
*  CASE7* y = (F*x^2/(48*E*I))*(3*L-4*x)
*  CASE8* y = (w*x^2/(24*E*I))/(L-x)^2
```

Figure 21–15. (Continued).

# CHAPTER 22

# Engineering Economics

If there is one factor that all fields of engineering have in common it is economics. Engineers are usually charged with the task of designing something that can be built at the lowest cost, that can be operated at the lowest cost, that will produce the best return on investments, etc. The words change, but the tune is always the same. The resources are always scarce.

Most engineering economics problems require that two or more alternatives be evaluated and compared. There may be two departments competing for limited funding for capital improvements projects. A client wants to know which of several systems has the best life-cycle cost. The manufacturing manager wants to know which of several designs for a product requires the lowest cost for tooling.

To complicate matters, the phrase, "other things being equal," never applies—things are never equal. Our first task in economics problems is usually to determine how the various alternatives can be compared. We need to establish a measure of "equivalence," a way to compare "apples to apples."

## CASH FLOW

Economics problems of interest to engineers are typically described in terms of cash flow, over a period of time. *Cash flows* consist of receipts and disbursements which occur at various times. A disbursement is made at the beginning of the time period to purchase a machine. Periodic disbursements are made for maintenance of the machine. Receipts occur periodically as the machine produces a product. At the end of the machine's useful life (or the period of the study), the machine can be sold for its salvage value. And, at regular intervals, the machine provides receipts in the form of depreciation taken.

This type of economics problem is illustrated with a cash flow diagram, as shown in Figure 22-1. The horizontal line is the time axis; it is marked off in equal intervals, corresponding to the number of periods in the life of the study. The most common interval, of course, is the year. The left end of the time axis is the origin, the beginning of the study, and the right end is the end of the study.

Disbursements are indicated by the arrows pointing down, and receipts by the arrows pointing up. We indicate two or more concurrent cash flows by superimposed arrows, with an arrowhead for each cash flow. We usually do not draw the arrows precisely to scale, but the lengths of the arrows should indicate the relative magnitude of the cash flows.

The left end of the time line is the beginning of the study, and an arrow, up or down, drawn at that point represents a cash flow at the beginning of the study. All other cash flows are assumed to occur at the end of a period.

## EQUIVALENCE OF CASH FLOWS

Alternatives typically differ in the amounts of cash flows, and the timing of cash flows, at the least. If you are to compare the amounts of two or more series of cash flows, you must find some measure of equivalence to account for the difference in timing. Certainly, $1000 income 5 years from now is not the same as $1000 income this week.

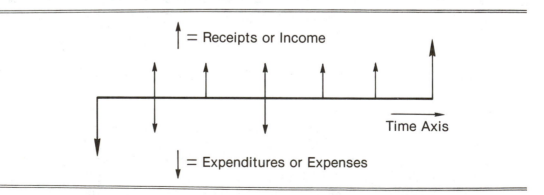

Figure 22-1. Cash Flow Diagram.

Alternative cash flows are usually compared on the basis of present worth; that is, you convert each future cash flow amount to an equivalent amount, as of the date of the beginning of the study. This concept can be illustrated with a simple example of a savings account.

If you deposit $1000 in a savings account drawing 5 percent interest (per year), it will be worth $1050 at the end of one year. The future worth of $1000, at 5 percent interest, at the end of one year, is $1050; the present worth of $1050 received one year from now, at 5 percent interest, is $1000. Thus the two amounts, $1000 received today and $1050 received one year from now, are equivalent.

The present worth of single future cash flows is calculated with the compound interest formula

$$P = F(1+i)^{-n}$$

*Where:* $P$ = present worth

$\phantom{Where:}$ $F$ = future worth, after $n$ periods

$\phantom{Where:}$ $i$ = interest rate per period

$\phantom{Where:}$ $n$ = number of periods

You can also calculate the future worth of cash flows expressed in current dollars with the reciprocal of the above formula

$$F = P(1+i)^n$$

Of course, not all cash flow problems consist of single cash flow amounts. You also have to consider the equivalence of series of equal annual amounts. For instance, you may need to know how large an amount to set aside each year to meet a future replacement cost—a sinking fund. You may need to compare two alternatives with unequal lives, by converting single cash flow amounts to annual amounts—a capital recovery factor problem.

# DISCOUNT FACTORS

Equivalence problems are typically solved with the aid of discount factors, which you can find in various handbooks and text books. The tables, of course, are produced on computers, using the formulas listed in Figure 22-2; the formulas are shown in TK!Solver form.

```
(5i) Input:                                                                    201/!
==================== VARIABLE SHEET ============================================
St Input       Name     Output   Unit     Comment
-- -----       ----     ------    ----     -------
               n                           NUMBER OF PERIODS
               FdivP                       SINGLE PAYMENT COMPOUND AMOUNT
               PdivF                       PRESENT WORTH
               AdivF                       UNIFORM SERIES SINKING FUND
               AdivP                       CAPITAL RECOVERY
               FdivA                       COMPOUND AMOUNT
               PdivA                       EQUAL SERIES COMPOUND AMOUNT
               PdivG                       UNIFORM GRADIENT
==================== RULE SHEET ================================================
S Rule
- ----
   FdivP = (1+i)^n
   PdivF = (1+i)^-n
   AdivF = (i/((1+i)^n-1))
   AdivP = (i*(1+i)^n/((1+i)^n-1))
   FdivA = ((1+i)^n-1)/i
   PdivA = ((1+i)^n-1)/(i*(1+i)^n)
   PdivG = ((1/i)-n/((1+i)^n-1))*((1+i)^n-1)/(i*(1+i)^n)
```

Figure 22-2.  Discount Factor Formulas.

With a little refinement, you could produce a complete set of discount factor tables, using the rules of Figure 22-2, but there is no need to do so. Instead, you can add four more rules, which will allow you to calculate not only the discount factors, but any three variables of interest rate, number of periods, present worth, future worth, annual amount, and gradient. The additional rules are shown in Figure 22-3.

The model of Figure 22-3 is adequate for cash flow problems in which the cash flow can be described in one of the categories listed—uniform annual amounts, single payments, gradients, or some combinations. Using the traditional method of looking up the factors in a table, you would attempt to structure the problem so that it could be solved by using one or more of the discount factors. But, real problems are not so neat; payments are not always uniform, perhaps not even uniform gradients.

With TK!Solver's List Solver, you can accommodate any schedule of payments, receipts or disbursements, and only use one rule, the first one mentioned, to convert each list element from future to present value, or vice versa. Then you add a rule to total the present worths, and one to calculate future worths, and you have a much more flexible tool. The model is shown in Figure 22-4.

With this model, you simply enter the cash flow diagram as the future worth list, enter the number of periods and the interest rate, and enter the List command. TK!Solver converts each future

```
(11r) Rule: "CONVERSION FORMULAS                                    200/!

==================== RULE SHEET =========================================
S Rule
- ----
  "DISCOUNT FACTORS
  "*****************
  FdivP = (1+i)^n
  PdivF = (1+i)^-n
  AdivF = (i/((1+i)^n-1))
  AdivP = (i*(1+i)^n/((1+i)^n-1))
  FdivA = ((1+i)^n-1)/i
  PdivA = ((1+i)^n-1)/(i*(1+i)^n)
  PdivG = ((1/i)-n/((1+i)^n-1))*((1+i)^n-1)/(i*(1+i)^n)

  "CONVERSION FORMULAS
  "*******************
  F = P*FdivP
  F = A*FdivA
  P = A*PdivA
  P = G*PdivG


(9c) Comment:                                                      200/!

==================== VARIABLE SHEET =====================================
St Input       Name    Output    Unit    Comment
-- -----       ----    ------     ----    -------
                                          DISCOUNT FACTORS
                                          ****************
   .1          i                          INTEREST RATE
               n        12                NUMBER OF PERIODS
   4000        F                          FUTURE WORTH
   1274.5233   P                          PRESENT WORTH
               A        187.05326         ANNUAL AMOUNT
               G        42.624457         GRADIENT

               FdivP    3.1384284         SINGLE PAYMENT COMPOUND AMOUNT
               PdivF    .31863082         PRESENT WORTH
               AdivF    .04676332         UNIFORM SERIES SINKING FUND
               AdivP    .14676332         CAPITAL RECOVERY
               FdivA    21.384284         COMPOUND AMOUNT
               PdivA    6.8136918         EQUAL SERIES COMPOUND AMOUNT
               PdivG    29.901220         UNIFORM GRADIENT
```

Figure 22–3. Discount Factors and Conversion Formulas.

worth value to its present worth and sums the present worth list on each pass. The value of the present worth total on the last pass is the value you are interested in. On the other hand, if you wish to evaluate the present worth of the cash flow series at any point in the series, you can associate a list with TOTAL__PW and maintain a running total of the value.

Of course, if you want to compare several alternatives, you can copy the rules with different variable names:

$$FW1 = PW1*(1 + i)^n$$

```
(4i) Input:                                                          201/!

==================== VARIABLE SHEET ===================================
St Input      Name      Output    Unit      Comment
-- -----      ----      ------    ----      -------
                                            PRESENT WORTH MODEL
                                            ******************
L   1         FW                            FUTURE WORTH
L             PW                            PRESENT WORTH
    .12       i                             INTEREST RATE, DECIMAL FRACTION
L   1         n                             NUMBER OF PERIODS
              I         12                  INTEREST RATE, PERCENT
              TOTAL_P   -1361.250           TOTAL PRESENT WORTH
L             ACCUM_F                       CUMULATIVE FUTURE WORTH


(1r) Rule: FW = PW*(1+i)^n                                           201/!

==================== RULE SHEET ===================================
S Rule
- ----
   FW = PW*(1+i)^n
   i = I/100

   TOTAL_PW = SUM('PW)
   ACCUM_FW = FW+ELEMENT('FW,ELEMENT()-1,0)
```

Figure 22–4. List-Oriented Present Worth Model.

$$FW2 = PW2*(1+i)\hat{}n$$
$$FW3 = PW3*(1+i)\hat{}n$$
•
•
•

And, you can present the alternatives as a table or a plot.

# DEPRECIATION AND TAXES

Depreciation is a method of allocating the cost of an asset over its useful life, or a major portion of it. Expenditures that have life of less than a year—supplies, small tools, direct material, etc.—are treated as expenses and are deducted from income for the purpose of calculating taxes. Major expenditures, for items which will contribute to income for several years, cannot be deducted in the year of purchase. Instead, the expenditure must be spread out over a period of several years.

Engineering economics problems involving depreciation must also involve income taxes. Analyses which do not account for taxes and depreciation may produce inaccurate conclusions. A project that is analyzed without consideration of income taxes may appear attractive, when in reality, it is not. The general principle is that both deductible expenses and revenues are reduced by the tax rate; if income taxes are $t$ percent, both deductible expenses and revenues are multiplied by the factor $(1 - t/100)$. Salvage values and initial purchase costs, on the other hand, are unaffected by the income tax rate.

Depreciation produces an increase in present worth, in the amount of $t$ percent of the depreciation allowance for the year. The depreciation allowance is deducted as an expense, but of course there is no actual disbursement.

Depreciation is normally calculated by one of four methods. The simplest method is the *straight line method*, which produces equal amounts of depreciation each year of the asset's life. You divide the quantity, cost-salvage value, by the number of years over which the asset is depreciated, to find the annual amount.

The second method is the *double-declining balance method*. This method does not account for salvage value. The annual amount is a fraction of the quantity, cost-accumulated depreciation. Each year the amount is less than that of the previous year; the amount would never drop to zero, regardless of the length of the time over which the asset is depreciated.

The third method is the *sum-of-the-years-digits method*. This method produces a high amount in the first year and smaller amounts in succeeding years. This method does account for salvage value and the book value is reduced to the salvage value at the end of the last year of the depreciation life.

The fourth method is the *sinking fund method*, which is not popular because it produces low amounts in the early part of the asset's life. All four methods are illustrated in the model shown in Figure 22-5.

```
(1i) Input:                                                    198/!

==================== VARIABLE SHEET ===================================
St  Input       Name     Output    Unit    Comment
--- -----       ----     ------    ----    -------
                                           DEPRECIATION
                                           ************
    10000       COST
    1000        SALVAGE
L   1           N
                                           ANNUAL DEPRECIATION AMOUNTS
L               Dsl      600               STRAIGHT LINE
L               Dddb                       DOUBLE DECLINING BALANCE
L               Daccum
L               Dsoyd                      SUM OF YEARS DIGITS
L               Dsf                        SINKING FUND
    .1          i                          INTEREST RATE FOR SINKING FUND DEPRECI
```

```
(5r) Rule: Dsf = (COST-SALVAGE)*i/((1+i)^MAX('N)-1)            198/!

==================== RULE SHEET ======================================
S Rule
- ----
  Dsl = (COST - SALVAGE)/MAX('N)
  Dddb = 2*(COST - ELEMENT('Daccum,ELEMENT()-1,0))/MAX('N)
  Daccum = ELEMENT('Daccum,ELEMENT()-1,0)+Dddb
  Dsoyd = (COST - SALVAGE)*(MAX('N)-ELEMENT()+1)/(MAX('N)*(MAX('N)+1)/2)
  Dsf = (COST-SALVAGE)*i/((1+i)^MAX('N)-1)
```

DEPRECIATION COMPARISON

| N | Dddb | Daccum | Dsoyd | Dsl | Dsf |
|---|------|--------|-------|-----|-----|
| 1 | 1333.33333 | 1333.33333 | 1125 | 600 | 283.263992 |
| 2 | 1155.55556 | 2488.88889 | 1050 | 600 | 283.263992 |
| 3 | 1001.48148 | 3490.37037 | 975 | 600 | 283.263992 |
| 4 | 867.950617 | 4358.32099 | 900 | 600 | 283.263992 |
| 5 | 752.223868 | 5110.54486 | 825 | 600 | 283.263992 |
| 6 | 651.927353 | 5762.47221 | 750 | 600 | 283.263992 |
| 7 | 565.003706 | 6327.47591 | 675 | 600 | 283.263992 |
| 8 | 489.669878 | 6817.14579 | 600 | 600 | 283.263992 |
| 9 | 424.380561 | 7241.52635 | 525 | 600 | 283.263992 |
| 10 | 367.796486 | 7609.32284 | 450 | 600 | 283.263992 |
| 11 | 318.756955 | 7928.07979 | 375 | 600 | 283.263992 |
| 12 | 276.256028 | 8204.33582 | 300 | 600 | 283.263992 |
| 13 | 239.421890 | 8443.75771 | 225 | 600 | 283.263992 |
| 14 | 207.498972 | 8651.25668 | 150 | 600 | 283.263992 |
| 15 | 179.832442 | 8831.08913 | 75 | 600 | 283.263992 |

Figure 22-5. Depreciation Model.

# APPENDICES

# TK!Solver for MS-DOS™ Computers

The text of this book describes the IBM PC version of the TK!Solver program, the first version of the program to become available. As mentioned in the text, the IBM PC version of the program is not portable to other computers—even those that run the MS-DOS operating system from which PC-DOS is derived—and is copy protected. However, the TK!Solver program is available for other computers that run the MS-DOS operating system.

The MS-DOS version of the TK!Solver program can be configured by the user to run on any of the following computers, using predefined configuration files on the distribution diskette:

Canon AS-100

Eagle 1600™ [1]

Grid Compass

Texas Instruments/Professional Computer

Toshiba T300

Zenith Z-100

The TK!Solver program is loaded with the command TK, as with the IBM PC version described in the text. However, the MS-DOS version signs on with the following message:

B:tk

This program runs on the following computers:

CANON
EAGLE

[1] Eagle 1600 is a trademark of Eagle Computer, Inc.

GRID
TI
TOSHIBA
ZENITH

If your computer is listed here, type the name in the field
below and press Return. If your computer is not listed
here, press Return.

Your computer's name (as shown above): \

**If your computer is listed, typing its name, and then pressing
ENTER will allow the TK!Solver to load and operate correctly.**
**Pressing ENTER without entering one of the names listed
produces the following message:**

\* \* \* \* \* \* \* \* \* \* \* \* \* \* \* \* \* \* \* \* \* \* \* \* \* \* \* \* \* \* \* \* \* \* \* \* \* \* \* \* \* \* \* \* \* \* \* \* \* \* \* \* \* \* \* \* \* \*

If your computer is not listed, the TK!Solver program diskette
does not contain a Configuration File for your computer. To run
the program, you must create your own Configuration File. To do
this you need:

1) a word processor or text editor
2) the CONFIG.SAI file, which is on the TK!Solver diskette
3) a technical reference manual for your computer

Copy the file CONFIG.SAI from the TK!Solver program diskette
to a diskette or hard disk and name it SOFTARTS.CFG. Use the
word processor or text editor to read the SOFTARTS.CFG file.
Follow the directions contained in the file to modify it for
your computer using the information in your technical reference
manual.

After you have modified the file, backup the TK!Solver program,
including the file SOFTARTS.CFG. You can then load the program
using the backup copy by typing TK and pressing Return. For
information about backing up and loading the TK!Solver program,
see Chapter 3 of the Instruction Manual.

Press Return to go back to DOS.

\* \* \* \* \* \* \* \* \* \* \* \* \* \* \* \* \* \* \* \* \* \* \* \* \* \* \* \* \* \* \* \* \* \* \* \* \* \* \* \* \* \* \* \* \* \* \* \* \* \* \* \* \* \* \* \* \* \*

**The MS-DOS version of the TK!Solver program differs from**

the IBM PC version in that it operates entirely through the operating system, whereas the IBM PC version makes some direct calls to the IBM ROM. Because the MS DOS version of the program operates strictly through the operating system, the program is portable to practically any running the MS DOS operating system.

The program can be configured to run on any MS-DOS computer using the procedure described in the disk file CONFIG.SAI, included on the TK!Solver diskette. The instructions in this file describe the procedure used to create a configuration file called SOFTARTS.CFG, which TK!Solver uses to communicate with the terminal of the MS-DOS computer. The program diskette contains a configuration file for each of the computers listed above.

The configuration file contains the character sequences TK!Solver uses to position the cursor on the screen, to erase all or part of the screen, and to perform other tasks required to provide the display of the various TK!Solver sheets on the screen. A typical configuration file appears below:

```
* This is a terminal configuration file for the Zenith Z-100

* Initialization...
init = / [ > /
* Termination...
terminate = ⌐[y5/

*Clearing this and that...
clear = ⌐[E/
cleol = ⌐[K/

* Screen Dimensions...
cols = 80
rows = 24

* Cursor Positioning...
cursor__ascii = FALSE
cursor__row__then__col = true
cursor__begin = /#1bY/
cursor__col__bias = 32
cursor__row__bias = 32

* Hiding the cursor...
cursor__disable = ⌐[x5/
cursor__enable = ⌐[y5/
```

* Highlighting...
high__on = /[p/
high__off = /[q/

* Keyboard Mappings...
key__abort = /C/
key__delete = /H/
key__delete = /#7f/
key__down = /[B/
key__enter = /#0d/
key__help = /?/
key__help = /[~/
key__left = /[D/
key__replot = /L/
key__right = /[C/
key__abort = /[|/
key__up = /[A/

When the TK!Solver program is loaded, it looks for a file named SOFTARTS.CFG. If it cannot find that file, it lists the names of the configuration files on the disk, and requests the user to enter the name of one of the files. If no name is entered, the program displays the message shown above, and returns to the operating system.

If your computer is not one of those for which a configuration file is provided, you have no choice but to create the SOFT-ARTS.CFG file as described in the CONFIG.SAI file. If your computer is one for which a configuration file is provided, you can speed the process of loading the TK!Solver program by copying the proper configuration file to the SOFTARTS.CFG file with the command

COPY name.CFG SOFTARTS.CFG

This command leaves the original configuration file unchanged, and creates a copy with the name SOFTARTS.CFG so that TK!Solver will automatically start up without asking for the name of your computer.

# THE CONFIG.SAI DISK FILE

```
*           TK!Solver Configuration File Template

*This file shows you how to create a TK!Solver Configuration
*File for your computer. You need to create a Configuration
*File only if there is no Configuration File named for your
*computer on the TK!Solver program diskette.

*Comment lines begin with an asterisk (*). Configuration
*fields begin with a keyword followed by an equals sign ( = ).
*Read the comments, then fill in the space after the  =  with the
*correct value for your computer, using either upper case or
*lower case characters. There are three types of values:

*Boolean Values—TRUE, FALSE
*Integer Values—any integer from 0 to 255
*Escape Sequences—any string of ASCII characters, hexadecimal
* values, or control characters. Hexademicals are preceded
* by a pound sign ( # ) and must be two digits in the range
* of 00 to FF. Control characters are preceded by a caret (^).
* Use one other character to define both the beginning and
* end of a sequence. For example, type one of these characters:
* / " ; \ before the sequence and use the same character
* at the end of the sequence.

*To determine the values for your computer's configuration, see
*the technical reference manual for your computer or contact
*your dealer. Before making any entries, copy this file, called
*CONFIG.SAI, to another diskette or to a hard disk and name it
*SOFTARTS.CFG. Make your entries in the SOFTARTS.CFG file. When
*you are finished, back up the TK!Solver program, including the
*SOFTARTS.CFG file, as described in Chapter 3 of the
*Instruction Manual. Then load the program using the backup
*copy by typing TK and pressing Enter.

*If any field below does not apply to your computer, change it
*to a comment line by typing an asterisk (*) before the keyword.

*           Terminal Configuration
*Sequence that sets up the characteristics of the screen and
*keyboard.
INIT =
```

*If you entered a setup sequence above, enter the termination
*sequence that restores the screen and keyboard to their
*original state.
TERMINATE =

*Sequence that clears the entire screen.
CLEAR =

*Sequence that clears a space on the screen from the current
*cursor position to the end of the same line.
*CLEOL =

*ROWS defines the number of horizontal lines of text on the
*screen. COLS defines the number of characters allowed per
*line. These keywords accept integer values only.
COLS =
ROWS =

*Sequence to cause a beep or ring a bell. HEX 07 is the default
*escape sequence for this keyword.
BELL =

*                    Cursor Positioning

*Boolean value of TRUE if the row and column numbers are
*represented in ASCII characters. If they are represented as
*integer byte values, enter FALSE.
CURSOR_ASCII =

*Boolean value of TRUE if the cursor position is listed
*by row and column. If listed by column and row, enter FALSE.
CURSOR_ROW_THEN_COL =

*Sequence to indentify the first character of an escape
*sequence that describes a cursor position.
CURSOR_BEGIN =

*The character inserted between the row and column values in
*the escape sequence that describes a cursor position.
CURSOR_MIDDLE =

*The last character in the escape sequence that describes
*a cursor position.
CURSOR_END =

*Integer values for the row and column of the upper left hand
*corner of the screen. The default is (0,0).
CURSOR__COL__BIAS =
CURSOR__ROW__BIAS =

*Sequence to make the cursor disappear from the screen (disable)
*and reappear (enable).
CURSOR__DISABLE =
CURSOR__ENABLE =

*               Highlighting
*To highlight characters on the screen, computers use either a
*high-order bit technique or an escape character sequence.
*Boolean value of TRUE below if high-bit technique used.
HIGH__BIT =

*Sequences if character sequences used.
HIGH__ON =
HIGH__OFF =

*If you have no highlighting attribute on your computer, you
*can surround the area to be emphasized with characters of your
*choice. Sequence for these characters:
WIDE__CURSOR__LEFT =
WIDE__CURSOR__RIGHT =

*               Keyboard Mappings
*The fields below define the keys on your keyboard to use
*with the TK!Solver program. You may define any key or
*series of keys on your computer in the fields below.

*Sequences for the up, down, right, and left arrow
*keys. (called the ARROW keys in the TK!Solver manual)
KEY__UP =
KEY__DOWN =
KEY__RIGHT =
KEY__LEFT =

*Sequence for the key that cancels or aborts the current operation.
*(called the BREAK key)
KEY__ABORT =

*Sequence for the key that deletes the character to
*the left of the cue. (called the BACKSPACE key)

KEY__DELETE =

*Sequence for the key that enters information (called the ENTER
*key).
KEY__ENTER =

*Sequence for the key to replot or refresh the screen. (optional)
KEY__REPLOT =

*Sequence for the key to call the Help Facility. (optional)
KEY__HELP =


The TK!Solver documentation describes commands in terms of TK!Solver keys or commands, which are available on all the computers listed above, although the keys may differ from the examples given in the text of this book. The actual key assignments of these computers are listed below.


## CANON AS–100

The following list shows how the TK!Solver names match the Canon AS-100 personal computer keyboard.

| TK!Solver name | Canon keys |
|---|---|
| Arrow keys | Press the CURSOR LOCK key, then use up, down, left, or right arrows |
| Backspace key | DEL |
| Break key | CANCEL |
| Enter key | RETURN |
| Edit command | Hold down CTRL and type E |

To load the TK!Solver program, boot MS-DOS and respond to any MS-DOS prompts. After the system prompt (A>), type AN-SISEQ1 ON and press the **ENTER** key. When this program is finished, replace the MS-DOS diskette with the TK!Solver program diskette. Change the default drive name from A to E. Then type TK and press **ENTER**. If you copy the program to a Canon diskette, load it from drive A.

## EAGLE 1600™

The following list shows how the TK!Solver names match the Eagle 1600 computer keyboard.

| TK!Solver name | Eagle keys |
|---|---|
| Arrow keys | up, down, right, left (BACKSPACE) arrows |
| Backspace key | DELETE |
| Break key | Hold down CONTROL and type C |
| Enter key | RETURN |
| Edit command | Hold down CONTROL and type E |

To load the TK!Solver program, boot MS-DOS and respond to any MS-DOS prompts. When the system prompt (A:) is displayed on the screen, replace the MS-DOS diskette with the TK!Solver program diskette. If you have two drives, you must also insert a blank diskette into drive B. Type TK and press the **ENTER** key. Note: To refresh a TK!Solver screen, hold down **CONTROL** and type L. Before printing, use the ASSIGN command. See the Eagle User's Guide.

## GRID COMPASS

The following list shows how the TK!Solver names match the Grid Compass keyboard.

| TK!Solver name | Grid Compass keys |
|---|---|
| Arrow keys | up, down, left, right arrows |
| Backspace key | BACKSPACE |
| Break key | Hold down CTRL and type C |
| Enter key | RETURN |
| Edit command | Hold down CTRL and type E |

To load the TK!Solver program, boot MS-DOS from the MS-DOS diskette. Respond to any MS-DOS prompts. When the system prompt (A>) is displayed on the screen, replace the MS-DOS diskette with the TK!Solver program diskette. Then type TK and press the **ENTER** key. **Note:** To refresh a TK!Solver screen, hold down **CTRL** and type L.

## TEXAS INSTRUMENTS PROFESSIONAL COMPUTER

The following list shows how the TK!Solver names match the TI Professional keyboard.

| TK!Solver name | TI Professional keys |
| --- | --- |
| Arrow keys | up, down, left, right triangular arrows |
| Backspace key | BACKSPACE |
| Break key | Hold down CTRL and type C |
| Enter key | RETURN |
| Edit command | Hold down CTRL and type E |

To load the TK!Solver program, boot MS-DOS and respond to any MS-DOS prompts. When the system prompt (A:) is displayed on the screen, replace the MS-DOS diskette with the TK!Solver program diskette. Then type TK and press the **ENTER** key. **Note:** To refresh a TK!Solver screen, hold down **CTRL** and type L.

## TOSHIBA T300

The following list shows how the TK!Solver names match the Toshiba T300 computer keyboard.

| TK!Solver name | Toshiba keys |
| --- | --- |
| Arrow keys | up, down, right, left arrows |
| Backspace key | BACKSPACE (large left arrow key) |
| Break key | BREAK |
| Enter key | ENTER (right angle arrow key) |
| Edit command | Hold down CTRL and type E |
| Help key | HELP |

To load the TK!Solver program, boot MS-DOS and respond to any MS-DOS prompts. When the system prompt (A>) is idsplayed on the screen, replace the MS-DOS diskette with the TK!Solver program diskette. Type TK and press the Enter key.

## ZENITH  Z–100/HEATHKIT  H–100

The following list shows how the TK!Solver names match the Zenith personal computer keyboard.

| TK!Solver name | Zenith keys |
|---|---|
| Arrow keys | up, down, left, right arrows |
| Backspace key | BACKSPACE |
| Break key | Hold down CTRL and type C |
| Enter key | RETURN |
| Edit command | Hold down CTRL and type E |
| Help key | HELP |

To load the TK!Solver program, boot MS-DOS and respond to any MS-DOS prompts. When the system prompt (A:) is displayed on the screen, replace the MS-DOS diskette with the TK!Solver program diskette. Type TK and press the **ENTER** key. **Note:** To refresh a TK!Solver screen, hold down **CTRL** and type L.

# DIF™ Technical Specification

## 1. INTRODUCTION

This document is the technical specification of DIF, a format for the exchange of data, developed by Software Arts Products Corp. It is a reference document and not a tutorial. It includes a description of the DIF file organization and structure, required items, and optional standard items. It also explains the use of the optional standard items by specific applications. The last section is an example of a DIF data file.

Programs should use defined standard items when possible. The DIF Clearinghouse will update this document to describe new items as they are defined and record their use in specific programs. Programmers developing new software that incorporates new optional items should inform the clearinghouse fully about them so that they can be standardized for common use by any program supporting the DIF format.

Programmers should remember that the program reading the data can be extremely simple. The program writing the data must handle it in such a way that it can be read by any program supporting DIF.

Within this text, upper case characters are actual values to be entered as shown and lower case characters name the value to be entered to a field. It is assumed that the ASCII character set is being used. See the section on Definitions for a discussion of character sets.

## 2. CONSTRAINTS OF THE FORMAT

The DIF format was designed for ease of use, and, for the sake of simplicity, certain constraints have been imposed on the format. Because DIF is not intended to be a universal representation for all data, one of these constraints is the representation of data in tables with rows of equal length and columns of equal length. A second constraint is that, because many users program in BASIC, the files must be compatible with BASIC programs. Programs written in another language, such as Pascal, can use a set of subroutines to read and write DIF files.

Below is a list of specific constraints on a DIF file.

1. Because some BASICs have only primitive facilities for reading and writing strings, the convention of keeping numbers and strings on separate lines has been adopted.

2. Two items, VECTORS and TUPLES, are required to support systems that require preallocation of space.

3. Because some systems do not allow programs to test for the end of a file, a special data value, EOD, provides graceful termination to a program.

4. To simplify programming, there are only two formats within the file, and all fields are predefined as character strings or numbers.

5. Strings must be enclosed in quotes if they contain characters other than alphanumerics.

6. The character set is restricted to the printable ASCII characters.

7. Although DIF places no explicit restriction on the length of data strings, some systems may impose restrictions.

Since the DIF format is not meant to meet all the needs for data representation, it may be necessary to use multiple DIF files or additional formats for some applications. A word processor, for example, would not use a DIF file to store text but could use DIF files for tables of values within a report.

===================== # 3. ORGANIZATION OF THE DIF DATA FILE

A DIF file is a text file using the standard printable character set of the host machine. The model for the data is a table. Fields are called vectors; records are called tuples. Data is organized into vectors of equal length. Each tuple consists of a row of corresponding values read across each vector. The user determines the specific groupings of vectors and tuples. Often vectors are treated as columns and tuples are treated as rows, but because DIF can transpose columns and rows, the terms *vectors* and *tuples* are used instead of the terms *columns* and *rows*.

The DIF file consists of two sections, a header section and a data section. The header section contains descriptions of the file and the data section contains the actual values.

===================== # 4. THE HEADER SECTION

The header section is composed of header items. There are four standard required header items and several standard optional header items.

## 4.1 The Header Item

The header items describe the data organization. Each header item consists of four fields arranged on three lines, as illustrated below. The first line is a token[1], the second line consists of two numbers, and the third line contains a string:

```
Topic
Vector Number, Numeric Value
"String Value"
```

### 4.1.1 The Topic

The first line of the header item is the Topic. It identifies the header item, and must be a token.

---

[1] A token is an upper case string of alphanumeric characters. It is usually short, 32 characters or less. See the Definitions section for more information.

### 4.1.2  The Vector Number

The first field on the second line is the Vector Number. If the header item describes a specific vector, the Vector Number specifies the vector being described. If the header item describes the entire file and not one specific vector, the Vector Number is zero (0).

### 4.1.3  The Numeric Value

The Value is an integer and occupies the second field of the second line, separated by a comma from the Vector Number. If the header item does not use a numeric value, the Value is zero (0).

### 4.1.4  The String Value

The String Value occupies the third line of the header item. The String Value is always enclosed in quotation marks. If it is not used, the line consists of a null string, a pair of quotation marks with no space between them.

## 4.2  Header Items

There are four required header items. The other header items described in this document are standard optional header items. The defined standard items should be used by new programs using DIF. If it is absolutely necessary, a new header item may be defined to meet the needs of a particular program. For details, see the section on Defining New Header Items.

A program may ignore all header items until it finds the header item DATA, described below.

The following four header items are required:

### 4.2.1  The First Header Item

```
          TABLE
          0,version
          "title"
```

The header item TABLE must be the first entry in the file. It identifies the file as a DIF file. The version number must be 1. The "title" is the title of the table and describes the data.

## 4.2.2 Vector Count

```
VECTORS
0,count
" "
```

The header item VECTORS specifies the number of vectors in the file.

**Note:** This header item must appear before header items that refer to vector numbers. Otherwise, it can appear anywhere within the header section.

## 4.2.3 Tuple Count

```
TUPLES
0,count
" "
```

The header item TUPLES specifies the length of each vector (the number of tuples). This can be used by a program to preallocate storage space for the data. This item may appear anywhere within the header section.

**Note:** Programs reading the data assume that the tuple count is correct. Some programs may be able to generate this information only after all the data has been generated. These programs must reread the DIF file to count the tuples, and rewrite the TUPLES item with the correct count.

## 4.2.4 The Last Header Item

```
DATA
0,0
" "
```

The header item DATA must be the last header item. It tells the program that all remaining data in the file are data values.

The following header items are optional. The programs that are known to use them are noted with the item. For detailed information on each program's specific use of the item, see the section below on Applications Programs.

### 4.2.5 Vector Label

LABEL
vector#,line#
''label''

The header item LABEL provides a label for the specified vector. The line number provides an option for labels that span more than one line, and can be ignored by a system that allows single line labels only. The values 0 and 1 are equivalent line numbers.
    **Note:** Some programs do not use the LABEL field. If the first vector in a tuple contains string values, the first data value in the tuple may be treated as a label.
    Used by the VisiPlot™ and VisiTrend/VisiPlot™ programs.

### 4.2.6 Vector Comment

COMMENT
vector#,line#
''comment''

The header item COMMENT is similar to LABEL. It provides an option to systems that allow an expanded description of a vector in addition to a label.
    Used by the VisiPlot and VisiTrend/VisiPlot programs.

### 4.2.7 Field Size

SIZE
vector,bytes
'' ''

The header item SIZE provides to programs such as data base systems the option to allocate fixed size fields for each value. Because SIZE is an optional item, programs using SIZE must be able to read files produced by programs unable to generate SIZE information.
    Used by the CCA/DMS program.

## 4.2.8  Periodicity

PERIODICITY
vector#,period
" "

The header item PERIODICITY provides the option of specifying a period in a time series.
 Used by the VisiPlot and VisiTrend/VisiPlot programs.

## 4.2.9  Major Start

MAJORSTART
vector#,start
" "

The header item MAJORSTART specifies the first year of a time series.
 Used by the VisiPlot and VisiTrend/VisiPlot programs.

## 4.2.10  Minor Start

MINORSTART
vector#,start
" "

The header item MINORSTART specifies the first period of a time series. Used by the VisiPlot and Visitrend/VisiPlot programs.

## 4.2.11  True Length

TRUELENGTH
vector#,length
" "

The header item TRUELENGTH specifies the portion of a vector that contains significant values.
 Used by the VisiPlot and VisiTrend/VisiPlot programs.

### 4.2.12 Units

> UNITS
> vector#,0
> "name"

The header item UNITS specifies the unit of measure for the values in the given vector. *Name* is the unit, for example *meters* or *ft*.
Used by the TK!Solver(tm) program.

### 4.2.13 Display Units

> DISPLAYUNITS
> vector#,0
> "Name"

The header item DISPLAYUNITS specifies the unit in which the values in the given vector should be displayed. This unit may be different from the one in the UNITS field. The values in the given vector are always stored in the unit specified in the UNITS field, and the application program is responsible for making the value conversion between the UNITS and DISPLAYUNITS.

For example, a vector might be stored in *km*, but displayed in the program in *miles*. The UNITS field would be *km*, the DISPLAYUNITS field would be *miles*, and the values in the vector would be in *km*. Any program using the vector would have to define the conversion between km and miles to display the values in *miles*.

Used in the TK!Solver program.

## 4.3  Defining New Header Items

If there is no standard optional header item to fulfill the specific need of a subsystem, a new header item may be defined. Because the DIF format is intended for common use, new optional header items should be standardized through the DIF Clearinghouse. They will then be added to this document.

To be accepted as standard items, new optional items must be consistent with existing conventions.

An optional item extends the format for a specific application. Any program reading the DIF file should be able to operate without optional items. If a reading program requires the information provided by an optional item, it should prompt the user to supply the missing information and not require the item iself.

# 5. THE DATA SECTION

The data section consists of a series of tuples. The Data Values within the tuples are organized in vector sequence.

Each Data Value represents one element of data in the file. The data may be either the actual data or one of the two Special Data Values that mark the beginning of a tuple (BOT) and the end of data (EOD) in the file.

Each Data Value consists of two lines. The first line consists of two fields containing numeric values, and the second line consists of one field containing a string value. The format is:

Type Indicator, Number Value
String Value

## 5.1 The Type Indicator Field

The Type Indicator is an integer that tells the program what kind of data is represented by this value. There are currently three possible values.

−1  The data is a Special Data Value, indicating either the beginning of a tuple or the end of data. The Number Value is zero (0) and the String Value is either BOT or EOD. See the description below of Special Data Values.

 0  The data is numeric. The Number Value field contains the actual value and the String Value field contains a Value Indicator. See the descriptions below of the Number Value and String Value fields.

 1  The data is a string value. The Number Value is zero (0) and the String Value field contains the actual string value.

## 5.2 The Number Value Field

When the Type Indicator is 0, the Number Value field contains the actual value. The value must be a decimal (base 10) number. It may be preceded by a sign (+ or −) and it may have a decimal point. It may be preceded or followed by one or more blanks. If the data value contains an exponent of a power of ten, the value is followed by the letter E and the signed or unsigned exponent power of ten.

**Note:** This is the only place where DIF allows a non-integer value. Some programs accept only integer values.

## 5.3  The String Value Field

The contents of the String Value field are dependent on the Type Indicator.

### 5.3.1  Special Data Value

If the Type Indicator is −1, the String Value is one of the two Special Data Values, BOT or EOD, and the Number Value is 0.

Each tuple begins with the Special Data Value BOT (Beginning of Tuple). If a program cannot generate a VECTORS header item before generating all data, it can use the Special Data Value BOT to determine the number of vectors in the file by counting the number of Data Values between BOTs when it rereads the file. A program can also verify its position in a file by using the BOT Special Data Value.

The Special Data Value EOD (End of Data) indicates the end in the file. The EOD occurs at the end of the last tuple in the file. If the program is unable to generate a TUPLES header item before generating all data, it can determine the number of tuples by counting the number of BOTs before the EOD when it rereads the file. A program can also use the EOD Special Data Value to detect the end of the file.

### 5.3.2  Numeric Value and Value Indicator

If the Type Indicator is 0, the data is numeric, and the String Value is one of the Value Indicators described below. The Value Indicator overrides the value.

A subsystem may define Value Indicators for its own needs. New Value Indicators should be registered with the DIF Clearinghouse.

The Value Indicators currently defined are:

V   Value. This is the String Value most commonly used with a numeric value. The Number Value contains the actual value.

NA   Not Available. The value is marked as not available. The Number Value is 0.

ERROR   The value represents the result of an invalid calculation. The Number Value is 0.

TRUE   Logical value. The Number Value is 1.

FALSE   Logical value. The Number Value is 0.

The String Value can be ignored in favor of the Number Value, or all values with a Value Indicator other than V can be considered nonexistent. Quotes are not permitted around the Value Indicator.

## 5.3.3 String Value

If the Type Indicator is 1, the String Value is the actual character string. If the value is a token, the quotation marks are optional. However, if there is a beginning quotation mark, there must be a terminating quotation mark.

# 6. DEFINITIONS

This section defines specific characteristics of DIF.

**Character Sets.** This document assumes use of the ASCII character set. The following characters are permitted:

```
 !"(#$%&'()* +,-./
0123456789:;< = >?
@ABCDEFGHIJKLMNO
PQRSTUVWXYZ[ \ ]^_
`abcdefghijklmno
pqrstuvwxyz{|}~
```

(The first character in this list is a space.)

There are 95 printable characters, including the space. If the host computer has more than 95 characters, the additional characters must be mapped into the 95 ASCII characters to transfer data to another machine.

Some computers permit only 64 characters. When data is transferred to these machines, lower case characters and the characters :{|}~ are mapped into their corresponding upper case characters. If these transformations affect the integrity of the data, associated documentation should specify the effect. Transfers between character sets should be transparent to most users. To assure compatibility, strings should not contain nonprinting characters.

**EBCDIC.** EBCDIC is a binary representation of characters and is used primarily for large IBM computers. An awareness of the representation used is not essential, but if files are transferred between machines they must be converted to the standard representation of the host machine. Because EBCDIC defines more than the 95 standard printable characters, users should avoid the additional characters when preparing data files on an EBCDIC machine.

**String Length.** Some programs place a limit length on strings that they read. This results in the truncation of long string values. Some systems also limit the length of lines in a data file. Programs should support a minimal string length of 64 characters, but longer ones are preferable.

**String Delimiters.** Some systems delimit strings with apostrophes instead of quotation marks. When files are transferred to or from these systems, appropriate changes must be made.

**Tokens.** A token is a string consisting of upper case alphanumeric characters. It should have a maximum length of 32 characters. Commonly, tokens may or may not be contained within quotation marks; however, a token that is a required string, such as a header item topic, must be represented without quotation marks.

**Floating Point Numbers.** A floating point number consists of an optional sign and a series of digits followed by an optional decimal point. The number may be followed by the letter E (exponent) and a signed decimal exponent.

**Note:** Some systems generate the letter D to indicate a double precision floating point number. This is not standard, but it can be read by compatible programs within a single system. When transferring data to other computers, the D must be converted to an E.

# 7. APPLICATIONS PROGRAMS

This section records the specific use of DIF by application programs that support it. Programmers who intend to interface with any of these programs should note the specifics listed here. Standardized optional items used by these applications are listed in the general section on Optional Header Items. However, if a program uses a header item that varies significantly from the conventions, it is not mentioned only in this section. The accuracy of this information is not guaranteed.

## 7.1 The CCA/DMS Program

Published and distributed by VisiCorp.
*Uses:* SIZE

## 7.2 The TREND-SPOTTER® Program

Published and distributed by Software Resources, Inc.
    The TREND-SPOTTER program requires that the DIF file contain either only one tuple or only one vector.

## 7.3 The VisiCalc® Program

Published and distributed by VisiCorp.
Program created and written by Software Arts™
    The VisiCalc program does not generate the LABEL items. Some programs interfacing to the VisiCalc program have adopted the convention of examining the first Data Value in a tuple, and, if it is a string value, treating it as a label.

## 7.4 TK!Solver

Published and distributed by Software Arts, Inc.
*Uses:* UNITS, DISPLAYUNITS

## 7.5 The VisiPlot and VisiTrend/VisiPlot Programs

Published and distributed by VisiCorp.

Early versions of the VisiPlot and VisiTrend/VisiPlot programs used the Number Value and String Value incorrectly, storing the Number Value in the String Value field. Programs exchanging data with these versions should check the String Value. If it is not null, the string must be converted and the Number Value computed.

*Uses:* LABEL, COMMENT, MAJORSTART, MINOR-START, PERIODICITY, TRUELENGTH

# I. SAMPLE DIF FILE

This is an example of a DIF data file. The data in the file is represented by the table below.

```
    PROFIT REPORT

YEAR |SALES |COST |PROFIT
-------------------------
1980 |  100 |  90 |     10
1981 |  110 | 101 |      9
1982 |  121 | 110 |     11
```

## The Data File

```
TABLE                 ------>
0,1                         >
"PROFIT REPORT"             >
VECTORS  ----> Header       >
0,4          >              >
""       ----> Item         >
TUPLES                      >
0,3                         >
""                          >
LABEL                       >
1,0                         >
"YEAR"                      > Header
LABEL                       >
2,0                         >
"SALES"                     >
LABEL                       >
3,0                         >
"COST"                      >
LABEL                       >
4,0                         >
"PROFIT"                    >
DATA                        >
0,0                         >
""                    ------>
```

```
-1,0                    ---------->
BOT                             >
0,1980                          >
V                               >
0,100                           >
V                               >
0,90                            >
V                               >
0,10                            >
V                               >
-1,0                            >
BOT                             >

0,1981          ------>         >
V                      >        > Data
0,110                  >        >
V                      >        > Part
0,101 ---> Data        > Tuple  >
V     ---> Value       >        >
0,9                    >        >
V                      >        >
-1,0                   >        >
BOT             ------>         >
0,1982                          >
V                               >
0,121                           >
V                               >
0,110                           >
V                               >
0,11                            >
V                               >
-1,0                            >
EOD                     ---------->
```

# II. SAMPLE BASIC PROGRAM THAT WRITES A DIF FILE

This program enters student records into a file by prompting the user for a student's name and test scores and copying the information into a DIF file.

```
100 REM - THIS PROGRAM CREATES A DIF FILE CONTAINING THE
110 REM - NAME AND TEST SCORES OF A GIVEN NUMBER OF STUDENTS.
120 REM - IT PROMPTS FOR A FILE NAME, THE TOTAL NUMBER OF
130 REM - STUDENTS, AND THE NUMBER OF TEST SCORES FOR
140 REM - EACH STUDENT.  IT THEN PROMPTS FOR A STUDENT'S
150 REM - NAME AND TEST SCORES, AND WRITES THEM TO THE
160 REM - FILE AS A TUPLE.

1000 PRINT "OUTPUT FILE NAME:"; :REM - GET FILE NAME.
1010 INPUT F$
1020 OPEN "O",1,F$            :REM - OPEN FILE FOR OUTPUT.
1030 PRINT "NUMBER OF STUDENTS:";
1035                          :REM - PROMPT FOR NUMBER OF
```

```
1040 INPUT NT                        :REM -     TUPLES.
1050 PRINT "NUMBER OF TEST SCORES PER STUDENT:";
1060 INPUT NV                        :REM - NUMBER OF VECTORS IS
1070 NV = NV + 1                     :REM -    NUMBER OF SCORES + 1.
1080 GOSUB 3000                      :REM - USE SUBROUTINE TO
1090                                 :REM -    OUTPUT DIF HEADER.

2000 FOR I = 1 TO NT                 :REM - OUTPUT A TUPLE FOR
2010                                 :REM -    EACH STUDENT.
2020    T = -1: V = 0: S$ = "BOT"
2025                                 :REM - OUTPUT BOT SPECIAL
2030     GOSUB 4000                  :REM -    DATA VALUE.
2040     PRINT "NAME OF STUDENT #";I;
2050     INPUT S$                    :REM - GET NAME OF THIS STUDENT.
2060     T = 1: V = 0                :REM - OUTPUT AS STRING DATA
2070     GOSUB 4000                  :REM -    VALUE.
2080     FOR J = 1 TO NV-1           :REM - PROCESS EACH SCORE.
2090       PRINT "SCORE #";J;
2100       INPUT V                   :REM - GET SCORE.
2110       T = 0: S$ = "V"           :REM - OUTPUT SCORE AS A DATA
2120       GOSUB 4000                :REM -    VALUE.
2130     NEXT J
2140 NEXT I
2150 T = -1: V = 0: S$ = "EOD"       :REM - OUTPUT EOD SPECIAL DATA
2160 GOSUB 4000                      :REM -    VALUE.
2170 CLOSE 1                         :REM - CLOSE THE OUTPUT FILE.
2180 STOP                            :REM - DONE.


3000                                 :REM - ROUTINE TO OUTPUT HEADER.
3010 PRINT#1,"TABLE":PRINT#1,"0,1":GOSUB 3500
3020 PRINT#1,"TUPLES":PRINT#1,"0,";NT:GOSUB 3500
3030 PRINT#1,"VECTORS":PRINT#1,"0,";NV:GOSUB 3500
3040 PRINT#1,"DATA":PRINT#1,"0,0":GOSUB 3500
3050 RETURN
3500                                 :REM - ROUTINE TO OUTPUT A
3510                                 :REM -    NULL STRING ("").
3520 PRINT#1,CHR$(34);CHR$(34)  :REM - PRINT 2 QUOTATION MARKS.
3530 RETURN


4000                                 :REM - ROUTINE TO OUTPUT A DATA
4010                                 :REM -    VALUE.  T IS THE TYPE
4020                                 :REM -    INDICATOR, V IS THE
4030                                 :REM -    NUMBER VALUE, AND S$
4040                                 :REM -    IS THE STRING VALUE.
4050 PRINT#1,T;",";V
4060 PRINT#1,S$
4070 RETURN
5000 END
```

# III. SAMPLE BASIC PROGRAM THAT READS A DIF FILE

This program uses the output DIF file from the previous sample program to calculate an average score and letter grade for each student.

```
100    :REM - THIS PROGRAM READS A DIF FILE CONTAINING THE
110    :REM - TEST SCORES OF A GROUP OF STUDENTS, CALCULATES
120    :REM - AN AVERAGE SCORE FOR EACH STUDENT, MATCHES THE
130    :REM - AVERAGE TO A LETTER GRADE, AND PRINTS THE
140    :REM - STUDENT'S NAME, AVERAGE, AND LETTER GRADE.


500 DIM T(100)          :REM - MAXIMUM OF 100 VECTORS.
510 DIM V(100)          :REM - T IS THE TYPE INDICATOR, V IS
520 DIM V$ (100)        :REM -    THE NUMBER VALUE, AND V$ IS
530                     :REM -    THE STRING VALUE OF EACH DATA
535                     :REM -    VALUE.
540 GOSUB 5000          :REM - INITIALIZATION SUBROUTINE.
550 GOSUB 6000          :REM - SUBROUTINE TO READ HEADER.
560 FOR I = 1 TO NT     :REM - FOR EACH TUPLE,
570    GOSUB 7000       :REM -    GET ALL VECTOR ELEMENTS IN
575                     :REM -    TUPLE.
580    M=0              :REM -    M IS THE SUM OF THE SCORES.
590    FOR J = 1 TO NV  :REM -    FOR EACH VECTOR VALUE,
600       IF T(J)=1 THEN PRINT V$(J)  :REM - PRINT NAME.
610       IF T(J)=0 THEN M = M+V(J)   :REM - ADD SCORES.
620    NEXT J
630    M = M/(NV-1): PRINT M :REM - PRINT STUDENT'S AVERAGE
640    IF M<=50 THEN PRINT "THIS STUDENT'S FINAL GRADE IS F"
650    IF M<=70 AND M>50 THEN PRINT "THIS STUDENT'S FINAL GRADE IS D"
660    IF M<=85 AND M>70 THEN PRINT "THIS STUDENT'S FINAL GRADE IS C"
670    IF M<=94 AND M>85 THEN PRINT "THIS STUDENT'S FINAL GRADE IS B"
680    IF M>94 THEN PRINT "THIS STUDENT'S FINAL GRADE IS A"
690 NEXT I
700 CLOSE 2
710 PRINT "FINISHED CALCULATING GRADES"
720 STOP


5000                    :REM - INITIALIZATION CODE.
5010 PRINT "FILE NAME";
5020 INPUT F$
5030 OPEN "I",2,F$      :REM - OPEN FILE FOR INPUT.
5040 NV = 0             :REM - INITIAL VECTOR COUNT.
5050 NT = 0             :REM - INITIAL TUPLE COUNT.
5060 RETURN
6000    :REM - READ HEADER. GET NUMBER OF VECTORS AND TUPLES.
6010 INPUT#2,T$         :REM - GET TOPIC.
6020 INPUT#2,S,N        :REM - GET VECTOR NUMBER AND VALUE.
6030 INPUT#2,S$         :REM - GET STRING VALUE.
6040 IF T$="VECTORS" THEN 6500:REM - CHECK FOR KNOWN HEADER
6050 IF T$="TUPLES" THEN 6600 :REM -    ITEMS.
6060 IF T$="DATA" THEN RETURN
6065                    :REM - "DATA" ENDS HEADER.
```

```
6070 GOTO 6010          :REM - IGNORE UNKNOWN ITEMS.
6500 NV = N             :REM - NUMBER OF VECTORS.
6510 IF NV<=100 THEN 6010 :REM - CHECK FOR 100 OR LESS VECTORS.
6520 PRINT "TOO MANY VECTORS. PROGRAM CAPACITY 100 VECTORS."
6530 CLOSE 2
6540 STOP
6600 NT = N             :REM - NUMBER OF TUPLES.
6610 GOTO 6010          :REM - GET NEXT HEADER ITEM.


7000    :REM - SUBROUTINE TO GET ALL VECTOR ELEMENTS IN A TUPLE.
7010 GOSUB 8000         :REM - GET NEXT DATA VALUE.
7020 IF T1<>-1 THEN 9000 :REM - MUST BE BOT, ELSE ERROR
7030 IF S$<>"BOT" THEN 9000
7040   FOR K = 1 TO NV :REM - GET EACH DATA VALUE.
7050      GOSUB 8000
7060      IF T1>1 THEN 9000
7070      T(K) = T1     :REM - SAVE TYPE INDICATOR.
7080      V(K) = V1     :REM - SAVE NUMBER VALUE.
7090      V$(K) = S$    :REM - SAVE STRING VALUE.
7100      NEXT K
7110 RETURN


8000    :REM - SUBROUTINE TO GET NEXT DATA VALUE.
8010 INPUT#2,T1,V1      :REM - GET TYPE INDICATOR, NUMERIC
8020 INPUT#2,S$         :REM -    VALUE, AND STRING VALUE.
8030 RETURN

9000    :REM - ERROR ROUTINE
9010 PRINT "ERROR IN FILE FORMAT"
9020 CLOSE 2            :REM - END PROGRAM
9030 STOP
9040 END
```

# IV. SAMPLE PASCAL PROGRAM USING A DIF FILE

**This program is a Pascal program that reads data from a DIF file into an array and displays the results on the terminal.**

```
{ This is a simple program which reads DIF file data into an array and
  displays the results on the terminal.  It makes use of a procedure
  called "get_dif_array" which handles only numeric data.  It is written
  for Apple Pascal 1.1 and may require modification to run on other systems. }

program dif_read;

const
  max_vector = 10;                          { maximum number of vectors }
  max_tuple  = 10;                          { maximum number of tuples }

type
  vector_index = 0..max_vector;
  tuple_index  = 0..max_tuple;
  dif_array    = array[1..max_vector, 1..max_tuple] of real;
```

```pascal
var
   in_file    : text;               num_vectors : vector_index;
   fname      : string[15];         num_tuples  : tuple_index;
   matrix     : dif_array;          code, i, j  : integer;


{ "Get_dif_array" reads a DIF file and returns the file data (currently
  only numeric) in an array.  Also returns number of vectors and tuples
  -- these must be specified in file header -- and an error code. }

procedure get_dif_array (var dif_file: text; var real_array: dif_array;
                         var nvectors: vector_index; var ntuples: tuple_index;
                         var return_code: integer);

   const                           { currently defined data types }
      special = -1; numeric =  0; char_string  = 1; other   = 2;

   type
      header_item  = record
                        topic       : string;
                        vector_num  : vector_index;
                        value       : integer;
                        string_value : string
                     end;
      data_value   = record
                        kind         : -1..2; { currently defined data types}
                        number_value : real;
                        string_value : string
                     end;

   var
      hdr_item    : header_item;
      data_val    : data_value;
      tuple, vector : integer;


{ "Read_integer" reads an integer terminated with a comma.  The routine
  is required because this Pascal dialect's "read" procedure recognizes
  only <space>, eoln and eof as delimiters of integer values. }

procedure read_integer (var number: integer);
   var
      sign, magnitude : integer;
      ch : char;
   begin
      sign := 1;  magnitude := 0;                  { initialize }
      read (difile, ch);                           { get 1st character }
      while ch <> ',' do                           { comma is delimiter }
         begin
            case ch of
               '-' : sign := -1;
               '0','1','2','3','4','5','6','7','8','9'
                   : magnitude := magnitude * 10 + ord(ch) - ord('0')
            end; { case }
            read (difile, ch)                       { get next character }
         end;
      number := sign * magnitude                   { return result }
   end; { read_integer }
```

```
{ "Read_string"  deletes leading and trailing blanks and strips the
  quotes from quoted strings. }

procedure read_string (var str: string);
   begin
      readln (difile, str);
      while str[1] = ' ' do                        { leading blanks }
         delete (str, 1, 1);
      if str[1] = '"'                              { strip quotes }
         then begin
               delete (str, 1, 1);
               delete (str, pos('"', str),
                       length(str) - pos('"', str) + 1)
            end
      else if pos(' ', str) > 0                    { trailing blanks }
         then delete (str, pos(' ', str),
                        length(str) - pos(' ', str) + 1)
   end;  { read_string }


procedure read_header_item (var item: header_item);
   begin
      read_string (item.topic);                    { get topic }
      read_integer (item.vector_num);              { get vector number }
      readln (difile, item.value);                 { get value }
      read_string (item.string_value)              { get string value }
   end;   { read_header_item }

procedure read_data_value (var value: data_value);
   begin
      read_integer (value.kind);                   { get data type }
      readln (difile, value.number_value);         { get number value }
      read_string (value.string_value)             { get string value }
   end;   { read_data_value }

begin   { get_dif_array }
   return_code := 0;                               { assume no problems }
   nvectors := 0; ntuples := 0;                    { initialize }
   repeat                                          { read header }
      read_header_item (hdr_item);
      if hdr_item.topic = 'VECTORS'
         then nvectors := hdr_item.value           { vector count }
      else if hdr_item.topic = 'TUPLES'
         then ntuples := hdr_item.value            { tuple count }
   until hdr_item.topic = 'DATA';

   if (nvectors = 0) or (ntuples = 0)              { check counts }
      then return_code := 1
      else begin
            for tuple := 1 to ntuples do           { read data }
               begin
                  read_data_value (data_val);           { BOT }
                  for vector := 1 to nvectors do
                     begin
                        read_data_value (data_val);
                        if data_val.kind = numeric
                           then real_array[vector, tuple] :=
                                       data_val.number_value
```

```
                    end
                end;
             read_data_value (data_val);              { EOD }
             if (data_val.kind <> special) or
                (data_val.string_value <> 'EOD')
                then return_code := 2
             end

end;   { get_dif_array }


begin    { dif_read }

    writeln;                               { get DIF file name }
    write ('DIF file name: ');
    readln (fname);
    reset (in_file, fname);                { open and point to BOF }

    get_dif_array (in_file, matrix, num_vectors, num_tuples, code);

    close (in_file);                       { close DIF file }
    case code of                           { display results }
      0: begin
            writeln;
            writeln ('"', fname:15, '"', ' contains ', num_vectors:3,
                       ' vectors and ', num_tuples:3, ' tuples.');
            writeln ('The data values follow in tuple order:');
            writeln;
            for i := 1 to num_tuples do
               begin
                 for j := 1 to num_vectors do
                    write (matrix[j, i]:10:2);
                  writeln
               end;
             writeln
         end;
      1: writeln ('Error. Tuple or vector count not found.');
      2: writeln ('Error. Data not properly terminated.')
      end    { case }

end.   { dif_read }
```

# DIF™ Program List

## EQUATION PROCESSING

| Product | Company | Comments |
|---------|---------|----------|
| TK!Solver | Software Arts, Inc.<br>27 Mica Lane<br>Wellesley, MA 02181<br>(617) 237–4000 | Solves simultaneous equations through direct and iterative methods. Provides units of measure conversions, list processing, and plotting. |

## SPREADSHEETS

| Product | Company | Comments |
|---------|---------|----------|
| MAGICALC | Artsci, Inc.<br>5547 Satsuma Ave.<br>North Hollywood, CA 91601<br>(213) 985–2922 | |
| VISICALC | VisiCorp<br>2895 Zanker Road<br>San Jose, CA 95134<br>(408) 946–9000 | Electronic spreadsheet that reads and writes in DIF files |
| VISICALC<br>(Advanced Version) | VisiCorp<br>2895 Zanker Road<br>San Jose, CA 95134<br>(408) 946–9000 | Advanced electronic spreadsheet version of VisiCalc that reads and writes data in DIF files |

| | | |
|---|---|---|
| VISICALC<br>(Business<br>Forecasting Model) | Visicorp<br>2895 Zanker Road<br>San Jose, CA 95134 | A program consisting of<br>seven interrelated VisiCalc<br>application worksheets |
| 1-2-3 | Lotus Development Corp.<br>55 Wheeler Street<br>Cambridge, MA 02139<br>(617) 492-7171 | Incorporates spreadsheet,<br>data base, and graphics<br>into one product |

## BUSINESS

| Product | Company | Comments |
|---|---|---|
| THE ACCOUNTANT<br>FINANCE DATA<br>BASE SYSTEM | Decision Support Software<br>1438 Ironwood Drive<br>McLean, VA 22102<br>(703) 241-8316 | |
| ACCTG | Accounting Systems, Ltd.<br>P.O. Box 1488<br>Boise, ID 83701<br>(208) 336-2281 | Accounting soft-<br>ware maintained on<br>General Electric<br>Information Ser-<br>vices Company's<br>Mark III remote<br>computing net-<br>work |
| BOOKKEEPER<br>INCOME STATEMENT<br>INCOME FORECASTER | SpreadSoft<br>P.O. Box 192<br>Clinton, MD 20735<br>(301) 856-1180 | Financial packages<br>for VisiCalc |
| DATA PLUS | Software Dimension, Inc.<br>6371 Auburn Blvd.<br>Citrus Heights, CA 95610<br>(916) 722-8000 | Converts Data<br>from Accounting<br>Plus II into DIF<br>format |
| INI Client<br>Write-up System | INI, Inc.<br>4013 Chestnut Street<br>Philadelphia, PA 19104 | General ledger<br>package and user-<br>definable report<br>generator |
| Investor's Interface | Marketware<br>P.O. Box 34647<br>Richmond, VA 23234<br>(804) 276-8577 | Stock data program |
| PEAR PLUS | Remote Computing Corp.<br>1044 Northern Blvd. | Financial package<br>converts PEAR |

| | Roslyn, NY 11576<br>(800) 645–3120<br>(212) 895–3810 (in-state) | (Portfolio Evalua-<br>tion and Reporting<br>System) to VisiCalc<br>then to the DIF<br>form |
|---|---|---|
| PEARFORMANCE | Remote Computing Corp.<br>1044 Northern Blvd.<br>Roslyn, NY 11576<br>(800) 645–3120<br>(212) 895–3810 (in-state) | Financial analysis<br>package |
| THE PERSONAL<br>INVESTOR | PBL Corporation<br>P.O. Box 559<br>Wayzata, MN 55391<br>(612) 471–7644 | Stock data program |
| SMART | Software Resources, Inc.<br>186 Alewife Brook<br>Parkway<br>Cambridge, MA 02138<br>(617) 497–5900 | Investment Tool |
| STOCK CRAFT | Decision Economics, Inc.<br>14 Old Farm Road<br>Cedar Knolls, NJ 07927<br>(201) 539–6889 | Stockmarket<br>system with port-<br>folio management |
| VISIMARK | Accounting Systems Ltd.<br>P.O. Box 1488<br>Boise, ID 83701<br>(208) 536–2281 | Converts data<br>maintained on<br>General Electric's<br>Mark III service<br>into the DIF<br>format |
| VISISCHEDULE | VisiCorp<br>2895 Zanker Road<br>San Jose, CA 95134<br>(408) 946–9000 | Job scheduling<br>program |

## DATA BASE/FILING PROGRAMS

| Product | Company | Comments |
|---|---|---|
| CCA/DMS<br>DIF Interface | F/S Associates<br>664 18th Street<br>Manhattan Beach,<br>CA 90266 | Sorts data, prints reports, and<br>labels |

| DB Master | Stoneware Microcomputer Products 50 Belvedere Street San Rafael, CA 94901 (415) 454-6500 | DB Master Utility Pak #1 is required to read and write DIF files |
|---|---|---|
| FILEMASTER | NF Systems, Ltd. P.O. Box 76363 Atlanta, GA 30358 (404) 252-3302 | Reads DIF files only |
| JINSAM | Jini Micro-Systems, Inc. P.O. Box 274 Kingsbridge Station Riverdale, NY 10463 | Data base manager, requires CALCPAK interface |
| List Handler | Silicon Valley Systems 1625 El Camino Real Belmont, CA 94002 (415) 593-4344 | |
| MDBS.QRS | Micro Data Base Systems, Inc. P.O. Box 248 Lafayette, IN 47902 | Query system/report writer |
| Micro-Analyst | Zacks Investment Research 135 South LaSalle St. Chicago, IL 60603 (312) 263-1464 | Extracts data from a number of data bases on time sharing systems and downloads to micros |
| MICROMAGIC/ MICROCOM | I.P. Sharp Associates 1200 First Federal Plaza Rochester, NY 14614 (716) 546-7270 | Communications package that accesses I.P. Sharp's data bases and downloads to DIF files |
| QTRAN (Qubit Transfer) | Compuserve 5000 Arlington Centre Blvd. P.O. Box 20121 Columbus, OH 43220 (614) 457-8600 | Data base management systems available on a time sharing basis through Compuserve. QTRAN converts DIF files into a format accepted by Qubit |
| TDM (The Data Machine) | Pascal Systems, Inc. 830 Menlo Avenue Suite 109 Menlo Park, CA 94025 (415) 321-0761 | |

| T.I.M. III | Innovative Software, Inc.<br>9300 W. 110th Street<br>Suite 380<br>Overland Park, KS 66210<br>(913) 383-1089 | Data base management |
| VISIFILE | VisiCorp<br>2895 Zanker Road<br>San Jose, CA 95134<br>(408) 946-9000 | |

# GRAPHICS

| *Product* | *Company* | *Comments* |
|---|---|---|
| APPLE BUSINESS<br>GRAPHICS | Apple Computer, Inc.<br>20525 Mariana Avenue<br>Cupertino, CA 95014<br>(800) 538-9696<br>(800) 662-9238 (in-state) | |
| BPS Business<br>Graphics | Business & Professional<br>Soft Inc.<br>143 Binney Street<br>Cambridge, MA 02142<br>(617) 491-3377 | |
| BIOSTATISTICS III | A2 Devices<br>P.O. Box 2226<br>Alameda, CA 94501<br>(415) 527-7380 | Combined statistical analy-<br>sis and graphical data<br>plotting software package |
| CHARTDIF and<br>PLOTDIF | The Consultant<br>414 Can-Data<br>Mt. Prospect, IL 60056 | Packages that either chart<br>or plot DIF files |
| CHARTMAN I, II,<br>III | Graphic Software Inc.<br>P.O. Box 367<br>Kenmore Station<br>Boston, MA 02215<br>(617) 491-2434 | |
| CHART-MASTER | Decision Resources<br>44 White Birch Road<br>Weston, CT 06883<br>(203) 222-1974 | |
| DIFBAR<br>DIFPIE<br>DIFPLOT | Strobe Incorporated<br>897-5A Independence<br>Avenue<br>Mountain View, CA 94043<br>(415) 969-5130 | Creates charts and linear<br>or semilog line graphs of<br>DIF data; only works with<br>Strobe Plotter |

| DIFMASTER | Starside Engineering<br>P.O. Box 18036<br>Rochester, NY 14618<br>(716) 461–1027 | Three-dimensional business graphics |
|---|---|---|
| DRAFTSMAN | Starware<br>1701 K Street, N.W.<br>Washington, D.C. 20006 | |
| FAST GRAPHS | Innovative Software<br>9300 West 110th Street<br>Suite 380<br>Overland Park, KS 66210<br>(913) 383–1089 | "Graphics Report" program |
| GIRAPH | Data Display<br>171 West 4th Street<br>New York, NY 10014<br>(212) 924–8167 | |
| Graph 'n' Calc | Desktop Computer<br>Software<br>303 Potreno St.<br>29/303<br>Santa Cruz, CA 95060 | Business graphics and calculation program |
| GRAFOX | Fox & Geller<br>604 Market Street<br>Elmwood Park, NJ 07407<br>(201) 794–8883 | Color graphics software |
| THE GRAPHICS GENERATOR | Robert J. Brady, Co.<br>Routes 197 and 450<br>Bowie, MD 20715<br>(301) 262–6300 | Business and technical graphics program |
| GRAPHMAGIC | International Software<br>Marketing<br>Suite 421, University<br>Building<br>120 East Washington St.<br>Syracuse, NY 13202<br>(315) 474–3400 | |
| GRAPHPOWER | Ferox Microsystems, Inc.<br>1701 N. Ft. Meyer Drive<br>Suite 611<br>Arlington, VA 22209<br>(703) 841–0800 | |

| | | |
|---|---|---|
| Micrograph by 2Y's | 2Y's Associates Ltd.<br>P.O. Box 6733<br>Station J<br>Ottawa, Ontario<br>Canada K2A 3Z4<br>(613) 232-3826 | Combines business<br>graphics and trend<br>analysis in one package |
| PFS;Graph | Software Publishing<br>Corp.<br>1901 Landings Drive<br>Mt. View, CA 94043<br>(415) 962-8910 | |
| THE PRIME<br>PLOTTER | Prime Soft Corp.<br>P.O. Box 40<br>Cabin John, MD 20818<br>(301) 229-4229 | Integrated statistics and<br>graphics package |
| TREND-SPOTTER | Software Resources, Inc.<br>186 Alewife Brook Pkwy<br>Cambridge, MA 02138<br>(617) 497-5900 | Produces trend graphics |
| ULTRAPLOT/DIF<br>DATA GRAPH<br>ULTRAPLOT/DIF<br>INTERFACE | Avant-Garde Creations<br>P.O. Box 30160<br>Eugene, OR 97403<br>(503) 345-3043 | |
| VERSAPLOT | Spectrasoft<br>350 N. Lantana Suite 775<br>Box 3000<br>Camarillo, CA 93010<br>(805) 987-6602 | |
| VISIPLOT | VisiCorp<br>2895 Zanker Road<br>San Jose, CA 95135<br>(408) 946-9000 | |
| VISITREND/<br>VISIPLOT | VisiCorp<br>1895 Zanker Road<br>San Jose, CA 95134<br>(408) 946-9000 | Combines graphics with<br>trend analysis in one<br>package |

# WORD PROCESSING

| Product | Company | Comments |
|---|---|---|
| THE EXECUTIVE SECRETARY | SOF/SYS Inc. 4306 Upton Avenue S. Minneapolis, MN 55410 (612) 929–7104 | Word Processing package that allows for integration of DIF files |
| MultiMate | Softword Systems Inc. 52 Oakland Avenue N. East Hartford, CT 06108 (203) 522–2116 | Reads DIF files only |
| WORDSTAR CONNECTION | Sofstar 13935 U.S. 1 Juno Square Juno Beach, FL 33408 (305) 627–5511 | Converts DIF files to Wordstar format |

# STATISTICS

| Product | Company | Comments |
|---|---|---|
| speedSTAT | SoftCorp International 229 Huber Village Blvd. P.O. Box 29765 Columbus, OH 43229 (800) 543–1350 (614) 890–2820 (in-state) | A line of business analysis tools that provides statistical projections and reports |
| STATISTICS WITH DAISY | Rainbow Computing, Inc. 19517 Business Center Dr. Northridge, CA 91234 (802) 423–5441 (213) 349–0300 (in-state) | |
| VISIBRIDGE/GL VISIBRIDGE/RPT | Solutions, Inc. Box 989 Montpelier, VT 05602 (802) 229–0368 | Report formatter for VisiCalc data |

# UTILITY PACKAGES

| Product | Company | Comments |
| --- | --- | --- |
| AEN Grading System | Apple Educators' Newsletter 9525 Lucerne Street Ventura, CA 93004 (805) 647–1063 | Allows transfer in support of the AEN Grading System |
| Bridge | Sun Microsystems, Inc. P.O. Box 1388 Ft. Lauderdale, FL 33302 (305) 368–8221 | Creates DIF files from PFS data files |
| CRS (COMPUSTAT RETRIEVAL SYSTEM)/DIF | Compuserve 5000 Arlington Centre Blvd. P.O. Box 20212 Columbus, OH 43220 (614) 457–8600 | Extracts data from COMPUSTAT or VALUELINE data bases and creates DIF files |
| Data Sorter | InterCalc P.O. Box 254 Scarsdale, NY 10583 (914) 472–0038 | Sorts VisiCalc print-to-disk files, creating DIF files |
| DATA*TRANS | ABT Microcomputer Software Dept. INF82A 55 Wheeler Street Cambridge, MA 02138 (617) 492–7100 | Provides telecommunications software program for Applesoft programs |
| LOADCALC | Micro Decision Systems P.O. Box 1392 Pittsburgh, PA 15219 (412) 276–2387 | Converts text files into DIF files |
| MAINLINE | Gregg Corporation 100 5th Avenue Waltham, MA 02254 (617) 890–7227 | Permits downloading between mainframe databases to microcomputers in the DIF format |
| MAKEFILE | Real Decisions Corp. 123 High Ridge Road Stamford, CT 06905 (203) 356–9200 | Creates DIF files from data drawn from COMPUSTAT II files |

| MERGECALC | Micro Decisions Systems<br>P.O. Box 1392<br>Pittsburgh, PA 15219<br>(412) 276–2387 | Manipulates and con-<br>solidates DIF or VisiCalc<br>files |
|---|---|---|
| Monte Carlo<br>Simulations (MCS) | Actuarial Micro<br>Software<br>3915 A Valley Court<br>Winston-Salem, NC 27106<br>(919) 765–5588 | Statistical distribution<br>analysis |
| PASCAL DOS 3.3<br>TRANSFER<br>UTILITY | Orange Software<br>293 Manley Heights<br>Orange, CT 06477 | Transfers selected files<br>between Apple DOS 3.3<br>and Apple Pascal |
| TEAK MANAGER | Open Systems, Inc.<br>430 Oak Grove<br>Minneapolis, MN 55403<br>(612) 870–3515 | Report writer allows data<br>from the Open Systems'<br>accounting applications<br>to be reformatted into the<br>DIF format |
| TRANSFER III | Mind Systems Corp.<br>P.O. Box 506<br>Northampton, MA 01601<br>(413) 586–6463 | Transfer data files<br>between Apple II DOS 3.3<br>diskette and Apple III |
| Versa-Prof | Rocking X Software<br>Company<br>1207 West Lee Street<br>Borger, TX 79007<br>(806) 273–3894 | Record keeping and grade<br>averaging program |
| VISITERM | VisiCorp<br>2895 Zanker Road<br>San Jose, CA 95135<br>(408) 946–9000 | Allows personal computer<br>to communicate with<br>other computers |
| VIZ.A.CON | Abacus Associates<br>Suite 240<br>65645 West Loop South<br>Bellaire, TX 77401<br>(713) 666–8146 | Combines multiple<br>"pages" of VisiCalc data<br>from a model for hier-<br>archical consolidations or<br>for summations over<br>periods of time |

# English-to-Metric Conversion Factors

| From | To | Multiply By | Add Offset |
|------|-----|-------------|------------|
| **ACCELERAT** | | | |
| ft/s^2 | m/s^2 | .2048 | |
| in/s^2 | m/s^2 | .0254 | |
| **ANGLE** | | | |
| deg__ang | radian | .01745329 | |
| min__ang | radian | .0002908882 | |
| sec__ang | radian | 4.848137E-6 | |
| **AREA** | | | |
| acre | m^2 | 4046.973 | |
| are | m^2 | 100 | |
| circ__mil | m^2 | 5.067075E-10 | |
| ft^2 | m^2 | .09290304 | |
| hectare | m^2 | 10000 | |
| in^2 | m^2 | .00064516 | |
| mi^2 | m^2 | 2589988 | |
| yd^2 | m^2 | .8361274 | |
| **MOMENT** | | | |
| dyne*cm | N*m | .0000001 | |
| kgf*m | N*m | 9.80665 | |
| ozf*in | N*m | .007061552 | |
| lbf*in | N*m | .1129848 | |
| lbf*ft | N*m | 1.355818 | |
| **ELECTRIC** | | | |
| abampere | ampere | 10 | |
| A | ampere | 1 | |
| abcoulomb | coulomb | 10 | |
| C | coulomb | 1 | |
| abfarad | farad | 1000000000 | |
| F | farad | 1 | |
| abhenry | henry | .000000001 | |
| H | henry | 1 | |

| | | |
|---|---|---|
| abmho | siemens | 1000000000 |
| S | siemens | 1 |
| abohm | ohm | .000000001 |
| abvolt | volt | .00000001 |
| V | volt | 1 |
| amp*hour | coulomb | 3600 |
| EMU(cap) | farad | 1000000000 |
| EMU(curr) | ampere | 10 |
| EMU(pot) | volt | .00000001 |
| EMU(ind) | henry | .000000001 |
| EMU(res) | ohm | .000000001 |
| ESU(cap) | farad | 1.11265E-12 |
| ESU(curr) | ampere | 3.3356E-10 |
| ESU(pot) | volt | 299.79 |
| ESU(ind) | henry | 898755400000 |
| ESU(res) | ohm | 898755400000 |
| faraday | coulomb | 96487 |
| gamma | tesla | .000000001 |
| gauss | tesla | .0001 |
| T | tesla | 1 |
| gilbert | ampere | .7957747 |
| maxwell | weber | .00000001 |
| Wb | weber | 1 |
| mho | siemens | 1 |
| oersted | A/m | .7957747 |
| ohm*cm | ohm*m | .01 |
| ohm*mil/f | ohm*m | 1.662426E-9 |
| statamper | ampere | 3.33564E-10 |
| statcoulo | coulomb | 3.33564E-10 |
| statfarad | farad | 1.11265E-12 |
| stathenry | henry | 898755400000 |
| statmho | siemens | 1.11265E-12 |
| statohm | ohm | 898755400000 |
| statvolt | volt | 299.7925 |
| unit__pole | weberv | 1.256637E-7 |

**ENERGY**

| | | |
|---|---|---|
| Btu(IT) | joule | 1055.056 |
| Btu(39F) | joule | 1059.67 |
| Btu(59F) | joule | 1054.8 |
| Btu | joule | 1059.67 |
| calorie | joule | 4.1868 |
| kilocalor | joule | 4186.8 |
| electronv | joule | 1.60219E-19 |
| erg | joule | .0000001 |
| ft*lbf | joule | 1.355818 |
| ft*pounda | joule | .04214011 |
| kW*h | joule | 3600000 |
| therm | joule | 105505600 |

| | | |
|---|---|---|
| W*hr | joule | 3600 |
| W*sec | joule | 1 |
| erg/(cm^2 | W/m^2 | .001 |
| W/cm^2 | W/m^2 | 10000 |
| W/in^2 | W/m^2 | 1550.03 |

**FLOW**

| | | |
|---|---|---|
| N | newton | 1 |
| dyne | newton | .00001 |
| dilogramf | newton | 9.80665 |
| kilopond | newton | 9.80665 |
| kip | newton | 4448.222 |
| ounceforc | newton | .2780139 |
| lbf | newton | 4.448222 |
| poundal | newton | .138255 |
| ton-force | newton | 8896.444 |

**FORCE/LGT**

| | | |
|---|---|---|
| lbf/ft | N/m | 14.5939 |
| lbf/in | N/m | 175.1268 |

**HEAT**

| | | |
|---|---|---|
| Btu*ft/h* | W/m*K | 1.730735 |
| Btu*in/h* | W/m*K | .1442279 |
| Btu/ft^2 | J/m^2 | 11356.53 |
| Btu/h*ft^2 | W/m^2 | 3.154591 |
| Btu/h*ft^2 | W/m^2*K | 5.678264 |
| Btu/lb | J/kg | 2326 |
| Btu/lb*F | J/kg*K | 4186.8 |
| Btu/ft^3 | J/m^3 | 37258.95 |
| cal/g | J/kg | 4186.8 |
| cal/g*C | J/kg*K | 4186.8 |
| clo | K*m^2/W | .2003712 |
| F*ft^2*h/ | k*m^2/W | .1761102 |
| F*ft^2*h/ | k*m/W | 6.933471 |

**LENGTH**

| | | |
|---|---|---|
| angstrom | m | .0000000001 |
| chain | m | 20.11684 |
| fathom | m | 1.828804 |
| foot | m | .3048 |
| inch | m | .0254 |
| lightyear | m | 9.46055E15 |
| microinch | m | .0000000254 |
| micron | m | .000001 |
| mil | m | .0000254 |
| mile | m | 1609.347 |
| nautmile | m | 1852 |
| rod | m | 5.02921 |
| yard | m | .9144 |

**LIGHT**

| | | |
|---|---|---|
| cd/in^2 | cd/m^2 | 1550.003 |

| foodcandl | lux | 10.76391 |
| footlambe | cd/m^2 | 3.426259 |
| lambert | cd/m^2 | 3183.099 |

**MASS**

| grain | kg | .00006479891 |
| gram | kg | .001 |
| hundredwt | kg | 50.80235 |
| ounce(av) | kg | .02834952 |
| ounce(tr) | kg | .03110348 |
| pound(av) | kg | .4535924 |
| pound(tr) | kg | .3732417 |
| slug | kg | 14.5939 |
| ton(assay) | kg | .02916667 |
| ton(lg) | kg | 1016.047 |
| ton(metric) | kg | 1000 |
| ton | kg | 907.1847 |
| tonne | kg | 1000 |

**MASS/AREA**

| oz/ft^2 | kg/m^2 | .3051517 |
| oz/yd^2 | kg/m^2 | .03390575 |
| lb/ft^2 | kg/m^2 | 4.882428 |

**MASS/LGTH**

| lbm/ft | kg/m | 1.488164 |
| lbm/in | kg/m | 17.85797 |

**MASS/TIME**

| perm | kg/Pa*s*m | 5.72135E-11 |
| perm-in | kg/Pa*s*m | 1.45322E-12 |
| lbm/h | kg/s | .0001259979 |
| lbm/min | kg/s | .007559873 |
| lbm/s | kg/s | .4535924 |
| lb/hp*h | kg/J | 1.689659E-7 |
| ton/h | kg/s | .2519958 |

**MASS/VOL**

| grain/gal | kg/m^3 | .01711806 |
| g/cm^3 | kg/m^3 | 1000 |
| oz/gal | kg/m^3 | 7.489152 |
| oz/in^3 | kg/m^3 | 1729.994 |
| lbm/ft^3 | kg/m^3 | 16.01846 |
| lbm/in^3 | kg/m^3 | 27679.9 |
| lbm/yd^3 | kg/m^3 | .5932764 |
| lbm/gal | kg/m^3 | 119.8264 |
| slug/ft^3 | kg/m^3 | 515.3788 |
| ton/yd^3 | kg/m^3 | 1186.553 |

**POWER**

| Btu/h | W | .2930722 |
| Btu/min | W | 17.58427 |
| Btu/s | W | 1055.056 |
| cal/s | W | 4.1868 |

| | | | |
|---|---|---|---|
| erg/s | W | .0000001 | |
| ft*lbf/h | W | .0003766161 | |
| ft*lbf/mi | W | .02259697 | |
| ft*lbf/s | W | 1.355818 | |
| hp | W | 745.6999 | |
| hp(boiler) | W | 9809.5 | |
| kilocal/h | W | 1.163 | |
| ton(ref) | W | 3516.853 | |

**PRESSURE**

| | | |
|---|---|---|
| atmosph | Pa | 101325 |
| bar | Pa | 100000 |
| cmHg | Pa | 1333.22 |
| cmH20 | Pa | 98.0638 |
| dyne/cm^2 | Pa | .1 |
| ftH20 | Pa | 2988.98 |
| gf/cm^2 | Pa | 98.0665 |
| inHg | Pa | 3386.38 |
| inH20 | Pa | 249.082 |
| kgf/cm^2 | Pa | 98066.5 |
| kgf/m^2 | Pa | 9.80665 |
| kgf/mm^2 | Pa | 9806650 |
| kip/in^2 | Pa | 6894757 |
| millibar | Pa | 100 |
| mmHg | Pa | 133.322 |
| poundal/f | Pa | 1.488164 |
| lbf/ft^2 | Pa | 47.88026 |
| lbf/in^2 | Pa | 6894.757 |
| psi | Pa | 6894.757 |
| torr | Pa | 133.322 |

**TEMPERATU**

| | | | |
|---|---|---|---|
| 'C | 'K | 1 | 273.15 |
| 'C | 'F | 1.8 | 32 |
| 'F | 'R | 1 | 459.67 |
| 'R | 'K | .5555555556 | |

**TIME**

| | | |
|---|---|---|
| day | s | 86400 |
| day(sider) | s | 86164.09 |
| hour | s | 3600 |
| hour(side) | s | 3591.7 |
| minute | s | 60 |
| minute(si) | s | 59.83617 |
| s(sideria) | s | .9972696 |
| year | s | 31536000 |
| year(sid) | s | 31558150 |
| year(trop) | s | 31556930 |

**VELOCITY**

| | | |
|---|---|---|
| ft/h | m/s | .00008466667 |
| ft/min | m/s | .00508 |

| | | |
|---|---|---|
| ft/s | m/s | 30.48 |
| in/s | m/s | .0254 |
| km/h | m/s | .2777778 |
| knot | m/s | .5144444 |
| mi/h | m/s | .44704 |
| mi/min | m/s | 26.8224 |
| mi/s | m/s | 1609.344 |
| **VISCOSITY** | | |
| centipois | Pa*s | .001 |
| centistok | m^2/s | .000001 |
| ft^2/s | m^2/s | .09290304 |
| poise | Pa*s | .1 |
| poundal*s | Pa*s | 1.488164 |
| lb/ft*h | Pa*s | .0004133789 |
| lb/ft*s | Pa*s | 1.488164 |
| lbf*s/ft^ | Pa*s | 47.88026 |
| lbf*s/in^ | Pa*s | 6894.757 |
| rhe | L/Pa*s | 10 |
| slug/ft*s | Pa*s | 47.88026 |
| stokes | m^2/s | .0001 |
| **VOLUME** | | |
| acre*ft | m^3 | 1233.489 |
| barrel | m^3 | 1.589873 |
| board-ft | m^3 | 2.359737 |
| bushel | m^3 | .03523907 |
| cup | m^3 | .0002365882 |
| fl__oz | m^3 | .00002957353 |
| ft^3 | m^3 | .02831685 |
| gallon | m^3 | .003785412 |
| gill | m^3 | .0001182941 |
| in^3 | m^3 | .00001638706 |
| litre | m^3 | .001 |
| peck | m^3 | .008809768 |
| pint | m^3 | .0004731765 |
| quart | m^3 | .0009463529 |
| stere | m^3 | 1 |
| tablespoon | m^3 | .00001478676 |
| teaspoon | m^3 | 4.928922E-6 |
| ton(reg) | m^3 | 2.831685 |
| yd^3 | m^3 | .7645549 |
| **VOL/TIME** | | |
| ft^3/min | m^3/s | .0004719474 |
| ft^3/s | m^3/s | .02831685 |
| gal/hp*h | m^3/s | 1.410089E-9 |
| in^3/min | m^3/s | 2.731177E-7 |
| yd^3/min | m^3/s | .01274258 |
| gal/day | m^3/s | 4.38126E-8 |
| gal/min | m^3/s | .0000630902 |
| GPM | m^3/s | .0000630902 |

# Properties of Vapors

| Name | Elements | Unit | Comment |
|------|----------|------|---------|
| | | | PROPERTIES OF VAPORS |
| | | | * * * * * * * * * * * * * * * * * * * * * * * * |
| VAPOR | 40 | | Name of Vapor |
| MOL__WT | 40 | | Molecular weight |
| BOILING__P | 40 | | Normal boiling point, 'K |
| CRIT__TEMP | 40 | | Critical temperature, 'K |
| CRIT__PRES | 40 | | Critical pressure, kPa |
| DENSITY | 40 | | Density, kg/m^3 |
| Cp | 40 | | Specific heat, J/kg*K |
| k | 40 | | Thermal conductivity, W/m*K |
| micro | 40 | | Viscosity, mmPa*s |

| Name | Domain | Mapping | Range | Comment |
|------|--------|---------|-------|---------|
| FMOLWT | VAPOR | Table | MOL__WT | Returns molecular weight |
| FBOILP | VAPOR | Table | BOILING__P | Returns boiling point |
| FCRTEMP | VAPOR | Table | CRIT__TEMP | Returns critical temperature |
| FCRPRES | VAPOR | Table | CRIT__PRES | Returns critical pressure |
| FDENSITY | VAPOR | Table | DENSITY | Returns density |
| FCp | VAPOR | Table | Cp | Returns specific heat |
| Fk | VAPOR | Table | k | Returns thermal conductivity |
| Fmicro | VAPOR | Table | micro | Returns viscosity |

PROPERTIES OF VAPOR

| VAPOR | MOL_WT | BOILING_PT | CRIT_TEMP | CRIT_PRES | DENSITY | Cp | k | micro |
|---|---|---|---|---|---|---|---|---|
| Alcohol_et | 46.07 | 351.7 | 516.3 | 6394 | | 1520 | .013 | 14.2 |
| Alcohol_me | 32.04 | 338.1 | 513.2 | 7977 | | 1350 | .0301 | 14.8 |
| Ammonia | 17.03 | 239 | 405.7 | 11300 | 7.72 | 2200 | .0221 | 9.3 |
| Argon | 39.948 | 87.4 | 151.2 | 4860 | 17.85 | 523 | .016 | 21 |
| Acetylene | 26.04 | 189.5 | 309.2 | 6280 | 1.17 | 1580 | .0187 | 9.34 |
| Benzene | 78.11 | 353.3 | 562.7 | 4924 | 2.68 | 1300 | .0071 | 7 |
| Bromine | 159.82 | 331.9 | 584.2 | 10340 | 6.1 | 230 | .0061 | 17 |
| Butane | 58.12 | 272.7 | 425.2 | 3797 | 2.69 | 1580 | .014 | 7 |
| Carbon_dio | 44.01 | 194.7 | 304.2 | 7384 | 1.97 | 840 | .015 | 14 |
| Carbon_dis | 76.13 | 319.4 | 552 | 7212 | | 599 | | |
| Carbon_mon | 28.01 | 81.7 | 132.9 | 3500 | 1.25 | 1100 | .023 | 17 |
| Carbon_tet | 153.84 | 349.7 | 556.4 | 4560 | | 862 | | 16 |
| Chlorine | 70.91 | 238.5 | 417.2 | 7710 | 3.22 | 490 | .008 | 12 |
| Chloroform | 119.39 | 334.9 | 536.5 | 5470 | | 528 | .014 | 16 |
| Ethyl_chlo | 64.52 | 285.5 | 460.4 | 5270 | 2.872 | 1780 | .00872 | 16 |
| Ethylene | 28.03 | 169.5 | 283.1 | 5120 | 1.25 | 1470 | .0176 | 9.6 |
| Ethyl_ethe | 74.12 | 307.8 | 465.8 | 3610 | | 2470 | | 11.3 |
| Fluorine | 38 | 86.2 | 144 | 5580 | 1.637 | 812 | .0254 | 37 |
| Helium | 4.0026 | 4.2 | 5.3 | 229 | .178 | 5192 | .142 | 19 |
| Hydrogen | 2.0159 | 20.1 | 33.2 | 1316 | .09 | 14200 | .168 | 8.4 |
| Hydrogen_c | 36.461 | 188.3 | 324.5 | 8260 | 1.64 | 800 | .0131 | 13.3 |
| Hydrogen_s | 34.08 | 212.4 | 373.5 | 9012 | 1.54 | 996 | .013 | 11.6 |
| Heptane | 100.21 | 371.6 | 539.9 | 2720 | 3.4 | 1990 | .0185 | 7 |
| Hexane | 86.18 | 340 | 507.9 | 3030 | 3.4 | 1880 | .0168 | 7.52 |
| Isobutane | 58.12 | 249.3 | 408.2 | 3648 | 2.47 | 1570 | .014 | 6.94 |
| Methyl_chl | 50.49 | 248.9 | 416.3 | 6678 | 2.307 | 770 | .0093 | 10.1 |
| Methane | 16.04 | 109.2 | 191.38 | 4641 | .718 | 2180 | .031 | 10.3 |
| Naphthalene | 128.18 | 55.2 | 742.2 | 3972 | | 1310 | | |
| Neon | 20.183 | 26.2 | 44.4 | 2698 | | 1030 | .0464 | 30 |
| Nitric_oxi | 30.01 | 121.2 | 180.3 | 6546 | | 996 | | 29.4 |
| Nitrogen | 28.01 | 77.4 | 126.3 | 3394 | | 1040 | .024 | 16.6 |
| Nitrous_ox | 44.01 | 184.7 | 309.5 | 7235 | | 850 | .01731 | 22.4 |
| Nitrogen_t | 92.02 | @ | 431.4 | 10133 | | 842 | .0401 | |
| Oxygen | 32 | 90.2 | 356 | 5077 | | 913 | .0244 | 19.1 |
| n_Pentane | 72.53 | @ | 469.8 | 3375 | | 1680 | .0152 | 11.7 |
| Phenol | 74.11 | 454.5 | 692 | 6130 | 2.6 | 1400 | .017 | 12 |
| Propane | 44.09 | 231.08 | 370 | 4257 | 2.02 | 1571 | .015 | 7.4 |
| Propylene | 42.08 | 225.45 | 364.9 | 4622 | 1.92 | 1460 | .014 | 8.06 |
| Sulfur_dio | 64.06 | 263.2 | 430 | 7874 | 2.93 | 607 | .0085 | 11.6 |
| Water_vapo | 18.02 | 373.2 | 647.3 | 22120 | .598 | 2050 | .0247 | 12.1 |

# Properties of Liquids

| Name | Elements | Unit | Comment |
|------|----------|------|---------|
| | | | PROPERTIES OF LIQUIDS |
| | | | * * * * * * * * * * * * * * * * * * * * * * * * * * * * * |
| LIQUID | 55 | | Liquids |
| BOILING__P | 53 | | Normal boiling point, 'K at 101 kPa |
| VENTHALPY | 53 | | Enthalpy of vaporization, kJ/kg |
| Cp | 55 | | Specific heat @ 293'K, J/kg*K |
| VISCOSITY | 55 | | Viscosity @ 293'K |
| FENTHALPY | 53 | | Enthalpy of Fusion, kJ/kg |
| DENSITY | 55 | | Density at 293'K, kg/m^3 |
| k | 55 | | Thermal conductivity at 293'K, W/m*K |
| VAP__PRESS | 53 | | Vapor pressure at 293'K, kPa |
| FREEZE__PT | 55 | | Freezing point, 'K |

| Name | Domain | Mapping | Range | Comment |
|------|--------|---------|-------|---------|
| FBOILING | LIQUID | Table | | Returns boiling point at 101 kPa |
| FVENTH | LIQUID | Table | | Returns enthalpy of vaporization |
| FCp | LIQUID | Table | | Returns specific heat at 293'K |
| FVISCOSIT | LIQUID | Table | | Returns viscosity at 293'K |
| Fk | LIQUID | Table | | Returns thermal conductivity |
| FDENSITY | LIQUID | Table | | Returns density at 293'K |
| FVAPPRESS | LIQUID | Table | | Returns vapor pressure at 293'K |
| FFREEZE | LIQUID | Table | | Returns freezing point |

Screen or Printer:     Printer
Title:     PROPERTIES OF LIQUIDS
Vertical or Horizontal:     Vertical

| List | Width | First | Header |
|---|---|---|---|
| LIQUID | 10 | 1 | |
| BOILING__P | 10 | 1 | |
| VENTHALPY | 10 | 1 | |
| Cp | 10 | 1 | |
| VISCOSITY | 10 | 1 | |
| FENTHALPY | 10 | 1 | |
| DENSITY | 10 | 1 | |
| k | 10 | 1 | |
| VAP__PRESS | 10 | 1 | |
| FREEZE__PT | 10 | 1 | |

| LIQUID | BOILING_PT | VENTHALPY | Cp | VISCOSITY | FENTHALPY | DENSITY | k | VAP_PRESS | FREEZE_PT |
|---|---|---|---|---|---|---|---|---|---|
| Acetic_acid | 391.7 | 405 | 2180 | 1222 | 195 | 1049 | .17 | 53.3 | 289.8 |
| Acetone | 329.4 | 532.4 | 2150 | 331 | 98 | 791 | .176 | 53.3 | 177.8 |
| Allyl_alcohol | 370.2 | 684.1 | 2740 | 1363 |  | 853.9 | .18 | 53.3 | 144.2 |
| n_Amyl_alcohol | 411.3 | 503.1 |  | 4004 | 112 | 817.9 | .16 | 13.3 | 194.2 |
| Ammonia | 240 | 1357 | 4601 | 266 | 332.4 | 696.8 | .5 | 53.3 | 195.4 |
| Alcohol_ethyl | 351.7 | 854 | 2840 | 1194 | 108 | 789.2 | .182 | 13.3 | 155.9 |
| Alcohol_methyl | 338.1 | 1100 | 2510 | 592.8 | 99.3 | 791.3 | .215 | 13.3 | 175.4 |
| Aniline | 457.5 | 434 | 2140 | 4467 | 114 | 1021 | .173 | 1.3 | 267 |
| Benzene | 353.3 | 394 | 1720 | 653 | 126 | 879 | .147 | 10 | 279 |
| Bromine | 331.9 | 185 | 448 | 988 | 66.3 | 3119 |  | 22 | 266 |
| n_Butyl_alcohol | 390.7 | 591.5 | 2350 | 2950 | 125 | 811 | .15 | .7 | 183 |
| n_Butyric_acid | 436.7 | 504.7 | 2150 | 1540 | 126 | 964 | .16 | .09 | 267 |
| Calcium_chloride_bri |  |  | 3110 | 2000 |  | 1180 | .574 |  | 257 |
| Carbon_disulfide | 319.4 | 346.1 | 1000 | 360 | 57.7 | 1260 | .16 | 39.3 | 162 |
| Carbon_tetrachloride | 349.9 | 195 | 842 | 967 | 29.8 | 1590 | .11 | 12 | 250.4 |
| Chloroform | 334.4 | 247 | 980 | 562 |  | 1489 | .13 | 21.3 | 209.9 |
| Decane_n | 447.2 |  | 2000 |  | 202 | 730 | .15 | .17 | 243.4 |
| Ethyl_ether | 307.6 | 351 | 2260 | 230 | 98.6 | 714.6 | .14 | 58.7 | 156.9 |
| Ethyl_acetate | 350.3 | 427.5 | 1950 | 451 | 119 | 838 | .175 | 9.6 | 190.8 |
| Ethyl_chloride | 285.5 | 385.9 | 1540 |  | 69.04 | 897.8 | .31 | 53.3 | 136.8 |
| Ethyl_iodide | 345.4 | 191 | 1540 | 9.9 |  | 1935.8 | .37 | 13.3 | 165.2 |
| Ethylene_bromide | 404.7 | 231 | 729 | 28.7 | 57.73 | 2179.3 |  | 1.3 | 282.7 |
| Ethylene_chloride | 356.7 | 356.8 | 1260 | 14 | 88.43 | 1235 |  | 8 | 237.8 |
| Ethylene_glycol | 471.2 | 800.1 |  |  | 181.1 | 1109 | .173 | .1 | 262.4 |
| Formic_acid | 373.9 | 502 | 2200 | 29.7 | 276.54 | 1219 | .18 | 5.3 | 281.5 |
| Glycerine | 454 |  |  | 17800 |  | 1261 | .195 | .1 | 293 |
| Heptane | 371.6 | 321 | 2220 | 409 | 140 | 684 | .128 | 4.73 | 182 |
| Hexane | 340 | 337 | 2250 | 320 | 150 | 658 | .125 | 16 | 178 |
| Hydrogen_chloride | 188.3 | 444 | 486 |  | 54.9 | 1190 |  |  | 158.4 |
| Isobutyl_alcohol | 381.2 | 579 |  | 3910 |  | 801 | .14 | 1.3 | 165.2 |
| Kerosene | 477 |  | 2000 | 2480 |  | 820 | .15 |  |  |

| | | | | | | | | | |
|---|---|---|---|---|---|---|---|---|---|
| Linseed_oil | | | | 42900 | | 920 | | | 249.3 |
| Methyl_acetate | 330.2 | 412 | 1950 | 389 | | 971 | .16 | 22.64 | 175 |
| Methyl_iodide | 315.7 | 192 | 1680 | 500 | | 2270 | | 42.7 | 206.7 |
| Naphthalene | 483.9 | 316 | 1700 | 901 | 151 | 976 | | .291 | 353.4 |
| Nitric_acid | 359.2 | 628 | 1450 | 910 | 166 | 1512 | .28 | .236 | 231.5 |
| Nitrobenzene | 484 | 330 | 2100 | 2150 | 93.69 | 1200 | 1.7 | .001 | 278.9 |
| Octane | 398.9 | 306.3 | 2500 | 562 | 180.7 | 703 | .15 | .056 | 216.7 |
| Petroleum | | 200 | | 10000 | | 820 | | | |
| n_Petane | 309.2 | 357.3 | 2330 | 226 | 117 | 626 | .11 | 56.7 | 143.4 |
| Propionic_acid | 414.3 | 413.6 | 1980 | 1102 | | 992 | .173 | .4 | 252.4 |
| Sodium_chloride_20% | 378 | | 3110 | 1570 | | 1150 | .583 | .076 | 256.8 |
| Sodium_chloride_10% | 375.1 | | 3260 | 1180 | | 1070 | .593 | .087 | 266.8 |
| Sodium_hydroxide_H20 | 374.8 | | 3610 | | | 1150 | | | 252.2 |
| Sulfuric_acid_100% | 560.9 | | 1400 | 22000 | | 1833 | | .001 | 283.7 |
| Sulfuric_acid_95% | 575 | | 1460 | 21000 | | 1836 | | .001 | 245 |
| Sulfuric_acid_90% | 533.2 | | 1600 | 25000 | | 1816 | .38 | .001 | 263.7 |
| Toluene | 383 | 363 | 1690 | 587 | 71.9 | 867 | .16 | .12 | 178.2 |
| Turpentine | 423 | 286 | 1700 | 546 | | 863 | .13 | | |
| Water | 373 | 2257 | 4180 | 988 | 333.8 | 998.2 | .602 | .045 | 273.2 |
| Xylene_ortho | 417 | 347 | 1720 | 831 | 128 | 881 | 1.6 | .026 | 248 |
| Xylene_meta | 412 | 342 | 1670 | 628 | 109 | 867 | 1.6 | .029 | 226 |
| Xylene_para | 411 | 340 | 1640 | 670 | 161 | 862 | | .03 | 286 |
| Zinc_sulfate_10% | | | 3700 | 1570 | | 1110 | .583 | | 271.9 |
| Zinc_sulfate_1% | | | 3300 | 1100 | | 1010 | .598 | | 273 |

# Properties of Solids

| Name | Elements | Unit | Comment |
|---|---|---|---|
| SOLID | 103 | | Name of solid |
| Cp | 102 | | Specific heat, J/kg*K |
| DENSITY | 102 | | Density, kg/m^3 |
| k | 102 | | Thermal conductivity, W/m*K |
| EMISSIVIT | 103 | | Emissivity ratio |

| Name | Domain | Mapping | Range | Comment |
|---|---|---|---|---|
| FCp | SOLID | Table | Cp | |
| FDENSITY | SOLID | Table | DENSITY | |
| Fk | SOLID | Table | k | |
| FEMISSIVI | SOLID | Table | EMISSIVIT | |

## PROPERTIES OF SOLIDS

| SOLID | Cp | DENSITY | k | EMISSIVITY |
|---|---|---|---|---|
| Aluminum__1100 | 896 | 2740 | 221 | .09 |
| Alum__bronze | 400 | 8280 | 100 | |
| Alundum | 779 | | | |
| Asbestos__fiber | 1050 | 2400 | .17 | |
| Asbestos__insul | 800 | 580 | .16 | .93 |
| Ashes__wood | 800 | 640 | .071 | |
| Asphalt | 920 | 2110 | .74 | |
| Bakelite | 1500 | 1300 | 17 | |
| Bell__metal | 360 | | | |
| Bismuth__tin | 170 | | 65 | |
| Brick__building | 800 | 1970 | .7 | .93 |
| Brass__red | 400 | 8780 | 150 | .03 |

| Material | | | | |
|---|---|---|---|---|
| Brass__yellow | 400 | 8310 | 120 | .033 |
| Bronze | 435 | 8490 | 29 | |
| Cadmium | 230 | 8650 | 92.9 | .02 |
| Carbon | 710 | | .35 | .81 |
| Cardboard | | | .07 | |
| Cellulose | 1300 | 54 | .057 | |
| Cement__portland | 670 | 1920 | .029 | |
| Chalk | 900 | 2290 | .83 | .34 |
| Charcoal | 840 | 240 | .05 | |
| Chrome__Brick | 710 | 3200 | 1.2 | |
| Clay | 920 | 1000 | | |
| Coal | 1000 | 1400 | .17 | |
| Coal__tars | 1500 | 1200 | .1 | |
| Coke__petroleum | 1500 | 990 | .95 | |
| Concrete__stone | 653 | 2300 | .93 | |
| Copper | 390 | 8910 | 393 | .072 |
| Cork | 2030 | 86 | .048 | |
| Cotton | 1340 | 1500 | .042 | |
| Cryolete | 1060 | 2900 | | |
| Diamond | 616 | 2420 | 47 | |
| Earth__dry | | 1500 | .064 | .41 |
| Felt | | 330 | .05 | |
| Fireclay__brick | 829 | 1790 | 1 | .75 |
| Fluorspar | 880 | 3190 | 1.1 | |
| German__silver | 400 | 8730 | 33 | .135 |
| Glass__crown | 750 | 2470 | 1 | .94 |
| Glass__flint | 490 | 4280 | 1.4 | |
| Glass__pyrex | 840 | 2230 | 1 | |
| Glass__wool | 657 | 52 | .038 | |
| Gold | 131 | 19350 | 297 | .02 |
| Graphite__powder | 691 | | .183 | |
| Graphite__Karbate | 670 | 1870 | 130 | .75 |
| Gypsum | 1080 | 1200 | .43 | .903 |
| Hemp | 1352.3 | 1500 | | |
| Ice__0__C | 2040 | 921 | 2.24 | .95 |
| Ice__neg__20__C | 1950 | | 2.44 | |
| Iron__cast | 500 | 7210 | 47.7 | .435 |
| Iron__wrought | | 7770 | 60.4 | .94 |
| Lead | 129 | 11300 | 34.8 | .28 |
| Leather | | 1000 | .16 | |
| Limestone | 909 | 1650 | .93 | .36 |
| Linen | | | .09 | |
| Litharge | 230 | 7850 | | |
| Magnesia__power | 980 | 796 | .61 | |
| Magnesia__carbonate | | 210 | .059 | |
| Magnesite__brick | 930 | 2530 | 3.8 | |
| Magnesium | 1000 | 1730 | 160 | .55 |
| Marble | 880 | 2600 | 2.6 | .931 |

| | | | | |
|---|---|---|---|---|
| Nickel | 440 | 8890 | 59.5 | .045 |
| Paint__wht__laq | | | | .8 |
| Paint__wht__ena | | | | .91 |
| Paint__blk__laq | | | | .8 |
| Paint__blk__shel | | 1000 | .26 | .91 |
| Paint__flt__blk | | | | .96 |
| Paint__alum | | | | .39 |
| Paper | 1300 | 930 | .13 | .92 |
| Paraffin | 2900 | 900 | .24 | |
| Plaster | | 2110 | .74 | .91 |
| Platinum | 130 | 21470 | 69 | .054 |
| Porcelain | 750 | 260 | 2.2 | .92 |
| Pyrites__cu | 549 | 4200 | | |
| Pyrites__iron | 569 | 4970 | | |
| Rock__salt | 917 | 2180 | | |
| Rubber__soft | 2000 | 1100 | .1 | .86 |
| Rubber__hard | | 1190 | .16 | .95 |
| Sand | 800 | 1520 | .33 | |
| Sawdust | | 190 | .05 | |
| Silica | 1320 | 2240 | 1.4 | |
| Silver | 235 | 10500 | 424 | .02 |
| Snow@32F | | 500 | 2.2 | |
| Steel | 500 | 7830 | 45.3 | .12 |
| Stone | 800 | 1500 | | |
| Tar__pitch | 2500 | 1100 | .88 | |
| Tar__bitum | | 1200 | .71 | |
| Tin | 233 | 7290 | 64.9 | .06 |
| Tungsten | 130 | 19400 | 201 | .032 |
| White__ash | | 690 | .172 | |
| American__elm | | 580 | .153 | |
| Hickory | | 800 | | |
| Mahogany | | 550 | .13 | |
| Sugar__maple | | 720 | .187 | |
| White__oak | 2390 | 750 | .176 | .9 |
| Black__walnut | | 630 | | |
| White__fir | | 430 | .12 | |
| White__pine | | 430 | .11 | |
| Spruce | | 420 | .11 | |
| Wool__fiber | 1360 | 1300 | | |
| Wool__fabric | | 220 | .05 | |
| Zinc__cast | 390 | 7130 | 110 | .05 |
| Zinc__hotrolled | 390 | 7130 | 110 | |
| Zinc__galvanizing | | | | .23 |

# Bibliography

American Society of Heating, Refrigerating and Air-Conditioning Engineers. *ASHRAE HANDBOOK 1981 FUNDAMENTALS.* American Society of Heating, Refrigerating and Air-Conditioning Engineers, Inc., 1791 Tullie Circle NE, Atlanta, GA 30329.

Bednar, Henry H., P. E. *Pressure Vessel Design Handbook,* Van Nostrand Reinhold, New York, NY, 1981.

Beil, Donald H. *The DIF File,* Reston Publishing Company, 11480 Sunset Hills Road, Reston, VA 22090.

Beil, Donald H. *The VisiCalc Book, for the IBM Personal Computer,* Reston Publishing Company, Reston, VA.

Bell, David A. *Fundamentals of Electric Circuits* (Second Edition). Reston Publishing Company, Reston, VA.

Beyer, William H. *CRC Standard Mathematical Tables.* CRC Press, Inc., Boca Raton, FL.

Burghardt, M. David. *Engineering Thermodynamics with Applications* (Second Edition). Harper & Row, New York, NY.

*Cameron Hydraulic Data.* Ingersoll-Rand Company, Woodcliff Lake, NJ.

Collier, Courtland, and Ledbetter, William B. *Engineering Cost Analysis.* Harper and Row, New York, 1982.

Coughanowr, Donald R., and Koppel, Lowell B. *Process Systems Analysis and Control.* McGraw-Hill, New York, NY.

Crocker, Sabin and King, Reno C., *Piping Handbook,* McGraw-Hill, New York, NY.

Davis, Don, & Davis, Carolyn. *Sound System Engineering.* Howard K. Sams, Indianapolis, IN.

*Disk Operating System (by Microsoft, Inc.) Reference Manual.* International Business Machines Corporation, Boca Raton, FL 33432.

Ewing, Richard. *VisiTrend/Plot Program Instruction/Reference Manual.* VisiCorp, 2895 Zanker Road, San Jose, CA 95134.

Ferziger, Joel H. *Numerical Methods for Engineering Application.* John Wiley & Sons, New York, NY.

Gebhart, Benjamin. *Heat Transfer.* McGraw-Hill, New York, NY.

*Guide to Operations.* International Business Machines Corporation, Boca Raton, FL 33432.

Harvey, John F. *Theory and Design of Modern Pressure Vessels* (Second Edition). Van Nostrand Reinhold, New York, NY.

Hickey, Harry E. *Hydraulics for Fire Protection.* National Fire Protection Association, 470 Atlantic Avenue, Boston, MA 02210.

Hicks, Tyler G. *Standard Handbook of Engineering Calculations.* McGraw-Hill, New York, NY.

Jennings, Alan. *Matrix Computation for Engineers and Scientists.* John Wiley & Sons, New York, NY.

John, James E. A., and Haberman, William L. *Introduction to FLUID MECHANICS* (Second Edition). Prentice-Hall, Englewood Cliffs, NJ, 1980.

Johnston, E. Russell Jr., and Beer, Ferdinand P. *Vector Mechanics for Engineers.* McGraw-Hill, New York, NY.

Kirk, Donald E. *Optimal Control Theory.* Prentice-Hall, Englewood Cliffs, NJ.

Kuo, Benjamin C. *Digital Control Systems.* Holt, Rinehart, and Winston, New York, NY.

Lindeburg, Michael R. *Mechanical Engineering Review Manual.* Professional Publications, San Carlos, CA 94070.

Maloney, Timothy J. *Industrial Solid-State Electronics—Devices and Systems.* Prentice-Hall, Englewood Cliffs, NJ.

Maron, Melvin J. *Numerical Analysis: A Practical Approach.* Macmillan, New York, NY.

Miller, Alan R. *FORTRAN Programs for Scientists and Engineers.* SYBEX Inc., 2344 Sixth Street, Berkeley, CA 94710.

Miller, David W., and Starr, Martin K. *Executive Decisions and Operations Research.* Prentice-Hall, Englewood Cliffs, NJ.

Oberg, Erik, Jones, Franklin D., and Horton, Holbrook L. *Machinery's Handbook*. Industrial Press, Inc., 200 Madison Ave., New York, NY 10016.

Olson, Reuben M. *Essentials of Engineering Fluid Mechanics*. International Textbook Company, Scranton, PA.

Parrish, A. *Mechanical Engineer's Reference Book*. Butterworths, Boston, MA.

Polya, G. *How to Solve It*. Princeton University Press, Princeton, NJ.

*Professional Engineering Examinations* (Volume I). The National Council of Engineering Examiners, Berkeley, CA.

Ralston, Anthony, and Reilly, Edwin, D., Jr. *Encyclopedia of Computer Science and Engineering* (Second Edition). Van Nostrand Reinhold, New York, NY.

Richmond, Samuel B. *Statistical Analysis* (Second Edition). The Ronald Press, New York, NY.

Schneider, Raymond K. *HVAC Control Systems*. John Wiley & Sons, New York, NY.

Schwartz, Abraham. *Analytic Geometry and Calculus*. Holt, Rinehart and Winston, New York, NY.

Steinberg, Dave S. *Cooling Techniques for Electronic Equipment*, John Wiley & Sons, New York, NY.

Taha, Hamdy A. *Operations Research—An Introduction*. Macmillan, New York, NY.

*TK!Solver Program Instruction/Reference Manual*. Software Arts, Inc., 27 Mica Lane, Wellesley, MA 02181.

Trost, Stanley R., and Pomernacki, Charles. *VisiCalc for Science and Engineering*. SYBEX Inc., 2344 Sixth Street, Berkeley, CA 94710.

Van Valkenburg, M. E. *Network Analysis* (3rd Edition). Prentice-Hall, Englewood Cliffs, NJ, 1974.

Van Wylen, Gordon J., and Sonntag, Richard E. *Fundamentals of Classical Thermodynamics*. John Wiley and Sons, Inc., New York, NY.

Wagner, Harvey M. *Principles of Operations Research with Applications to Managerial Decisions*. Prentice-Hall, Englewood Cliffs, NJ.

Walverton, Van. *VisiCalc Program Instruction/Reference Manual.* VisiCorp, 2895 Zanker Road, San Jose, CA 95134.

Wylie, C. R. Jr. *Advance Engineering Mathematics.* McGraw-Hill, New York, NY.

Yarbrough, Raymond B. *Electrical Engineering Review Manual.* Professional Publications, San Carlos, CA 94070.

Zill, Dennis G. *A First Course in Differential Equations.* Prindle, Weber & Schmidt, Boston, MA.